Geology of the country between Aberystwyth and Machynlleth

The country between Aberystwyth and Machynlleth provided inspiration to two of Wales' pioneer geologists, O. T. Jones and his student W. J. Pugh. Both occupied successively the chairs of geology at Aberystwyth and Manchester. The former rose to become Woodwardian Professor at Cambridge and the latter was knighted as Director of H. M. Geological Survey.

The district is a scenic hilly part of mid-Wales. It includes the remote upland area around Plynlimon and seaside resorts such as Aberdovey in the Snowdonia National Park. It has been carved from sedimentary rocks of Ordovician and Silurian ages formed in the Lower Palaeozoic Welsh Basin. The rocks are mainly turbidites with mudstone predominating. Hemipelagites are dispersed through the turbiditic sequence and these reflect an alternation of oxic and anoxic conditions of deposition.

The macrofossils preserved in these rocks are almost entirely the remains of graptolites, marine animals which had a planktonic existence. By means of these remains the rocks have been divided biostratigraphically into Zones and since the completion of the writing of the memoir the 27th International Geological Congress in Moscow has used these zones to redefine the base of the Silurian System at the base of the *Parakidograptus acuminatus* Zone. Thus the sharp lithological contrast between the Bryn-glâs Formation and the Cwmere Formation no longer signifies the Ordovician – Silurian boundary. The boundary now lies within the Cwmere Formation, but cannot be recognised.

The structural fabric of the rocks, the folds, faults and cleavage, stimulates debate. The memoir is unable to provide an unequivocal exposé of the genesis of this fabric. Rather did the survey encourage belief that many structures were not the product of an end-Silurian event, but more the result of a continuum of movement and that the fabric had evolved during the accumulation of the whole sedimentary prism. The folding of the sediments may thus have taken place while they were in various states of dewatering. Bedding-parallel decollement and dextral deviation of cleavage strike compared with fold trends are two common elements of this fabric.

Frontispiece   View of Drosgol with Nant-y-môch Reservoir in foreground. Bench features on the slope mark the outcrops of bedded sandstone (Pl.7)

BRITISH GEOLOGICAL SURVEY

R. CAVE and B. A. HAINS

CONTRIBUTORS

Geophysics
R. B. Evans

Palaeontology
D. E. White

Petrography
G. E. Strong and B. R. Young

Stratigraphy
D. I. Jackson and R. A. Waters

Geology of the country between Aberystwyth and Machynlleth

Memoir for 1:50 000 geological sheet 163
(England & Wales)

Natural Environment Research Council

LONDON: HER MAJESTY'S STATIONERY OFFICE   1986

First published 1986

ISBN 0 11 884394 X

Bibliographical reference

CAVE, R. and HAINS, B. A. 1986. Geology of the country between Aberystwyth and Machynlleth. *Memoir of the British Geological Survey*, Sheet 163 (England & Wales).

Authors

R. CAVE, PhD
B. A. HAINS, PhD
British Geological Survey, Aberystwyth SY23 4BY

Contributors

R. A. Waters, BSc, PhD
British Geological Survey, Aberystwyth

D. I. Jackson, MA
British Geological Survey, Edinburgh

R. B. Evans, MSc, DIC, G. E. Strong, BSc, D. E. White, MSc, PhD and B. R. Young, MSc
British Geological Survey, Keyworth

Other publications of the Survey dealing with this district and adjoining districts

BOOKS
Memoirs
Lead and Zinc. The mining district of north Cardiganshire and west Montgomeryshire (out of print)
British Regional Geology
North Wales (3rd edition)
South Wales (3rd edition)
MAPS
1:625 000
Solid geology (South sheet)
Quaternary geology (South sheet)
1:100 000
Central Wales Mining Field

Printed in the United Kingdom for Her Majesty's Stationery Office
GP 398/2. Dd238944 C20 2/87 C.C.N. 12521

CONTENTS

1 **Chapter 1   Introduction**
Geological history   2
Geological research   3
Turbidites   3
Hemipelagite   4
Cleavage   5

6 **Chapter 2   Ordovician—general account**
Nant-y-Môch Formation   8
Drosgol Formation   9
 Pencerrigtewion Member   10
Bryn-glâs Formation   14

16 **Chapter 3   Ordovician—details**
Nant-y-Môch Formation   16
Drosgol Formation   20
Bryn-glâs Formation   36

41 **Chapter 4   Silurian—general account**
Cwmere Formation   42
 Mottled Mudstone Member   42
 Strata above Mottled Mudstone Member   43
 Conditions of deposition   44
Derwenlas Formation   44
 Conditions of deposition   45
Cwmsymlog Formation   46
 Dark grey mudstone 'member'   47
 Green mudstone 'member'   47
 Red mudstone 'member'   47
 Arenite facies   47
 Conditions of deposition   48
Devil's Bridge Formation   49
 Fauna   50
 Conditions of deposition   50
Borth Mudstones Formation   50
 Fauna   51
 Stratigraphical relationships and sedimentation   51
Aberystwyth Grits Formation   53
 Lithologies   54
 Distribution of lithologies   57
 Faunas   57
 Conditions of deposition   57
Dolwen Mudstones Formation   58
Blaen Myherin Mudstones Formation   58
Biostratigraphy   58

64 **Chapter 5   Silurian—details**
Cwmere Formation   64
Derwenlas Formation   68
Cwmsymlog Formation   82
Devil's Bridge Formation   91
Borth Mudstones Formation   95
Aberystwyth Grits Formation   96
Dolwen Mudstones Formation   97

99 **Chapter 6   Structure and geophysics**
Folds   99
Faults   100
Cleavage   100
Minor structures   100
Curious markings   103
Suggested sequence of structural events   103
Geophysics   104
Details   105

111 **Chapter 7   Quaternary—general account**
Boulder Clay   112
Morainic Drift   112
Head and Scree   113
Glacial Sand and Gravel   113
Glacial Lake Deposits   113
Diatomaceous Earth   113
River Terraces, Alluvial Fans and Alluvium   115
Peat   116
Marine and Estuarine deposits   117
General relationships   117
Summary of Late Devensian to Recent events   118

119 **Chapter 8   Quaternary—details**
Boulder Clay   119
Morainic Drift   120
Head   126
Scree   126
Glacial Sand and Gravel   126
Submerged Forest   126
Storm Gravel Beach Deposits   127
Marine and Estuarine Alluvium   127
Marine Beach Deposits and Tidal Flats   127
River Terraces   128
Alluvial Cones or Fans   128
Alluvium   128
Blown Sand   128
Made Ground   129
Landslip   129

130 **Chapter 9   Mineral products**
Metalliferous minerals   130
Building stone and slate   132
Sand and gravel   132
Peat   132

133 **References**

137 **Appendices**
 1 List of boreholes   137
 2 List of Geological Survey photographs   138
 3 List of non-graptolitic fossil localities and
   faunas   140

141 **Index of fossils**

143 **General index**

PLATES

Frontispiece   View of Drosgol with Nant-y-môch
   Reservoir in foreground
1   Phosphatic concretions (black) in the Mottled
   Mudstone Member. Slabs from the Cardiganshire
   Slate Quarry   5
2   Nant-y-Môch Formation, Craig y Dullfan subfacies,
   Maesnant stream   8
3   Nant-y-Môch Formation, Maesnant subfacies, Maesnant stream   9
4   Loaded-ripples, Bryn-glâs Formation at Bryn-yr-
   afr   15
5   Craig y Dullfan subfacies. Convolute lamination in
   arenitic layer, Craig y Dullfan   18
6   Craig y Dullfan subfacies. Turbiditic sand occupying a
   contemporaneously loaded scour pocket, Craig y
   Dullfan   18
7   Drosgol Formation. Base of a massive turbidite sandstone containing intraclasts of mudstone. Nant-y-môch
   Dam   21
8   Carn Owen. Quarry A in evenly bedded massive sandstones of the Pencerrigtewion Member   28
9   Pseudonodules on a bedding-plane surface of an inverted fallen block. Quarry A, Carn Owen   29
10   Drosgol Formation. Pillow-form sandstone at foot of
   the incline. Carn Owen   30
11   Cwmere Formation. Layers of hemipelagic and turbiditic mudstone. Nant Rhyddlan   43
12   Devil's Bridge Formation, mainly mud turbidites with
   arenites subordinate. Coed Rhiw-lwyfan   50
13   Aberystwyth Grits. Harp Rock (Craig y Delyn)   55
14   Aberystwyth Grits. Harp Rock Type 2 turbidite   56
15   Flute moulds, Cwmsymlog Formation. Morben
   Quarry   89
16   Aberystwyth Grits. Evenly bedded bi-modal turbiditic
   sandstones. Cliffs below Constitution Hill   97
17   Cusp-and-furrow structure, Devil's Bridge Formation.
   Ynys Edwin   102
18   Fold, east of the Glandyfi Tract   105
19   Fold, east of the Glandyfi Tract   106
20   Bedding-parallel slide, Devil's Bridge Formation,
   Forge   107
21   Fold, west of the Glandyfi Tract. Anticline, cusp-
   shaped and axially ruptured, in the Aberystwyth
   Grits   107
22   Bedding-parallel slide, Aberystwyth Grits; foot of Allt-
   wen looking N   108
23   Bedding-parallel slide, Aberystwyth Grits; foot of Allt-
   wen looking S   109
24   Two bedding-parallel slides (arrowed) with a slightly
   discordant package of beds between; Devil's Bridge
   Formation. Near Ynys Edwin   110
25   Two small listric normal faults dipping northward into
   the lower of the bedding-parallel slide planes of
   Pl.24   110
26   Two-tier moraine left by late-glacial corrie-glacier.
   Near Rhiw-gam   124
27   Gravelly Morainic Drift overlying glacial-lake clays.
   Hengwm   124

FIGURES

1   Locality map and index to areas covered by other
   figures within this memoir   1
2a   The five terms erected by Bouma for the intervals
   making up the complete sequence in a turbidite from
   Ta at the bottom to Te at the top   4
2b   The lithological theme of mud turbidites   4
3   Ordovician sequences of the Machynlleth and
   Plynlimon inliers   7
4   Distribution of rock-types, Pencerrigtewion Member;
   a possible sedimentational model of a westward prograding 'Pencerrigtewion Event'   12
5   Facies distribution within the Pencerrigtewion
   Member around Plynlimon   13
6   Conjectured palaeogeography of late-Ordovician (Hirnantian) times   14
7   The geology of the centre of the Plynlimon
   Pericline   17
8   Section through parts of the Nant-y-Môch and
   Drosgol formations 400 m WSW of Dynyn   19
9   Locality map for Drosgol Formation (lower part);
   Nant-y-môch Dam   20
10   Drosgol Formation on Drosgol   22
11   Section through east flank of Drosgol from the summit to the Nant-y-môch Reservoir   23
12   Location diagram for outcrops on the south side of
   Llyn Llygad Rheidol   24
13   Map of the Pencerrigtewion Member at Pencerrigtewion   25
14   Drosgol Formation on Banc Llechwedd-mawr   26
15a   Map of Carn Owen area   27
15b   Diagrammatic section through Carn Owen on a
   traverse c.70 m north of the Hafan Fault   28
16   The Drosgol and Bryn-glâs formations between Ogof
   Morris and Bwlcheinion   31
17   The Drosgol and Bryn-glâs formations between
   Bwlcheinion and the Llyfnant Fault   32
18   Section across Craig Caerhedyn   33
19   Channel-margin at the base of the Drosgol Formation, Craig Caerhedyn   33
20a   The Drosgol Formation between Moel Hyrddod and
   the Llyfnant Valley   34
20b   Section across channel-fill near Dynyn. Tectonic dip
   c.10° towards 280°. (Loc.1, Fig.20a)   35
21   Distribution of sandstone within the Bryn-glâs Formation near Bwlchystyllen   37
22   The geology of the southern extremity of the
   Machynlleth Inlier, 2.5 km east of Talybont   39
23   Silurian sequence and graptolite zones   41
24a   Bodies of arenaceous turbidites, Derwenlas Formation   45
24b   Channel sandstones at the base of the Devil's Bridge
   Formation   46
24c   The encroachment eastwards during the mid-Silurian
   of northward delivered turbidite sands   46

24d Plan of the Navy Fan and Navy Channel superimposed on a map of mid-Wales 47
25 The Cwmsymlog Formation with its sandstones and the relationship to the overlying Devil's Bridge Formation and its basal sandstones 48
26 Facies formations in part of the upper Llandovery 54
27 A sedimentary model explaining the disposition and relationships of the Devil's Bridge, Borth Mudstones and Aberystwyth Grits formations 54
28 Some characteristic Silurian graptolites 62
29 Sketch-map of the Craigypistyll area 65
30 Cardiganshire Slate Quarry, near Cymerau. A section through the top of the Bryn-glâs Formation and the Mottled Mudstone Member of the Cwmere Formation 66
31a Geological sketch of Fridd Cae-crŷdd 67
31b Succession estimated from the outcrop near Fridd Cae-crŷdd 68
32 Composite section of the Derwenlas Formation in the Rheidol Gorge-Eisteddfa Gurig area 69
33 Folding in the Derwenlas and adjoining formations, south of Disgwylfa Fâch 70
34 Geological sketch-map of the Pen y Graig-ddu area 71
35 Comparative sections in the Derwenlas and Cwmsymlog formations in the area from Capel Tabor to the Afon Lluestgota 72
36 Sketch-map of the Cwmere area 78
37a Sketch-map of the Coed y Fedw area 79
37b Sketch-map of the Tyn y Garth area 79
38 Exposures along the old coach road, Derwenlas, type-section of the Derwenlas Formation 80
39 Sections through the Derwenlas Formation; railway cuttings, north side of Dovey Estuary 81
40 The Llyn Ieuan Cwmsymlog Formation inlier 83
41 Sketch section through slope at Y Chwareli 85
42 Comparative sections of the basal Devil's Bridge Formation, Cwmsymlog Formation and uppermost Derwenlas Formation in the Cwmere area 88
43 Lower quarry, Craig yr Hesg; lower part of the Devil's Bridge Formation 92
44 Section through the base of the Aberystwyth Grits Formation at Harp Rock 96
45 Main structural features of the Aberystwyth district 101
46 A generalised structural model of the area 104
47 Small ramp of strata produced by movement on a bedding-parallel slide, between Aberdovey and Trefeddian 109
48 Quaternary deposits; generalised distribution 111
49 Drift deposits in Hengwm 114
50 Drift deposits near Glaspwll 115
51 Ditch section through the Drift deposits near Glaspwll 116
52 'Incised meanders', Ponterwyd 120
53 Drift deposits in Cwm-byr and near Rhiw-gam 123
54 Drift deposits of Cwmrhaiadr 125

TABLES

1 Graptolites from the Glan-fred Borehole 52
2 Graptolite faunas in the Aberystwyth Grits 53
3 Zonal distribution of Silurian graptolites in BGS collections, Aberystwyth district 59

NOTES

In this book, the word 'district' means the area depicted on 1:50 000 Geological Sheet 163 (Aberystwyth) (Figure 1).

Numbers in square brackets are National Grid references within the 100-km square SN. Single grid references of sections that have an appreciable lateral extent refer to the base of the section.

Numbers preceded by the letter A refer to the British Geological Survey Photograph Collection.

Numbers preceded by the letter E refer to the Sliced Rock Collection of the Survey.

Numbers preceded by the letter MR refer to the Museum Reserve Collection of the Survey.

The authorship of fossil species is given in the index.

PREFACE

Primary six-inch to one mile (1:10 560) geological surveying around the university town of Aberystwyth started in 1965. This ended the long absence of the British Geological Survey from the Lower Palaeozoic heartland of Wales. At this time Professor O. T. Jones, who made such a vigorous input into Welsh geology, was still active and the authors had the privilege of discussing the proposed survey with him. The ensuing survey was greatly indebted to his pioneer work in the district with Dr W. J. Pugh, who eventually became Director of the Geological Survey.

The first map of the district by the Geological Survey was surveyed on the scale of one-inch to one-mile (1:63 360) and published in 1848. The primary six-inch survey was conducted between 1965 and 1975 by Drs R. Cave and B. A. Hains, later joined by Mr D. I. Jackson (1969–74) and Dr R. A. Waters (1973–75). The District Geologists were, successively, Dr G. A. Kellaway until 1968 and then Dr J. R. Earp; the 1:50 000 map (Solid Edition) was published in 1984. A large collection of fossils, mainly graptolites, was made by Dr D. E. White and colleagues of the Palaeontology Unit, to supplement those of the surveyors. Graptolite identifications and reports on them were provided initially by Drs P. Toghill and A. W. A. Rushton, but the main effort came later when graptolite determinations were made by Dr R. B. Rickards of the Sedgwick Museum, Cambridge. Graptolite biostratigraphy based on this information has been compiled by Dr White who also identified the benthic faunas.

Reports on the sedimentary petrography of some Silurian and Ordovician rocks were provided by Mr G. E. Strong, while XRD analyses of clay minerals in some of the mudstones were undertaken by Mr B. R. Young, both of the Petrology Unit. A geophysical assessment of the district has been made by Mr R. B. Evans of the Applied Geophysics Unit, and photography in the district was undertaken by Mr C. Jeffrey of the Photographic Department. This memoir has been edited by Dr R. A. B. Bazley and Mr. W. B. Evans.

Place names used in this memoir are those shown on the 1:50 000 map. Where this has been impracticable recourse has been made to the 1:10 560 maps. We are grateful for the assistance of the landowners and tenants in granting free access to all areas. though reference herein to places of interest does not impute public accessibility.

G. Innes Lumsden, FRSE
Director

British Geological Survey
Keyworth, Nottingham

1st August 1986

SIX-INCH MAPS

The following is a list of six-inch geological maps included wholly or in part within the land area of 1:50 000 Geological Sheet 163 with the names of the surveying officers and the date of survey for each map. The officers are R. Cave, B. A. Hains, D. I. Jackson and R. A. Waters. The manuscript copies of the six-inch maps are deposited for public reference in the library of the Aberystwyth Office of the British Geological Survey. Those sheets marked * have been surveyed only in part.

*SN 57 NE	(Southgate) Waters	1973
SN 58 SE	(and part 58 NE) (Aberystwyth) Waters	1973 – 75
SN 59 NE	(Rhowinar) Cave	1973
*SN 67 NW	(Capel Seion) Waters	1973
SN 67 NE	(Glan Rheidol) Jackson and Waters	1970, 1975
SN 68 NW	(and part 58 NE) (Llandre, Borth (southern)) Cave, Hains and Waters	1973 – 75
SN 68 NE	(Talybont, Bontgoch) Cave, Hains and Jackson	1970 – 71, 1973 – 74
SN 68 SW	(Penrhyncoch, Llanbadarn Fawr (eastern)) Jackson and Waters	1970, 1973 – 74
SN 68 SE	(Goginan, Pen-bont Rhydybeddau) Jackson	1969 – 71
*SN 69 NW	(Aberdovey) Cave	1972 – 74
SN 69 NE	(Eglwysfach, Gogarth) Cave	1965 – 66, 1971 – 74
SN 69 SW	(Ynyslas, Cors Fochno) Hains	1965 – 66, 1972
SN 69 SE	(Tre'r-ddôl, Foel Goch) Cave and Hains	1965 – 66, 1972 – 73
*SN 77 NW	(Bwa-drain) Hains	1971 – 72
*SN 77 NE	(Afon Myherin) Hains	1971 – 72
SN 78 NW	(Carn Owen, Craigypistyll) Cave and Hains	1968 – 71
SN 78 NE	(Nant-y-môch dam, Plynlimon) Cave and Hains	1967 – 71
SN 78 SW	(Ponterwyd, Syfydrin) Hains	1968 – 71
SN 78 SE	(Bryn-glas, Eisteddfa Gurig) Hains	1967 – 68, 1970 – 71
*SN 79 NW	(Llyfnant Valley) Cave	1965 – 67, 1971
*SN 79 NE	(Talbontdrain) Cave	1967, 1972
SN 79 SW	(Anglers' Retreat, Moel-y-Llyn) Cave	1966 – 67, 1970 – 71
SN 79 SE	(Uwchygarreg) Cave	1967, 1970, 1972

x

CHAPTER 1

Introduction

This Memoir describes the geology of Aberystwyth and the district close to the north and east (Figure 1) of that town. It encompasses the southern extremity of the Snowdonia National Park, the Dovey Estuary and twenty kilometres of Cardigan Bay coast. Inland, much of the district is high moorland, and on the eastern margin is Plynlimon (752 m). The drainage flows mainly southwards and westwards from the high land as Afon Rheidol, but other minor rivers flow westward and northward to join the River Dovey. Husban-

dry is devoted mainly to forests and sheep, while manufacturing is negligible and the once-flourishing metalliferous mines are derelict. Electricity is generated from Afon Rheidol by a scheme that involved the construction in 1957–62 of tunnels, impressive dams and reservoirs, which add to the tourist attractions of the district. Aberystwyth is the centre of population and administration, housing the University College of Wales, the National Library of Wales and several Colleges of Further Education.

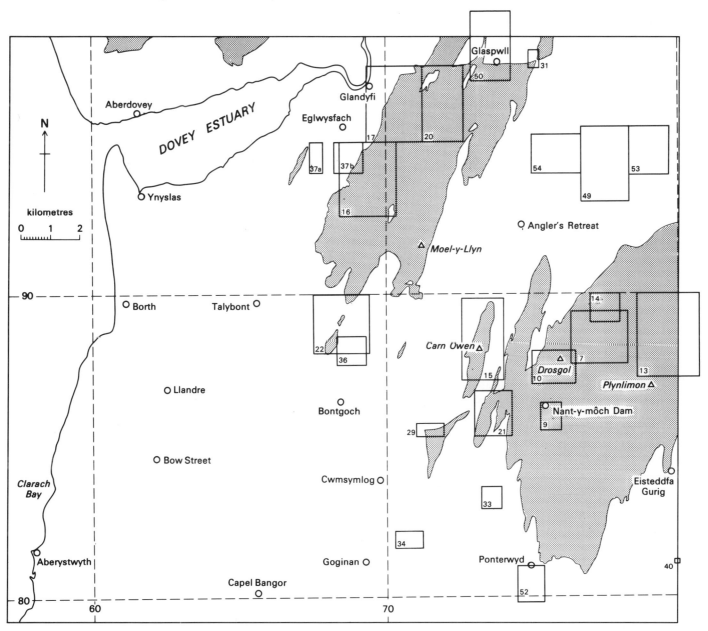

Figure 1 Locality map and index to areas covered by other figures within this memoir

GEOLOGICAL HISTORY

The rocks of the district preserve its history in two widely separated periods only; one near the start of Phanerozoic time, the other during the Quaternary. The Phanerozoic rocks are included in the youngest series (Ashgill) of the Ordovician and the oldest series (Llandovery) of the Silurian. They have been subdivided into zones by the use of graptolite faunas; only the highest zone of the Ordovician, that of *Dicellograptus anceps*, is present; the Silurian zones range from the basal *Glyptograptus persculptus* Zone to the *Monograptus turriculatus* Zone (Figure 23). The rocks are almost entirely sedimentary, though their constituent grains and clasts were commonly derived from igneous rocks. Exceptions are two very thin fine grained layers containing montmorillonite, one in the Derwenlas Formation and one in the Aberystwyth Grits, which are almost certainly volcanogenic. The predominant macrofossils are graptolites. Shelly faunas are sparse and restricted to a few local lags at the bases of some turbidites.

The Phanerozoic sediments were muds and sands which accumulated in a marine basin that may have been many hundreds of metres deep; this was the Welsh Basin, defined on the east and south by a slope rising to the shelf-sea areas of the Welsh Borderland and South Wales. To the north the contemporary basin margin probably extended north-westwards towards the North Wales coast. Evidence for the existence of this shelf-sea from Cambrian times through to the end of the Silurian (Bridges, 1975; Greig and others, 1968) is found in the Welsh Borderland.

The earliest rocks that crop out within the district, of late Ordovician (Ashgill) age, reveal that it then lay to the west of a submarine slope that separated the shelf along the Borderland from deeper water to the west (Figure 6). A prism of siliciclastic sediment accumulated at the foot of this slope and fanned out westward across the district (Cave, 1979). The glaciation in the latter part of the Ordovician, so evident in some other parts of the world, seems to have indirectly, but quite abruptly, terminated this orderly pattern of basinal sedimentation, for the Ordovician rocks contemporary with this glaciation reflect a rapid accumulation of sediment showing penecontemporaneous disorder. It is reasonable to presume that the glaciation engendered marine regression across, and possibly complete withdrawal of the sea from, the shelf to the east. Detritus might then have been introduced more directly to the slope and thence to the basin, so that the deposits within the district probably accumulated near to the base of the slope and were in part emplaced by mass-movement.

At the beginning of the Silurian this episode of deposition, and the associated marine regression, ceased as abruptly as it had commenced. The succeeding Llandovery history of the district is one of basinal sedimentation, not unlike that of present continental rises, at a time when marine transgression at the basin edge was dominant. In early Llandovery (Cwmere Formation) times the area lay largely out of the path of high energy turbidity currents and beyond the fringes of submarine fans or lobes. Indeed, in this early phase of the transgression, influx of detritus into the basin as a whole appears to have been low, and iron sulphide-bearing muds of high pelagic content accumulated. Later came incursions from the east or south-east of more turbidite-dominated sediments, which were coarser and were more sharply differentiated into alternations of sand and mud. Periodically, lithological changes occurred rather abruptly, and these may be related to pulses in the shelf-sea transgression.

More fundamental was the change in direction of transportation of sediment which took place in the upper part of the Llandovery, within *Monograptus turriculatus* times, when sediment began to arrive largely from the south or south-south-west and also became coarser; this change resulted in the deposition of the Aberystwyth Grits. Here the sedimentary record ends within the district, although evidence from adjacent districts shows that this influx from the south persisted and progressively affected areas to the east and north.

The Ashgill and Llandovery rocks all show signs of deformation. They have a complex history of dewatering, compaction, mineralogical alteration, jointing, cleavage formation, folding and faulting. It can be argued that a good deal of the structural fabric of mid-Wales had its origin while the sediments were still wet and only partially lithified, and may have been related initially more to depositional slopes, active faults, shifts in the sediment pile, etc., than to a later single phase of crustal compression. Sediment deformation commenced immediately upon deposition; indeed, in the case of some turbidite arenites, it occurred during deposition. By the advent of the next epoch, the Wenlock, major folds were already present in the older sediments, and folding had affected all the rocks of mid-Wales by the end of the Silurian.

During the Silurian, the sediment pile was fractured on several transverse lines trending ENE–WSW and NNW–SSE, but the precise date of the faulting is uncertain. The faults barely reach the Wenlock outcrop to the east, but where they do they seem to have less effect on the Wenlock than they do on the underlying rocks; it is thus possible that fracturing commenced within the Llandovery epoch with a limited continuation of movement into the Wenlock epoch. Mineralised fluids invaded the fracture planes, possibly even assisting in their formation. As a result, quartz, carbonates, sulphides and baryte were deposited in places. Again the date of this mineralisation is not known precisely.

The effects of crustal plate movements are not readily readable in the rocks of the district. By late-Ordovician times volcanic activity in the Welsh Basin had largely ceased. The 'accretionary prism' of the Southern Uplands was still growing, and the basin in Central Wales, as part of a Southern Britain plate, was still closing with a Northern Britain plate, and continued to do so until the end of the Silurian (Dewey, 1982, p.405). By this time the Welsh Basin had filled, early Přídolí sediments of a very shallow marine, or brackish water environment having spread across the whole region.

Much speculation has been exercised on later events but for much of the time the district was part of a dominantly positive region that was subjected to long intervals of erosion. The only post-Silurian sediments are the superficial deposits of the Quaternary. During the Pleistocene, ice moved southwards along Cardigan Bay and, at least in the Devensian, locally generated ice moved south-westwards and westwards across the district. In places tills were deposited. Later, lake clays formed in some valleys, followed by gravels. Some of the latter probably resulted from the growth and retreat of very late Devensian valley-glaciers. Other valley deposits in mid-Wales, such as those in the Dovey and Rheidol, appear to be kame terraces subsequently benched by melt-waters.

At the end of the glaciation sea level was some 30 m lower than it is now; consequently, at the coast, so were the valley floors. Melting of the ice brought about a rise in sea level which inundated the river mouths and the low coastal land, and initiated a depositional episode, while a cliff line was eroded back into higher land. Thus were created the fine cliff sections north of Aberystwyth, and the extensive Flandrian infill of the Dovey Estuary and the tract to the south.

GEOLOGICAL RESEARCH

The pioneer advances in the geology of North and South Wales were made by Adam Sedgwick and Roderick Impey Murchison in the early nineteenth century. The rocks of mid-Wales lay between their areas of study, and the classifications of the sequences there provoked controversy. Did they belong to Sedgwick's Cambrian System of North Wales (1852, p.151) or Murchison's Silurian of South Wales (1839, p.195)? Lapworth (1878, 1879) resolved the argument by devising an Ordovician System that included strata formerly considered variously as Upper Cambrian and Lower Silurian. This led to the solid rocks within the Aberystwyth district being assigned to the Ordovician and Silurian systems.

Geological maps of the district at the scale of 1 inch to 1 mile were published by the Geological Survey of England and Wales in 1848 as parts of Sheets 57 NE (Aberystwyth) and 59 SE (Machynlleth). The survey was conducted by A.C. Ramsay, joined on the latter sheet by A.R. Selwyn, while the mineral lodes were mapped by W.W. Smyth and Sir Henry de la Beche. The solid rocks were depicted as being of Silurian (Lower Llandovery) age, and the more arenaceous parts of the sequence were shown by a stipple. Thus were outlined crudely the distribution at surface of those beds now known as the Drosgol Grits, the Rhuddnant Grits and the Aberystwyth Grits. (The term 'Grits' is of very loose definition, deriving from a period in geological research when it was fashionable to refer to any arenaceous rock within the Lower Palaeozoic sedimentary pile as a grit). More revealing of the high quality of the work done by Ramsay are the Horizontal Sections published by the Geological Survey in 1845. Section 1 of Sheet 6 reaches, at its northwestern end, the southern edge of the Aberystwyth district and gives a good interpretation of the structure.

In 1848, Hunt and Smyth respectively wrote a history and described the operation of local mining, but the first major study specifically devoted to the rocks of the Aberystwyth district was probably that of Keeping (1878, 1881), for Ramsey made only passing references to Aberystwyth in his writings on North Wales (1866, 1881). It was, therefore, unfortunate that Keeping misinterpreted the major structure of the district, for this resulted in him portraying the rocks in inverse order. A clue to the truth was afforded by Lapworth (*in* Keeping, 1881, p.167) from examination of Keeping's graptolites but it was left to Marr (1883, p.48) to correct the error.

In directing his attention to the district at the turn of the century, O.T. Jones initiated a major advance in geological understanding of the district, and indeed of the whole of Lower Palaeozoic Wales. Still on the staff of the Geological Survey of Great Britain, he was versed in detailed mapping and it was this practice which unravelled the stratigraphy and revealed the intricacies of the structure. His account (1909), presented with a map of the SE part of the district, records in detail the geology along several river traverses. In 1916, then as professor at Aberystwyth, and in conjunction with his student W.J. Pugh, he produced an account of the geology of a northern part which became the model for a host of works on the geology of Central Wales by many other authors. It also laid the foundations for the recent survey, for although O.T. Jones, W.J. Pugh and R.M. Jehu, a student of Pugh at Aberystwyth, produced further accounts in the inter-war years (Jehu, 1926; Jones and Pugh, 1935a), these are only extensions or syntheses of earlier work.

The metalliferous lodes and mines of north Cardiganshire and west Montgomeryshire were described in a memoir by O.T. Jones in 1922. It would be difficult to better this study, for it was the product of a special investigation by one who had knowledge of several of the mines while they were still in production. Mining has long since ceased and the mines have fallen into decay or become obliterated in reclamation schemes. Several recent publications supplement or broaden the scope of the 1922 memoir. Lewis (1967) has given a history and sociology of lead mining in Wales; Bick (1974–8) has produced a series of booklets each devoted to the mines of a portion of the mining field; Ball and Nutt (1976) have prepared a preliminary report on a mineral reconnaissance of Central Wales, which includes a 1:100 000 geological map of the Central Wales mining field.

After the 1939–45 war geological research veered away from outcrop mapping and interpretation into sedimentary and structural studies; the statistical approach also entered the field, and probably the most important work stemming from this is that of Wood and Smith (1959) on the sedimentation and history of the Aberystwyth Grits. Sudbury (1958) made a study of triangulate monograptids from the Rheidol Gorge, and Price (1962) described the tectonics of the Aberystwyth Grits. The cleavage, generally a rather open 'fracture cleavage' affecting most of the mudstones of the district, has been described by Davies (1980). No account of geological research in the district would be complete without mention of John Challinor who has recorded so many important phenomena since describing 'curious markings' with K. E. Williams (1926).

TURBIDITES

Most of the Ordovician and Silurian sediments around Aberystwyth were introduced by turbidity currents. It is consequently apposite to mention briefly some of the recent researches into turbidity, for much of it is relevant, though not specific to the district.

The terminology used in this text for the detailed description of turbidite beds follows Bouma (1962) (Figure 2a). Since so many of the turbidites are mud-dominated it is amplified by further dividing the pelitic *Te* interval (Cave,

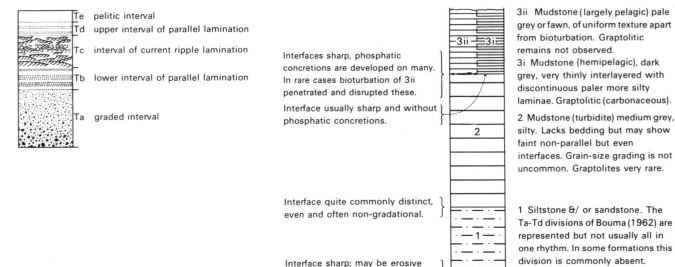

Figure 2a. The five terms erected by Bouma for the intervals making up the complete sequence in a turbidite from Ta at the bottom to Te at the top

b. The lithological theme of mud turbidites

1979, p.517) (Figure 2b). The grain size of intervals *Ta-d* is commonly silt and sand, usually generally described as arenite in this memoir. The *Tc* and *Td* intervals are expressions of traction currents, and where there is no evidence of the lower intervals doubt may exist that the deposits are indeed turbidites. In many such cases *Te* interval mudstone overlies *Tc* and *Td* in one of the forms depicted in Figure 2b. In these cases *Tc* and *Td* are seen as clastic sediments tractionally worked in the base of a mud turbidity flow of insufficient energy to entrain grains above clay grade.

The basinal turbidite environment has recently been studied in detail both in the geological sequence and in modern ocean basins (Rupke, 1978). These accounts are very relevant to the basin in Central Wales, but local influences are also highly significant. The size of a basin—and the Welsh basin is small in comparison with those covered by recent turbidite studies—the nature of the terrain supplying detritus, and the composition and size of the detritus available to nourish turbidity currents, can each be locally of over-riding importance. Rather than large turbidite fans with an ordered arrangement of component members, in the style of Mutti and Ricci Lucchi (1972), it is easier in Wales to envisage smaller lobes of turbidites and perhaps several source loci along the shelf margin. One of these was undoubtedly that near Rhayader described by Kelling and Woollands (1969), but the district is so separated from the basin margin by younger rocks that it is not possible to be more precise.

HEMIPELAGITE

Hemipelagic mud was the 'background' sediment of the basin into which the turbidites were interjected. It consists of fine silt, mud and organic debris which descended through the water column as 'rain'. This facies is represented by two distinctive types of rock depicted in Figure 2b as 3i and 3ii. Type 3i accumulated under anoxic water, and type 3ii was formed in more oxic bottom conditions. Each type occurs commonly as thin interturbidite layers in thinly multilayered turbidite sequences. Type 3i is a dark grey very finely laminar silty mudstone. Thin, uneven laminae of black, carbonaceous detritus are separated by equally thin laminae of pale grey silt which have suffered disruption or segregation into lenticles and ovoid blebs. The latter are thought to be the product of very frequent pulses of fall-out from suspended nepheloid layers which possibly travelled out over the basin on pycnoclines. A deposit of this nature would have suffered a great deal of compaction, and thus thinning of the laminae: it may have been at this stage that the silt laminae were disrupted. The resulting rock was described by Cummins (1959, p. 162, pl. 14) as Laminated Muddy Siltstones. He found it common in the Nantglyn Flags (Wenlock) of Montgomeryshire, but it occurs widely in the Welsh Basin from Cambrian (e.g. Clogau Shales) through the Ordovician (e.g. Nant-y-Môch Formation) and Silurian (e.g. Cwmere Formation (Plates 1, 3 and 11) and, in the east, Lower Ludlow Shales). Because of the anoxicity of the conditions of deposition the rock is not normally bioturbated. This facies is associated in places with non-ferrous mineralisation.

Type 3ii is a pale grey mudstone, commonly unlayered and homogeneous, but commonly bioturbated. Because of the bioturbation it is not possible to assess how much of the mudstone is truly hemipelagic fall-out and how much was incorporated from turbidite mud below. The mudstone is commonly underlain by a distinctive dark grey apatite-enriched layer of very early diagenetic origin (Figure 2b. Plate 1, p.5).

CLEAVAGE

Most of the mudstones of the district are cleaved. The intensity of cleavage varies in accord with factors such as grain size, clay mineral content, fold style, position within a fold, and perhaps even geographical position. Roofing slates were not a noted product of the district, though a few roofs have been 'tiled' with thin Division 1 layers of turbidite siltstone. Slate slabs, however, have been quarried from some of the particularly argillaceous parts of the succession. Generally the cleavage is a rather open, spaced cleavage, commonly with a spacing of between 2 mm and 10 mm (Davies, 1980, p.14). In many of the unweathered rocks such as the Borth Mudstones and mudstones of the Devil's Bridge Formation, the cleavage is very difficult to discern or reveal. Clearly weathering affects the fabric of the rock, opening it up and making cleavage obvious only in natural exposures.

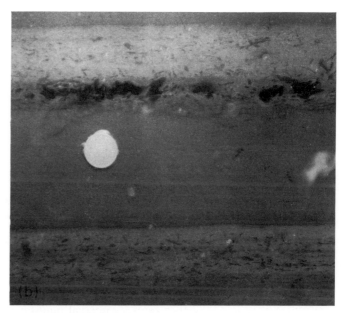

Plate 1 Phosphatic concretions (black) in the Mottled Mudstone Member. Slabs from the Cardiganshire Slate Quarry [6991 9595]. The concretions underlie layers of pale grey mudstone 3(ii). In (a) the concretions have remained undisturbed below the level of bioturbation. In (b) bioturbation of the pale mudstone is intensive and has penetrated the phosphatic concretions. Diameter of penny is 2 cm

CHAPTER 2

Ordovician – general account

The rocks assigned to the Ordovician System are sedimentary, consisting of silty mudstones and subordinate greywackes (arenites of sand and silt grades). Characteristically the majority of the mudstones are unbedded, dark blue-grey, and of splintery fracture due to the unevenness of their cleavage. A further characteristic of parts of these mudstones is a small content of extra-basinal pebbles, usually matrix supported, while all contain a scatter of large and small inclusions of balled-up synbasinal sandstone and siltstone. These arenitic bodies have shapes and contorted bedforms suggestive of large-scale penecontemporaneous disturbance. Only the basal third of the Ordovician rocks are pervasively well bedded, consisting of thinly multilayered mudstones and arenites, and these rocks have yielded a few graptolites proving them to be of Ashgill age.

The Ordovician rocks occur mainly in two inliers, one around Plynlimon and the other south of Machynlleth (Figure 1), and in two smaller areas north of the River Dovey. In addition several small periclinal outcrops occur at Carn Owen [732 882] 8 km east of Talybont, Banc Lletty-Evan-hen [718 852], Cwmere [683 884], near Alltgoch-ymynydd [707 882], Penrhyngerwyn [670 942] and near Bwlch-glâs [701 874]. This last area is too small to show on the 1:50 000 geological map.

It is estimated that about 700 m of Ordovician rocks are exposed in the Plynlimon Inlier, where the oldest beds crop out, though the equivalent beds at the northern margin of the district are probably even thicker (Jehu, 1926). The large thicknesses and the monotonous lithological nature of the constituent formations, and the rarity of good marker beds, inhibit direct measurements of thickness, and only allow estimates based on outcrop width to be made.

The Ordovician rocks have been divided into the Nant-y-Môch Formation, the Drosgol Formation (which includes the Pencerrigtewion Member), and the Bryn-glâs Formation, largely following O.T. Jones (1909). The basis of this classification is the distinction between a set of rhythmically layered rocks in the lowest parts and massive mudstones above, a set of sandstones (the Pencerrigtewion Member) dividing the massive mudstones into two. The summarised sequence is:

	Thickness m
Bryn-glâs Formation	
Mudstone, medium to dark grey; massive ill-cleaved; balled inclusions of sandstone and some thin-bedded siltstones in places	30–195
Drosgol Formation	
Mudstone, medium to dark grey; massive, ill-cleaved; commonly feldspathic and in places pebbly; divided by several sandstones up to 10m thick; includes	290–405

PENCERRIGTEWION MEMBER
(0–180m) at top, of mudstone, medium grey siltstone and sandstone in various proportions and states of disturbance (includes 'Drosgol Grits' of authors)

Nant-y-Môch Formation
Mudstone, medium grey, with thinly interbedded sandstones and siltstones; subordinate thinly bedded dark grey hemipelagic mudstone   seen 280

This stratigraphy and the correlations that support it (Cave and Hains, 1967, pp.65–66; 1968, p.79), differ from that propounded by James (1972, pp.291–292 and Fig. 1) who considers that only the Bryn-glâs and Drosgol formations are present within the Machynlleth Inlier, whereas all three formations have now been recognised there (Figure 3). The sedimentational model he builds upon his stratigraphy is, therefore, somewhat different from the one that we envisage though ours offers less precision.

The Bryn-glâs and Drosgol formations are, in the main, pelitic synbasinal mélanges. The mixing mechanism is not fully understood, but prelithified or exotic clasts are not common except in parts of the Drosgol Formation. Most clasts are of contemporary arenite, and were not lithified at the time of mixing. Simple inverse density loading may have been a very important factor in causing disturbance, as was probably downslope translation by gravity.

The argillites are cleaved; in those without sign of bedding or in which the bedding has been disturbed, this cleavage is very uneven and phacoidal—the double-cleavage of Jones (1909, p.470). The high-matrix arenites possess a similarly trending foliation while all rocks show evidence of regional metamorphism to Chlorite 1 grade of the greenschist facies (Strong, 1979).

Faunal remains are sparse. Graptolites of the *Dicellograptus anceps* Zone have been found in the Nant-y-Môch Formation, both in the Plynlimon (Jones, 1909, p.469) and Machynlleth inliers, but the next higher graptolite horizon is about a metre above the base of the overlying Cwmere Formation where the basal Silurian zonal form *Glyptograptus persculptus* occurs. Only one or two linguloid brachiopods have been recorded from the Drosgol and Bryn-glâs formations. The Ordovician–Silurian boundary thus lies somewhere within the large thickness of barren rocks separating these two graptolite horizons. As an expedient the boundary has been placed at the abrupt change from the unlayered argillites of the Bryn-glâs Formation to the bedded argillites of the Cwmere Formation. No depositional break has been detected in the sequence (Cave, 1979, p.521), but

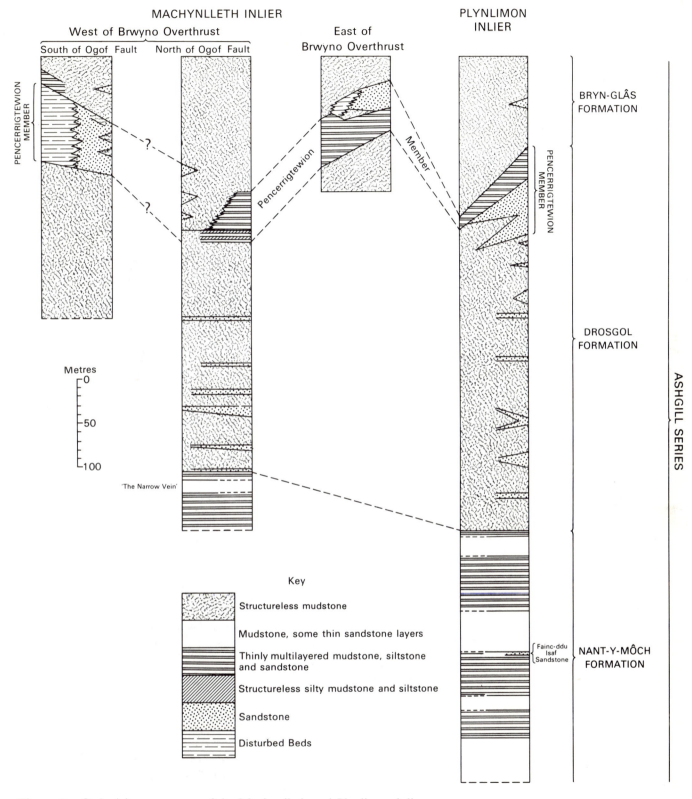

Figure 3 Ordovician sequences of the Machynlleth and Plynlimon inliers

the lithological change maintains its position at about a metre below the *G. persculptus* horizon over wide areas and so appears to be almost synchronous across the basin. It is, moreover, a boundary that is eminently mappable (p.41), and was specifically defined by Jones and Pugh (1916, p.351) in the Machynlleth Inlier. Although Jones (1909, fig. 11 and p.482) had earlier placed the Ordovician–Silurian boundary some 16 m below the *G. persculptus* horizon in the Plynlimon Inlier and had stated that it was difficult to find 'a definite baseline to the Eisteddfa Beds' [basal Silurian], we have found that the boundary as defined above is equally applicable to the Plynlimon area.

NANT-Y-MÔCH FORMATION

The formation is distinctive within the Ordovician rocks of the district in being their only part which is essentially a rhythmically and thinly bedded turbiditic sequence of sandstone and siltstone (arenite) and mudstone (argillite) couplets (Figure 2b), albeit with hemipelagic intervals in parts. It was first defined (Jones, 1909, p.468) as the Nant-y-Môch Group and also as the Nant-y-Môch Beds, the term formation being introduced later (James, 1971a).

In the Machynlleth Inlier the formation has been taken to extend upwards to the base of a thick sandstone. It has a narrow outcrop between Foel Goch [695 928] and Craig Caerhedyn [711 969]. In this inlier the Brwyno Overthrust has truncated the sequence, leaving only the highest parts exposed.

The main outcrop of the formation occurs in the core of the Plynlimon Inlier where the oldest beds crop out at the centre. It is estimated that 280 m of beds occur in this inlier. The top of the formation is sharp, its characteristic thin bedding being absent from the overlying rocks.

In this inlier the sequence has been separated into seven informal members (Figure 3) on the basis of the relative proportions of arenite to argillite in the couplet (Figure 2b).

Three of these are composed of couplets in which the arenite part (Division 1) is well represented; such sequences are here called the Craig y Dullfan subfacies from the crag of that name [772 885]. The other four informal members are dominantly argillaceous, though the topmost 20 m of the formation contains many thin arenites. The couplet in these more argillaceous parts consists largely or wholly of Divisions 2 and 3ii, and this type is called the Maesnant subfacies from a stream [775 879] in which it is well exposed.

In the **Craig y Dullfan subfacies** couplets are commonly between 4 and 12 cm thick, with arenitic bases usually between 1 and 4 cm thick, but ranging up to 30 cm (Plate 2). The rest of the couplets consist of argillite up to four times the thickness of the basal arenitic units. Most of the argillite is homogeneous and medium grey (Division 2 of Figure 2b). A thin layer of pale grey argillite caps some individual couplets (Division 3i), and some of the couplets are sufficiently mature to possess the concretionary phosphatic layer. In others, however, the pale layer is missing; either it never formed or it was eroded before the deposition of the next couplet. Small aggregates of iron sulphide, now pyrite, grew at the top of some pale argillite layers, at the interface with overlying turbiditic arenite. The only deviant from this description is a bed of sandstone, about 1 m thick, the Faincddu Isaf Sandstone, which caps the middle member of those that exhibit the Craig y Dullfan subfacies.

The various bedforms of the arenite divisions of this subfacies have previously been detailed (James, 1971a, p.180). In the terms of Bouma they are mostly base-missing sequences in which parallel and cross-lamination predominates (*Tc-Td* intervals); coarse sand of the *Ta* interval is sparse, occurring particularly in the fining upward fill of a few small scour channels. Penecontemporaneous loading effects accompanied the infilling of these channels, as can be determined from the convergence and warping of immediately underlying layers (Plate 6).

In the **Maesnant subfacies**, the rhythms are commonly

Plate 2 Nant-y-Môch Formation, Craig y Dullfan subfacies, Maesnant stream [7751 8791]. Lens cap 5.5 cm diameter

Plate 3 Nant-y-môch Formation, Maesnant subfacies, Maesnant stream [7755 8790]. Pyrite occurs at the base of thin arenite layers. Polished section perpendicular to bedding, × 1.0

2–7 cm thick. Where present, the arenaceous Division 1 is usually a pyritous siltstone no more than a few millimetres thick (Plate 3). In places aggregates of pyrite have detached themselves from the bases of arenite layers and have sunk into the mud below. The rest of the rhythm is argillite, in approximately equal parts of Divisions 2 and 3i. In parts of the sequence the thin arenite divisions are uneven (wavy), lenticular and even discontinuous. Internally the arenite laminae are contorted (as distinct from convoluted) and wispy. Such a bed form is likely to have arisen almost penecontemporaneously with deposition, from high frequency emplacement of sand layers on particularly wet mud substrates, with consequent loading and fluidisation of the sand. Bottom structures are not diverse; trace fossils are present as casts, but current scour casts and tool-mark casts are uncommon.

The formation has an even cleavage which is most intense in the more argillaceous subdivisions, especially that at the top of the formation in the Machynlleth Inlier which is considered to be an extension of the Narrow Vein (Y Faen gul) (Pugh, 1923). This part of the formation has been quarried for slabs and slate (p.132).

The alternation of these two mainly turbidite subfacies represents repeated waxing and waning of the turbidity currents reaching the district; this can be accounted for by repeated events affecting the basin margin to the east or, more likely, the shifting of the depocentre of turbidite lobes.

DROSGOL FORMATION

The Drosgol Formation was first defined (Jones, 1909, pp.468–469) as the Drosgol Group (also Drosgol Grits and Drosgol Beds) from the hill of that name [760 878] in the Plynlimon Inlier. It consists mainly of dark grey, pale brown or buff weathering, massive, homogeneous mudstones and is 290–405 m thick. About eight sandstones, each with a maximum thickness of about 10 m, divide these mudstones into units up to 80 m thick. In most places the top part of the formation is an impersistent heterogeneous assemblage of bedded arenite and penecontemporaneously disturbed siltstone and mudstone, the Pencerrigtewion Member.

The base of the formation is conformable and sharp, and because it is considered to mark a regional change in deposition is thought to be synchronous across the district. In the Plynlimon Inlier non-layered mudstone, in places with an abundance of small quartz pebbles, rests on the multilayered arenites and argillites of the Nant-y-Môch Formation. In the Machynlleth Inlier, the exact position of the base is less easily justified. Here two sandstones similar to those in the Drosgol Formation sandwich some 20 m of rocks of impersistent bedforms just above the base of the formation which is drawn at the base of the lower sandstone.

The mudstones that make up the greater part of the formation are composed largely of chlorite and sericite with parts charged with feldspar, quartz, chert and rock-fragment sand. More rarely, well rounded pebbles and some cobbles of quartz, chert, quartzite and rock fragments occur, some in clusters, but mostly disseminated. In places there is a liberal scatter of pyrite as individual cubes with edges up to 1 cm. Additional petrographic data on the clasts are given by James (1971b, p.265) with subsequent remarks on the matrix.

The feldspar sand, most noticeable by reason of its pale colour in a dark grey matrix, is not randomly distributed. It seems to occur most commonly in the lower parts of the thick sheets of mudstone above the greywacke-sandstone layers. The sedimentological implications of this arrangement have not been resolved. If only for reason of scale it seems unlikely that this crude grading of the particulate matter means that the whole of one of these sheets of mudstone some tens of metres thick was emplaced during a single event. On the other hand the lithologies presented are typical of mass flow deposits, so that there is little doubt that the basin was experiencing rapid mass emplacement of sediment at the time.

Sporadic rafts of arenite and rudite from penecontemporaneous bedded sequences occur within the mudstone, and these may also have been derived during mass emplacement. More common are wispy inclusions of fine arenite, mainly fine sandstones and siltstones, and balled-up internally contorted masses of arenite, some many metres across and some possessing relicts of multilayering. In places, and for short distances, these inclusions seem to occur at preferred levels almost as a crude bedform, and this is thought to be attributable to reversed density gradients within thick piles of rapidly deposited, wet sediments. The disruption of a spread of arenite into aggregated masses which then descended into a wet mud medium may have produced this crude bedform characteristic. How much these fluidised sediments also suffered translatory movement down an incline as slumps is difficult to determine.

The sandstone units which divide the mudstone are up to 10 m thick and include thinly bedded as well as massive sandstones. They commonly crop for 1.5 km or more but probably are of much wider extent than exposure reveals. Indeed, the number and spacing of the sandstones in the two inliers are so similar as to suggest a direct correlation between the outcrops; if this is so they did not suffer intense sedimentary deformation. These sandstones are medium

grey and massive, and are not normally foliated except in some associated thinly multilayered sequences in the Machynlleth Inlier. Macrobedding is evident, but there are few other internal primary (sedimentary) structures. The bases of the sandstones are commonly coarse grained; fining upwards occurs, but a continuously graded profile is not characteristic. In grain size they range from coarse sand to silt. Grain to matrix ratios vary, giving rock types ranging from greywackes through subgreywackes to quartzose arenites. Dark grey mudstone clasts, probably ripped up from the substrate, are common in individual sandstone beds; some occupy a preferred position towards the top of the bed. Deformation structures resulting from reversed density gradients are common, and include pseudo-nodular irregularities and entrapped 'flames' of underlying mud.

Pencerrigtewion Member

This member of the Drosgol Formation is a packet of greywacke sandstones and usually subsidiary thinly multilayered arenites and mudstones that occurs at the top of the formation. Its composition is, however, very varied across the district, the greywackes ranging from comparatively mature siliceous sandstones, massive and well bedded, to high matrix siltstones which are unbedded. There is a high degree of contemporary disturbance in places which further increases the heterogeneity of the member. The type locality is at Pencerrigtewion [798 882], 1.5 km NE of Plynlimon. Its areal distribution is shown in Figure 4; its thickness ranges from nil to 180 m.

Within the Plynlimon Inlier its depositional fabric is largely undisrupted; variations in thickness, and the spatial distribution of facies within this inlier, are illustrated by the ribbon diagram Figure 5. At Pencerrigtewion, some 42 m of massively bedded sandstones of greywacke type are overlain by about 18 m of thinly multilayered arenites and argillites. At the top of the sequence bedding is vague, and there is probably passage upwards into unlayered mudstone of the Bryn-glâs Formation. Southward and westward of Pencerrigtewion both the lower, massively bedded sandstone (Drosgol Grits of authors, e.g. Jones, 1909, pl.xxiv, *non* p.251) and the upper, thinly bedded elements of the member are thinner. In places, such as between Banc Llechwedd-mawr and the summit of Drosgol, they are totally absent; in others, the massive sandstones only are absent, whilst in southern parts of the inlier there are intercalations of unlayered mudstone in the sandstone.

The Pencerrigtewion Member is less orderly in other outcrops. In the Carn Owen periclinal inlier it consists largely of fine- to medium-grained sandstone. The top part is usually well bedded and massive, but various proportions of the lower part have suffered disturbance, and in places this extends throughout the whole thickness. The disturbance takes various forms. In places the sandstone appears homogeneous, in self-accommodating pillow-form or, more commonly, with sporadic rafts of well layered sandstone. In other places the well-layered more massive sandstone is thoroughly disrupted and contorted on a very large scale and forms a 'suspensate' within mudstone (Cave, 1967, pl.1). The well layered massive sandstones are generally quartzose with some subgreywacke. They are well jointed and, though commonly of structureless appearance, some show fine parallel laminations. The unlayered sandstones have a rather finer grain, contain appreciable matrix, and are subgreywackes. Mapping of the Carn Owen pericline has revealed that these two types of sandstone are laterally equivalent, and that the change from bedded in the east to largely unbedded in the west coincides approximately with the axis of the major pericline. A similar disruption of the strata occurs on the western side of the anticlinal outcrop on strike to the south at Banc Lletty-Evan-hen [717 851]. It may be that the folding of the strata and the rearrangement of their sedimentological fabric at these two localities is related, though alternatives are discussed later (p.29).

In the Machynlleth Inlier the Pencerrigtewion Member contains much high-matrix sandstone of homogeneous appearance, with mud clasts; well bedded massive arenite is rather uncommon. Grain size is generally finer than in the Plynlimon Inlier; siltstones and fine sandstones are common, though coarse feldspathic sandstone also occurs in places. The homogeneous arenite is discontinuous, with bodies of it, tens of metres thick, terminating abruptly along the strike. Over wide areas it is totally absent. Such bodies can be construed as channel-fills. In other places its stratigraphical position is occupied by balled-up and contorted sandstone and siltstone, recorded on the map as 'Disturbed Beds'.

The inlier is divided in two by the Brwyno Overthrust. To the east of this the massive arenitic element of the Pencerrigtewion Member is underlain by a laterally persistent, thinly multilayered sequence of sandstones or siltstones and mudstones. This sequence is very like that which overlies the massive arenites in the Plynlimon Inlier, and provides the reason for including such a variety of rocks in a single member. By so doing it is possible to view the member as a sequence of bedded arenites and mudstones in which the arenite is dominant in the lower parts in the Plynlimon Inlier and in the upper parts in the Machynlleth Inlier. Again this arrangement can be construed as representing channel fills and overbank deposits (levée and interchannel). The massive arenite of the Pencerrigtewion Member is responsible for some of the main eminences of the Machynlleth Inlier: Moel-y-Llyn [712 916]; Moel Hyrddod [722 953]; the western flank of Tarren Tyn-y-maen [722 968]; and the quarries at Parc [758 997] (Sheet 149). The Disturbed Beds 'facies' crops out, along with massive sandstone, in the crags on the northern side of Foel Einion [717 937], along Pencarreg-gopa [723 947], and in the core of a minor pericline at the southern end [7288 9417] of Bwlch Corog. The underlying thinly multilayered sequence is present in all these places; it is usually between 8 and 22 m thick, though in the NE at Parc it is reduced to 2 or 3 m.

West of the Brwyno Overthrust massive arenite dominates the member in the south where it has given rise to two more of the higher hills, Moel y Garn [692 912] and Garn Wen [691 928]. Between Garn Wen and Bwlcheinion [6967 9466] the arenite has been dismembered and occurs as masses, some over 100 m across, surrounded by mudstone indistinguishable from that of the Bryn-glâs Formation. This is interpreted as a channel margin along which there has been sediment failure. Northwards these masses are progressively sparser and smaller, and north of Bwlcheinion as far as the

Llyfnant Fault, which terminates this outcrop, they are absent. Northwards along the strike of the arenite there is a broad hollow with an incompletely exposed succession, some 45–50 m thick, mostly of unlayered splintery mudstone but including some thinly and evenly bedded siltstones and mudstones especially at the bottom and top. Presumably these rocks are interchannel deposits laid down north of the Garn Wen channel.

Northward, beyond the Llyfnant Fault, the member is again exposed. Although lying west of the overthrust, the lithologies are identical with those east of the thrust but south of the Llyfnant Fault. From this it is inferred that differences in the successions west and east of the overthrust are sedimentological and not a result of the tectonic introduction of an exotic facies.

North of the Dovey Estuary, the Pencerrigtewion Member occupies the core of a large anticline plunging SW, and the base is not exposed. It is composed of massive arenites of generally fine grain, most being poorly bedded, commonly with a high matrix and mud clasts: in places there are coarser and feldspathic sandstones, and in all the facies is similar to the member at Garn Wen.

The environmental model to explain the deposition of the heterolithic Pencerrigtewion Member must accommodate the accumulation in juxtaposition of bodies of sand and mud, sediment failure, and marine currents and sediment transportation from an easterly quarter. Since the arenitic rocks are generally not associated with conglomerates bearing extra-basinal clasts, the environment is assumed to be basinal, and not immediately at the base of the slope though a base of slope can be identified to the east along the Towy Anticline. Here the crucial evidence is in the late-Ordovician rocks (Roberts, 1929), where [9969 6261] a mud-dominated sequence is divided by a package of mass-flow rudites and arenites of a thinning- and fining-upwards nature. The latter is typical of proximal channel-fill, and clearly is intimately related to the more distal deposits of the Drosgol Formation (Figure 6). Late-Ordovician times thus saw a rapid influx of mud and silt during the marine low, punctuated by an event which introduced a large quantity of sand from a shelf to the east. This sand spread out into the basin to fill channel systems with massive sand bodies and form finer thin-bedded silt/mud interchannel sequences. The major sand body of the Plynlimon Inlier is consequently interpreted as a channel-fill thinning south-westwards along Banc Llechwedd-mawr and Drosgol, where it is replaced by interchannel deposits (Figures 4 and 5). The sand bodies of Banc Lletty-Evan-hen and Carn Owen can be explained as another, slightly later, but perhaps related, channel-fill which has suffered sediment failure of one margin. It is not possible to relate the sand bodies exposed in the Machynlleth Inlier to the channel system farther east in the Plynlimon Inlier because of major interruption of outcrops, but they are probably also channel-fills, though more distal and probably slightly younger than those to the east. The fact that they overlie a more thinly bedded sequence implies progradation westwards of the sand during this event. Disturbed beds, seen north of Garn Wen and around Pemprys and Pencarreg-gopa, represent sediment failures along channel margins as a result of reversed density gradients, though there was minor slumping locally. The influence of contem-

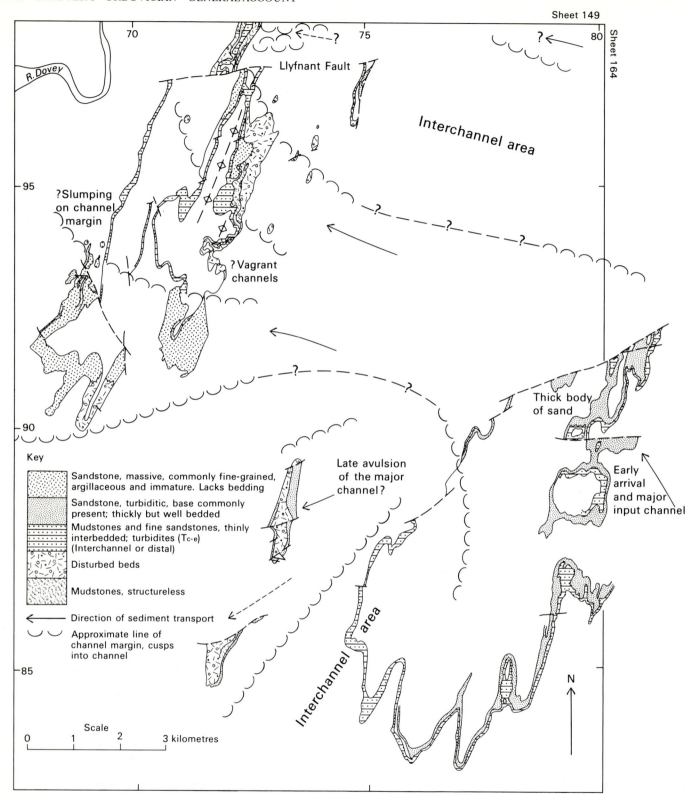

Figure 4 Distribution of rock-types, Pencerrigtewion Member; a possible sedimentational model of a westward prograding 'Pencerrigtewion Event'

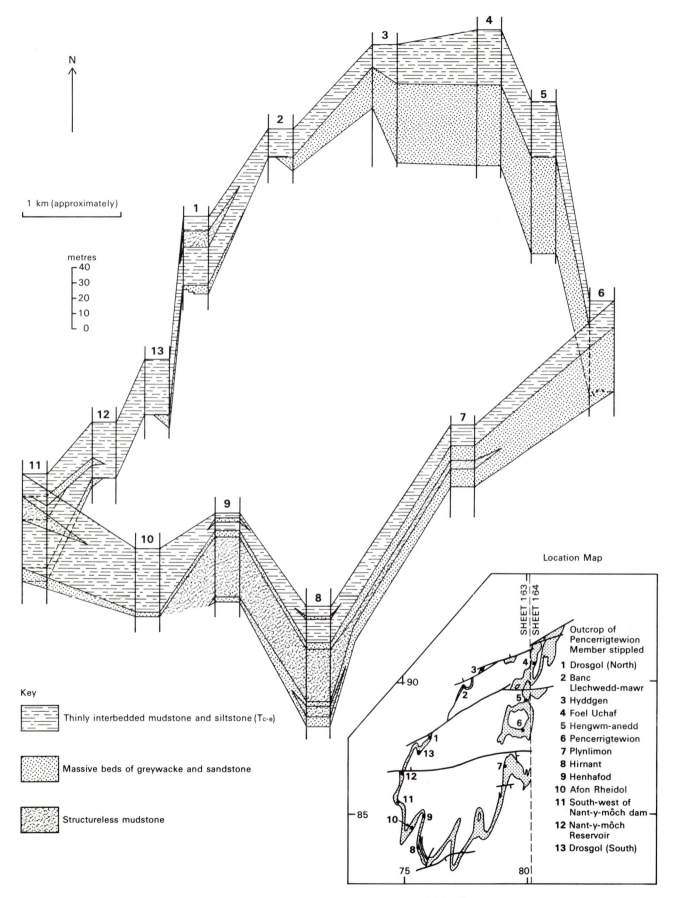

N

1 km (approximately)

metres
- 40
- 30
- 20
- 10
- 0

Key

Thinly interbedded mudstone and siltstone (Tc-e)

Massive beds of greywacke and sandstone

Structureless mudstone

Location Map

SHEET 163
SHEET 164

Outcrop of
Pencerrigtewion
Member stippled

1 Drosgol (North)
2 Banc
  Llechwedd-mawr
3 Hyddgen
4 Foel Uchaf
5 Hengwm-anedd
6 Pencerrigtewion
7 Plynlimon
8 Hirnant
9 Henhafod
10 Afon Rheidol
11 South-west of
   Nant-y-môch dam
12 Nant-y-môch
   Reservoir
13 Drosgol (South)

Figure 5 Facies distribution within the Pencerrigtewion Member around Plynlimon

poraneous basement faulting on sedimentation in this basinal area (James, 1972) remains hypothetical.

BRYN-GLÂS FORMATION

The Bryn-glâs Formation was defined by Jones (1909, pp.468, 471–472) under the names Bryn-glâs Group and Bryn-glâs Mudstones. His definition, with a minor modification to the upper limit of the formation (p.8), has been adopted in the present account. The formation makes sharp lithological contrast with the Silurian rocks above. In consequence the tough mudstones of the Bryn-glâs Formation form a bold dip surface rising away from the younger rocks (Jones and Pugh, 1916, p.351).

It is rather uniform consisting mostly of massive mudstone very similar to that of the Drosgol Formation, namely dark grey, silty, unevenly cleaved and thus splintery, and largely devoid of layering. It differs in that only very rarely does the Bryn-glâs Formation embrace massive arenite, pebbly mudstone or feldspathic mudstone. A bulk assessment of the rocks is that they are of finer grain than those of the Drosgol Formation.

Within the mudstone there are sporadic rafts of thinly multilayered arenite, mostly of fine grain, and argillite. Some are tens of metres long, within which the beds may be but slightly disturbed and approximately concordant. Near their edges they are commonly much more disordered. The small bodies of arenite are usually much more contorted, and form rounded lumps and wisps; their margins are always smooth and rounded.

In the south-west, as for instance near Cwm Ceulan and Cwmere [684 882], relatively undisturbed thinly multilayered arenite/argillite sequences are much more common. This may reflect a regional trend, for cliff sections [3137 5510] near Carregifan, Llangranog, display a well ordered stratigraphy in which bundles of bedded strata, each several metres thick, are very common and are evenly interstratified

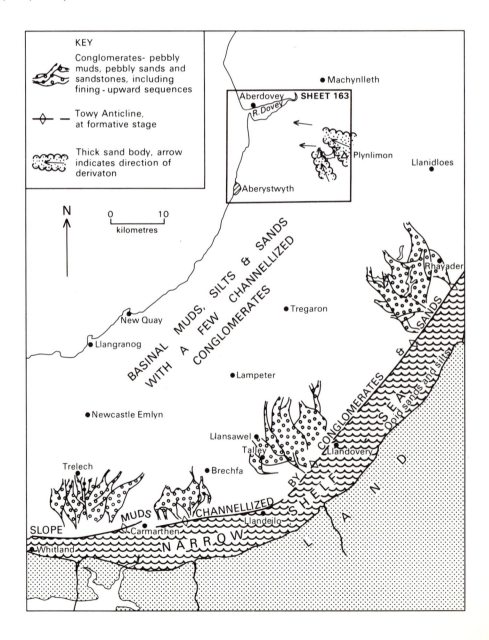

Figure 6 Conjectured palaeogeography of late-Ordovician (Hirnantian) times

Plate 4 Loaded-ripples, Bryn-glâs Formation at Bryn-yr-afr [745 878]. One loaded ripple is completely detached from its parent layer as a pseudonodule. The arrow indicates the direction of the current. Polished section perpendicular to bedding, × 0.6

with unlayered, nearly homogeneous, mudstones containing siltstone pseudonodules and channellised mass-movement deposits.

The thickness of the formation varies greatly across the district, from about 180 m in the SE to as little as 30 m in the centre and as much as 195 m in the NW. These figures are, however, based upon outcrop widths; although no folding is apparent, since the formation is almost non-layered, it may well be present and, if so, must distort apparent thicknesses.

No obvious conclusion can be drawn from the variations in thickness. It is possible (p.11) that early folding, for instance possibly at Carn Owen, may have induced sediment disturbance. It is equally plausible that such movements, created local unevennesses of bathymetry, and that these influenced deposition by promoting inequalities in sediment thickness. The marked attenuation of the Derwenlas Formation along the trend of the Carn Owen pericline (pp.45,71) supports this view of early folding. In a wider context the formation is thinner in western mid-Wales than are the equivalent beds to the east (300 m) near Llanidloes (Jones, W.D.V., 1945, p.314) and to the north (Pugh, 1923; Jehu, 1926).

The small-scale disruption and contortion of arenite in the formation is due probably to reversed density gradients within wet unlithified sediments. Occurrences of piled load-casted ripples (Plate 4) and other pseudonodules are common indicators of such a mechanism, but any original depositional fabric of the formation has been so widely disordered or obliterated that additional mass movements of a lateral nature need to be invoked, though the magnitude of such lateral movement is unknown. The rafts and clasts are not exotic, so they are not olistoliths. Neither are the deposits clearly divisible into thick layers in which there is abundant coarse detritus or grading, such as would indicate mass-flow deposits or slumps. It is concluded that these rocks were probably mass-movement muds interbedded with thin sequences of current-laid sediments penecontemporaneously disrupted in reversed density gradient systems. This type of mass-movement is much more clearly displayed in the contemporary rocks farther south [3137 5510] near Llangranog, which display bedding-parallel slides, clastic dykes, and probable slump folds of westward vergence.

CHAPTER 3

Ordovician—details

NANT-Y-MÔCH FORMATION

Plynlimon Inlier

All the exposed rocks in the southern part of the inlier are of the Craig y Dullfan subfacies, the best sections being in a line of crags [7693 8595 – 7708 8652] situated between two hollows each some 50 m wide, which parallel the eastern margin of the inlier as far south as Nant Maesnant-fach [769 860], and in crags and an old quarry [7699 8729] immediately south of Brynybeddau. In Nant Maesnant-fach [7656 8608 – 7649 8616], a rather atypical section shows arenites up to 30 cm thick in the top 3m; arenites up to 22 cm thick occur in a broad anticline [7682 8619-7691 8618] some 350 m to the east. These two occurrences appear to be at about the same horizon, not far below the base of the second-highest division of the Maesnant subfacies and, therefore, the approximate equivalent of the Fainc-ddu Isaf Sandstone to the north (p.17). The junction between the Nant-y-Môch and Drosgol formations can easily be traced along a well marked feature on the eastern side of the inlier north of Drum Peithnant [769 857], on the western side on the slopes of Drosgol, and also around the subsidiary fold west of the Nant-y-môch dam. Graptolite traces were found [7638 8676] in Maesnant-fach, but a fossil locality (Jones 1909, p.468, Locality F1) just to the north is now usually flooded by the reservoir. In the north of the inlier it has been possible to map seven informal members within the formation (Figure 3; p.7); each member is described separately below, grouped under the two subfacies.

MAESNANT SUBFACIES

a. The basal member occupies the core of the pericline. There is a good exposure of about 32 m of the succession in the Maesnant [7737 8813] (Figure 7, Loc.1). Arenite interbeds, weathering pale grey and up to 2.5 m thick, occur in the higher strata, and some contain clusters of pyrite crystals near their bases. The member is also exposed in the Rheidol [7764 8872] (Figure 7, Loc.2).

b. The second member of the Maesnant subfacies is exposed during low-water in a hollow [7703 8815] (Figure 7, Loc.3) separating the island from the eastern shore of the reservoir. There are thin arenite layers, commonly 0.5 cm or less, in 'units' of up to 10 cm thick. A sequence 30 to 35 m thick is well exposed (Loc.4) between Craig y Dullfan [7722 8852] and the Rheidol. In general this is an arenite thinning-upwards[1] sequence, for near the bottom the arenite layers reach a thickness of about 2 cm, in the middle only about 1 cm, while the top 3 to 4 m consist of mudstones (of Divisions 2 and 3i) with arenites less than 0.5 cm thick or absent. The best exposure is in the Maesnant [7755 8790] (Figure 7, Loc.5) just below the southern end of Fainc-ddu Isaf. It reveals about 10 m of mudstone, mainly as alternating layers, 2 to 10 cm thick, of medium grey homogeneous mudstone (Division 2) and dark grey laminar mudstone (Division 3i) containing *Dicellograptus* cf. *anceps*, *Orthograptus truncatus* s.l. and *Pseudoclimacograptus?*. Siltstone or fine sandstone layers are present at the bases of some of the mudstones (Division 2), and small crystals or granular clusters of pyrite are present in many of these. There are 'units' in which the mudstone types deviate from this pattern and indicate that oxic bottom conditions prevailed at times, in particular at the top of the member near the

---

[1] This term denotes progressive upward thinning of arenite layers in a multilayer turbidite sequence. Converse: thickening-upwards.

foot of the water cascade, where bioturbated mudstone of Division 3ii is present above the mudstone of Division 2. Graptolite fragments have also been found in the Rheidol [7798 8892] to the north (Figure 7, Loc. 6).

c. The third member is best exposed when the water level is low, in the old bed of the Rheidol [7675 8790] (Figure 7, Loc.7). There, the siltstones are pale grey, cross-bedded, up to 1 cm thick with overlying mudstones up to 2 cm thick and showing rusty tints on weathered fractures.

d. The highest member of the subfacies is 25 to 30 m thick. It is exposed on the steep eastern slope of Drosgol [7642 8755], where well cleaved dark grey mudstones up to about 10 cm thick are interbedded with arenites up to 0.5 cm. A more active input of detritus is indicated in the topmost strata of the member, for the junction with the overlying Drosgol Formation is sharp near here [7650 8771] (Figure 7, Loc.8), with unlayered medium grey silty mudstone of the Drosgol Formation, containing in parts an abundance of small pebbles, mainly quartz, overlying thinly bedded homogeneous mudstones with arenites up to 10 cm thick. The lithological contrast between the formations clearly records a sharp event in the depositional history of the basin. The junction is also visible at the western end [7874 8961] of Banc Lluestnewydd, but there, although the junction between the two types of lithology is well defined, the contrast is not so great. Cleaved mudstones interbedded with thin arenites 0.5 to 3 cm thick give way abruptly to splintery unlayered mudstone containing arenite knots and lumps.

CRAIG Y DULLFAN SUBFACIES

a. The lowest member forms a low ridge on the eastern limb of the pericline from Brynybeddau to the Rheidol [7781 8889] (Figure 7, Loc.9), where some 3 cm of beds are exposed with arenites 2 to 3 cm thick and interbeds of mudstone of divisions 2 and 3ii type up to 7 cm thick. It is best exposed in the Maesnant [7751 8791] (Locality 10), and in folded strata near the footbridge where some 12 m of thinly interbedded arenites and mudstones are visible in the stream. The arenites are siltstone and fine sandstone, commonly about 2 cm thick, and making up about 60 per cent of the strata. They are pale grey, commonly lenticular and ruptured; both top and bottom interfaces are sharp, but uneven as a result of loading, water escape, and possibly bioturbation. They are mainly unevenly laminated to cross-laminated with oversteepened dips, and the laminae have been disturbed. Some arenite layers have coarser sand at the bases, commonly in scour pockets, and are unlaminated though graded *(Ta)*. Pyrite cubes and clusters are present in some layers, usually in the basal and coarser parts. The uneven, disrupted nature of the arenites was thought sufficiently unusual for James (1971a) to consider that these arenites formed a marker band: such characteristics are, however, not unique to this level. The mudstones are medium grey, about 1.5 cm thick and bioturbated, so that only Division 3ii is present in many layers. In the western limb the member, about 40 m thick, is exposed in rapids of the Rheidol [7733 8844] (Figure 7, Loc.11). The arenites there, containing some pyrite, are up to 2.5 cm thick, with mudstones up to about 5 cm. At the base of the exposed sequence the mudstones are darker and rust-stained, and probably belong to the top of the underlying member.

b. The second member of the subfacies is well exposed in the Maesnant [7758 8787] (Figure 7, Loc.12) and at Craig y Dullfan [7717 8839] (Figure 7, Loc.13). It also produces the bold scarp of Fainc-ddu Isaf in which it is estimated there are about 40 m of

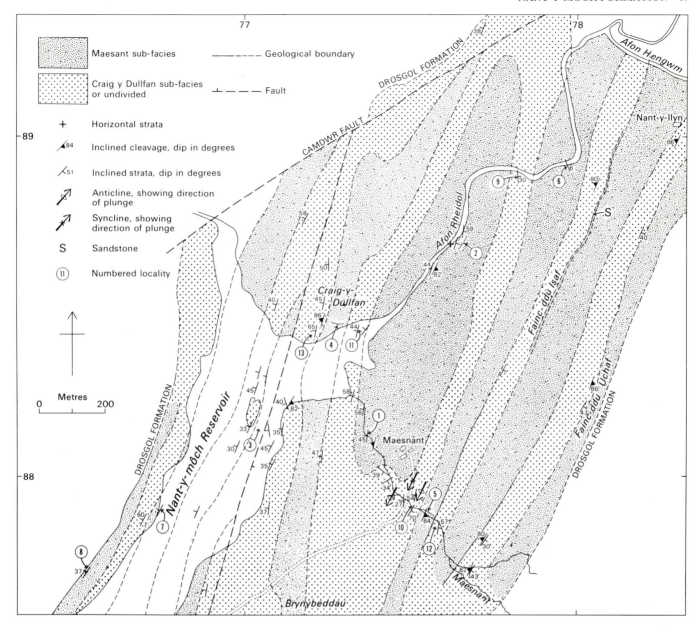

Figure 7 The geology of the centre of the Plynlimon Pericline

strata. A strike fault affects the western limb of the pericline, so that the outcrop of the second member is repeated. Here it forms a ridge, of which the island [7702 8820] and the outcrops [7710 8850] west of Craig y Dullfan are part. The topography of most of these ridges is etched by hollows into five or so subsidiary ridges, indicating that the arenites are probably separable into four or five bundles within the member.

At the top of the cascade in the Maesnant (Figure 7, Loc.12) some 8 m of strata are well exposed, consisting of interbeds of pale grey arenite, mainly fine sandstone, and medium grey homogeneous mudstone of Divisions 2 and 3i. The arenites are of uneven thickness here, usually between 2 and 3 cm thick. Some are lenticular and disrupted; others amalgamate. Internally they are laminar and commonly cross-laminated, mainly base-missing Bouma sequences. Arenites in the lower part of the member, exposed just downstream in the gorge, are thinner and more evenly bedded,

while the mudstones are thicker. There is thus some indication of a thickening-upwards sequence. Indeed each couplet of a Maesnant subfacies overlain by a Craig y Dullfan subfacies can be interpreted in the same way, culminating perhaps in the Fainc-ddu Isaf Sandstone. This sandstone is not visible in the Maesnant but is exposed about 1150 m to the NNE where it is about 1.2 m thick and is composed of two 'leaves'. Internally it contains a few thin partings of dark grey mudstone, and these are convoluted with the bedding lamination. The few metres of beds overlying this sandstone contain dark grey mudstones and thin arenites; they represent the base of member c of the Maesnant subfacies.

Palaeocurrent indicators were observed in two beds in the crags [7748 8783] just south of the Maesnant. The foresets of a cross-laminated fine-grained sandstone face due west, and the surface of a grey mudstone is scoured by flutes indicating a current flow from 100°.

At the southern end of Craig y Dullfan (Figure 7, Loc.13), best exposed at low water, some 14 m of very evenly bedded strata are well exposed. In the top third, between 60 and 70 per cent is arenite; there is less below, again suggestive of minor thickening-upwards in the sequence. The arenites commonly are about 3 cm thick, but one or two reach 20 cm thick. They show minor cross-lamination, ripple lamination and parallel lamination; some of the thicker ones show convolute lamination (Plate 5) and others thicken into loaded scour-pockets up to 65 cm deep and over 1 m wide. In the base of these pockets the sand is coarse, fining upwards, representing perhaps a very local *Ta* division of the Bouma sequence, but most of the arenites are base-missing turbidites. Several loaded scour-pockets are exposed both here (Plate 6) and on the shores of an 'island' immediately to the south. A number of bedding surfaces on this 'island' show flute scours, indicative of current directions from the SSE.

c   The highest member of the subfacies forms the bold scarp of Fainc-ddu Uchaf, where its thickness is estimated at 60 to 65 m. Nant y Llyn [7846 8900] exposes some 3 m of the division where the arenites are fairly thin, about 0.5 to 1 cm, separated by medium grey homogeneous mudstones up to 10 cm. During low water there are exposures [7651 8759] on the opposite limb of the pericline at the foot of Drosgol. There the mudstones are up to 2.3 cm thick, and the arenites, showing small-scale cross-lamination, convolute lamination, and pyrite in the basal parts, are up to 2 cm thick, though they swell to 4 cm where they have overloaded the mud substrate.

Machynlleth Inlier

The outcrop extends along the western limb of the inlier, immediately above the Brwyno Overthrust, between the Ogof Fault

Plate 5   Craig y Dullfan subfacies. Convolute lamination in arenitic layer, Craig y Dullfan [7717 8839]. Scale 15.2 cm

Plate 6   Craig y Dullfan subfacies. Turbiditic sand occupying a contemporaneously loaded scour pocket, Craig y Dullfan [7717 8839]. Scale 15.2 cm

[696 920] in the south and the Llyfnant Fault [711 973] in the north. The best exposures occur where the outcrop crosses the valley of the Einion and at the confluence [707 964] of the Dynyn and Brwyno. The most southerly exposure is 110 m SSW of Coronwen [6984 9389] where well cleaved mudstones contain thin, small scale cross-laminated and parallel-laminated arenites. In the higher part of the exposure the arenites become thinner and sparser; these beds correlate with part of the quarried sequence 0.5 km to the north and represent the Narrow Vein of Pugh (1923).

Immediately behind Coronwen and northwards, a small scarp marks the outcrop of some 7 m of evenly interbedded thin arenites and thicker mudstones. They underlie the well cleaved mudstone of the Narrow Vein, and a quarry [6994 9420] some 300 m NNE of Coronwen may have worked them although the mapping suggests it is in beds a few metres lower. The workings form a trench along the strike. The bottom 4 m contain arenites commonly 2.5 to 7.5 cm thick, with some up to 15 cm; at the base the arenites are uneven and of inconstant thickness. They are overlain by about 2 m of dark grey mudstone with thinner arenites (up to 2.5 cm), and then by some 2.5 m of beds, obscured in the trench, on the western side of which there is a small strike fault, possibly a thrust. Above, in the foot of the western wall of the trench, is a 3 m sequence, dominantly of arenites up to 10 cm thick with c.1 cm mudstone partings. Overlying this are some 10 m of cleaved medium and dark grey mudstones with very thin siltstone 'stripes'. Nearby [6993 9419] graptolites were obtained from dark grey mudstone bands (Division 3i, Figure 2b) underlying the thin siltstone 'stripes'. The fauna includes *Orthograptus truncatus* cf. *intermedius*, *O. truncatus* s.l. and *O. truncatus abbreviatus?*

A quarry in the Narrow Vein [6993 9440] 430 m SE of Bwlcheinion provides good exposure. Below the quarry, and forming a line of small crags diagonally up the slope to the NE, are some 30 m of interbedded thin arenites and cleaved mudstones. The arenites are up to 6 cm thick, and in parts constitute some 60 per cent of the rock. The individual beds become thinner upwards, especially towards the top, and are more widely spaced. Flute-casts consistently reflect currents from between 100° and 110°. The quarried beds, used for slabs and slate, are some 12 m thick and constitute the upper part of a fining-and thinning-upward sequence of thin arenites in cleaved, medium to dark grey mudstones. Like the Narrow Vein they have not yielded any graptolites though trace fossils are present on the bases of the arenites, which are up to 0.5 cm thick and 6 to 40 cm apart. Near the top of the quarry thicker arenites, up to 14 cm, are present, and in the overlying 1 to 2 m sandstones up to 20 cm, with convolute lamination, become dominant heralding the change of sedimentological pattern which produced the overlying Drosgol Formation.

In the slope some 400 m WSW of Dynyn [7090 9568] the following sequence has been recorded (Figure 8):

		Thickness m
Drosgol Formation		
9	Sandstone in trench with overlying thinly bedded arenites and mudstones	seen
8	Mudstone, splintery, with dispersed inclusions of arenite, small in upper part, larger balls and lumps in lower part	c.40
7	Sandstone, pale greyish green, base not seen, top sharp and flat	c. 5
6	Arenites and mudstones, thinly interbedded	c. 1
	Obscured	c. 4
5	Mudstone, arenaceous and feldspathic with ball-like inclusions of finer, low-matrix arenite; some coarse feldspathic sandstone near base. Mostly non-layered, traces of bedding towards the bottom	c.16
4	Sandstone, pale grey, hard, flat top	at least 1.5
Nant-y-Môch Formation		
3	Mudstone, grey, cleaved, with thin arenite interbeds up to 4 cm	3.5
2	Mudstone, grey, cleaved, with thin silty stripes, mostly concealed	13
1	Mudstones, grey, cleaved; close interbeds of arenite	seen c.10

Another locality where an extensive sequence can be seen is in the angle of confluence [7070 9635] of the Brwyno and Dynyn. The basal sandstone of the Drosgol Formation, some 2.5 m thick, is visible on the west of the road to Dynyn. Its base is not exposed, but below its position lies 1 m of well cleaved mudstone with widely spaced arenites. These beds are the top of the Narrow Vein as seen to the south. Below them, a further 8 m of well cleaved mudstones with sparse and thin arenites are exposed, while the underlying 5.5 m are obscured. These 13.5 m of beds must contain the whole of the quarried sequence seen 430 m SE of Bwlcheinion, including the Narrow Vein. Some 20 m of well cleaved mudstones, with thicker and more numerous arenite layers, also crop out immediately to the east.

The top few metres of the formation are exposed at several places in Craig Caerhedyn [709 966] where the arenite layers thicken slightly upwards and are capped by a thick massive bed of greywacke sandstone taken as the base of the Drosgol Formation.

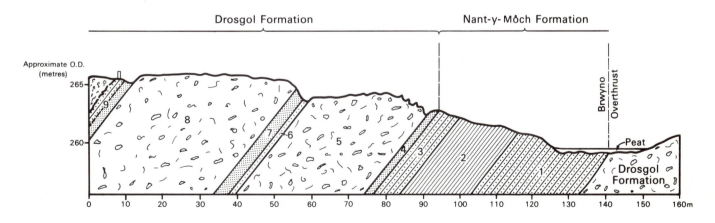

Figure 8 Section through parts of the Nant-y-Môch and Drosgol formations 400 m WSW of Dynyn

DROSGOL FORMATION

Plynlimon Inlier

STRATA BELOW THE PENCERRIGTEWION MEMBER

The lower part of the formation is best displayed in a number of sections in and near the Rheidol downstream of the Nant-y-môch Dam (Figure 9). Mudstones below the lowest sandstone can be seen in the river at Loc.1. They are dark grey, silty and massive, with an ir-regular cleavage and apparent absence of layering. A variable content of feldspar sand is present in places, and wisps and small contorted masses of arenite (p.10) are uncommon. Similar mudstones at higher levels are seen in crags (Figure 9, Loc.2), in the Rheidol (Figure 9, Loc.3) and in a long roadside section (Loc.4). In addition there is a discontinous gorge section (Figure 9, Loc.5), mainly in mudstone, which extends for more than a kilometre downstream [7536 8553 to 7517 8450] to the base of the Pencerrigtewion Member.

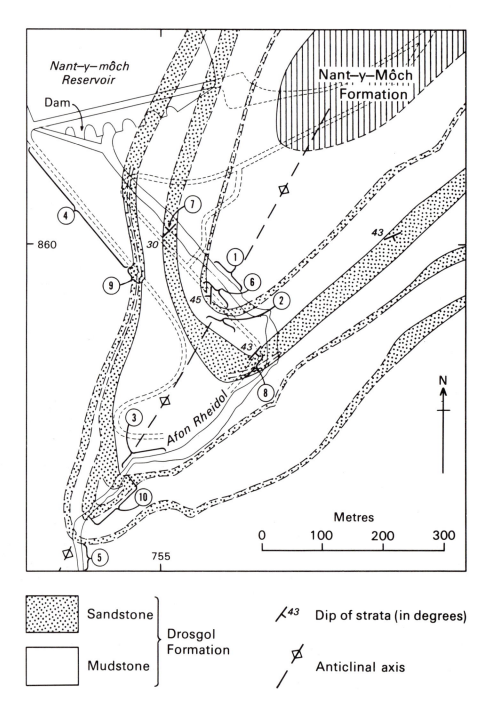

Figure 9   Locality map for Drosgol Formation (lower part); Nant-y-môch Dam

The lowest sandstone unit is well exposed in the roadside (Figure 9, Loc.6):

	Thickness m
Sandstone, grey, massive, quartzose, fine- to medium-grained	2.10
Interbedded mudstone and sandstone	0.30
Sandstone, grey, argillaceous (subgreywacke); some angular clasts of dark grey mudstone (cf. Plate 7)	0.15
Sandstone, grey, massive	0.75
Mudstone, grey, finely laminated	0.15
Sandstone, dark grey, argillaceous, feldspathic (greywacke)	0.60

A succeeding and thicker sandstone unit is exposed in the river (Figure 9, Loc.7) and, more clearly, in the roadside (Figure 9, Loc.8):

	Thickness m
Sandstone, grey, massive, medium-grained	c.1.20
Mudstone, dark grey, splintery, sandy, feldspathic, unlayered	c.3.05
Sandstone, grey, massive, medium-grained	c.9.15

In the sandstone crags to the NW of this last section there is extensive quartz veining in the core of the anticline. A third sandstone unit in this part of the sequence is well exposed in a roadside section (Figure 9, Loc.9), where 4.3 m of grey medium-grained sandstone are overlain by 3.65 m of dark grey unlayered mudstone with a locally high content of feldspar sand, in turn overlain by a further 3.65 m of grey medium-grained sandstone with some thin beds of sandy feldspathic mudstone. The lower part of this sandstone unit is also seen (Figure 9, Loc.10) in the river where it includes a 2.4 m bed of mudstone. All these sections show that the mapped sandstone units are commonly multiple, including beds of mudstone of varying thickness. The third sandstone unit splits into two separate mappable units south of the Nant-y-môch Dam (Figure 9) with an increasing thickness of mudstone between them.

The eastern and southern slopes of Drosgol (Figures 10 and 11) contain the outcrop of the entire formation and afford a liberal scatter of exposures throughout its thickness. Most reveal medium grey structureless mudstone, possessing a very uneven cleavage fabric imparting a splintery fracture. Small macroscopic inclusions of paler grey siltstone and fine sandstone are common. These are usually of uneven shapes, and are highly contorted by soft sediment deformation. Some of the mudstone has a high content of white feldspar sand as scattered grains, while abundant rounded pebbles up to 3 cm in diameter are scattered and locally form pockets.

The interbedded sandstone units, with massive sandstones, are poorly exposed and crop out along topographic benches. The massive sandstones are largely medium- to fine-grained and weakly graded, though rip clasts of mudstone are locally abundant at their bases. There are good sections along the shore of Nant-y-môch Reservoir, where surfaces of both mudstones and sandstones show a great variety of sedimentary types and structures. In particular, one section [7569 8696 – 7560 8697] shows a gradual westward transition from grey unlayered silty mudstone to grey unlayered sandy feldspathic and pebbly mudstone, which is overlain by 1.8 m of massive fine- to medium-grained sandstone with internal ball-and-pillow structure. Above the sandstone are similar sandy pebbly mudstones but with incorporated discrete sandstone rolls, and some impersistent layers of pseudo-nodules.

About 1.5 km to the east, to the west of Craig y Fedw [790 890], the outcrops of two persistent sandstone units traverse the lower slopes; the lower, near the base of the formation, is visible in Nant y Llyn [7853 8877] where it is 10 m thick and divided into massive beds up to 2 m thick showing parallel laminae. The second is some 45 m higher. Both persist northwards into the western end of Banc Lluestnewydd, where the lower is about 6 m thick.

The crops of several massive sandstones cross the hillside of Pumlumon Fach, while exposures of pebbly and feldspathic mudstone are common, more so in the lower part of the formation than near the top. Within the mudstone outcrops there are some very pronounced strike ridges, an example being displayed [7815 8810] 80 m SE of the track to Llyn Llygad Rheidol. The rocks that form them are tough, unevenly cleaved, grey, structureless mudstones, pebbly in parts and containing a scatter of siliciclastic clasts of synbasinal origin. Such strike features indicate that the mudstone intervals are crudely layered, each probably recording an individual mass-flow event. Masses of sand-matrix-supported conglomerate and lenses of mud-matrix-supported pebbles occur sporadically and may represent small channel fills—possibly subsequently disturbed by loading processes. The pebble bodies are

Plate 7 Drosgol Formation. Base of a massive turbidite sandstone containing intraclasts of mudstone [7522 8620]. Nant-y-môch Dam. Polished section perpendicular to bedding, × 0.6

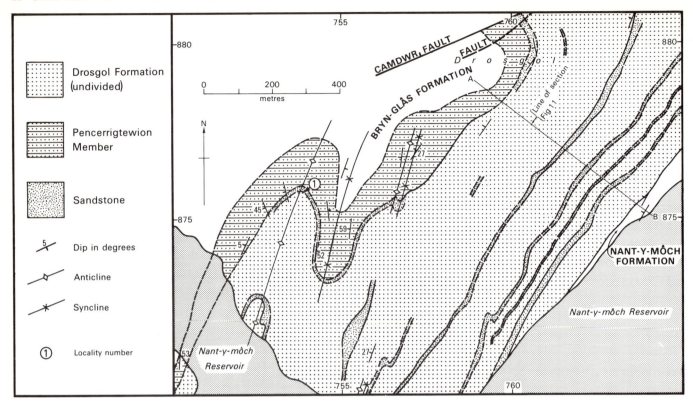

Figure 10 Drosgol Formation on Drosgol

small, in the order of 5 m across at most, and have diffuse margins passing into pebbly mudstone. The pebbles are almost all exotic, mainly of white vein-quartz and igneous rocks. One such mass is exposed adjacent to a track [7807 8820] and another, with mud-matrix support, occurs 70 m SE of the track [7801 8897].

Three quarries used in the construction of this track also expose some of these rocks. One [7797 8798] is very near the base of the formation and exposes about 6 m of mudstone, massive, structureless, and fawn to khaki when weathered, but probably grey where unweathered. It contains pyrite in isolated cubes. At the base of the quarry is sandstone, grey and khaki, very feldspathic and hard. Southwards the sandstone persists, and is visible in a 3 m cut. Some 10 m to the south it is very conglomeratic, beyond which all 3 m of the face are occupied by pebbly, sandy feldspathic mudstone in which there are also contorted inclusions of uncontaminated mudstone and rounded inclusions of coarse sandstone. The relationships between these different exposures are not obvious, but the rocks appear to be the product of mass-flow, and may include channel-fill debris. Another quarry [7845 8844] shows 1 m of mudstone, structureless and splintery, on 3 m of grey, argillaceous and hard sandstone, with mudstone clasts and layers, on 2.6 m of blue-grey, fine-grained, massive, cross-bedded sandstone, on massive sandstones (seen for 2.6 m). The third quarry [7861 8847] shows 6 to 7 m of mudstone, grey, tough and structureless, weathering to khaki, and containing large isolated cubes of pyrite.

On the south side of Llyn Lygad Rheidol there is a rock face and steep slope in which several minor folds plunge steeply northward (Figure 12). Massive mudstones predominate. They are splintery, and in part coarsely feldspathic and pebbly. A small thickness of thinly interbedded mudstones and siltstones is present in the sequence, and is visible along the axis of a syncline [7905 8750] (Figure 12, Loc.1); it is probably equivalent to similar beds

on the southern side of the E–W fault some 250 m south of the lake (Figure 12, Loc.2), and also to beds near the margin of the lake visible in the limbs of another syncline [7923 8754] (Figure 12, Loc.3). At this last locality the thin-bedded sequence overlies pebbly feldspathic mudstones, in which the scattered pebbles are mainly of quartz or quartzite, and are up to 2 cm in diameter. Nearby [7940 8750] the pebbles are up to 8 cm across. In places the pebbles are aggregated as impersistent layers or lenses of conglomerate in the pebbly mudstone. Balled-up 'clasts' of siltstone are also present within the mudstones. Overlying the thin-bedded sequence, the mudstones are again structureless though largely free of feldspathic sand and pebbles. They do, however, contain irregular shaped clasts and wisps of siltstone, the whole sequence having the appearance of a mass-flow deposit, probably channelised and containing impersistent, thinly multilayer, overbank turbidites.

Mudstones just below the Pencerrigtewion Member are well displayed [794 872 – 797 875] to the SE of Llyn Lygad Rheidol, and in crags [786 857 – 788 865] along the western side of the Plynlimon ridge. The mudstones are again almost entirely unlayered with a variable content of feldspar sand. At the northern end of the latter locality [7875 8632 – 7877 8655], there is a thick, but laterally limited, development of high matrix feldspathic sandstone within the mudstone.

PENCERRIGTEWION MEMBER

There are two main areas of outcrop: first, the type area of Pencerrigtewion and northwards to Foel Uchaf; secondly, the outcrop from Plynlimon through Hirnant to Drosgol. Facies distribution and thickness variations (18–100 m) are shown on the ribbon diagram (Figure 5).

In the type area the member is well displayed in natural crags. Its

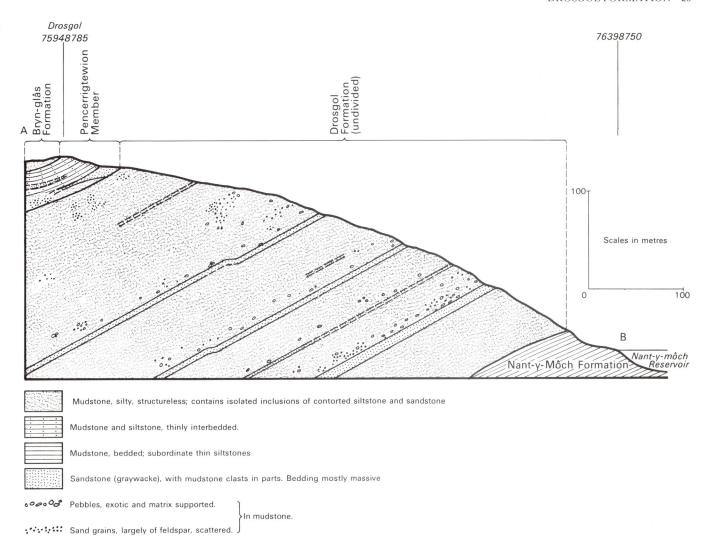

Figure 11 Section through east flank of Drosgol from the summit to the Nant-y-môch Reservoir

lower half consists of massively bedded sandstones [7950 8796] (Figure 13, Loc.1) resting on pebbly and splintery mudstones. The sandstones are some 42 m thick, and form beds up to 3 m thick near the base and 2 m near the top [7958 8811] (Figure 13, Loc.2), where they are overlain by 2 to 5 cm thick interbeds of mudstone and 1 cm beds of siltstone. The sandstones are coarse grained, especially near the base (E46558 and E46559), and intra-basinal mud clasts are present near the top. Bedding is generally evident, with some cross-lamination and some convolute-lamination near the top. Loading irregularities occur on the bases of some beds, and there is a current rippled surface near the top with ripple crests striking at about 015° and with a periodicity of about 8 cm.

Between Pencerrigtewion and Lluest y Graig the massive sandstones are well exposed and about 30 m thick. At the base, however, poorly sorted argillaceous khaki feldspathic sandstone (E42594) appears to replace dark grey mudstone, laterally, so the overall thickness is variable. This sandstone is probably the initial channel deposit, and is similar to sandstone occurring in other nearby outcrops. The overlying sandstones are grey, better sorted and feldspathic; near the top a current-rippled surface shows ripples striking at 020°.

The overlying thin-bedded sequence here is at least 26 m thick. It is a thinning- and fining-upward sequence, hence interpreted as interchannel turbidites of waning energy; its top fades vaguely into the unlayered splintery mudstone of the Bryn-glâs Formation. In the basal part [7989 8911] (Figure 13, Loc.3), mudstone layers are 2 to 30 cm thick with siltstone interbeds up to 2.5 cm. A few thicker layers of poorly sorted sandstone, up to 30 cm thick, contain contorted rip-up clasts of mud and show uneven scour-cast bases; they are presumably the products of the waning channel as it evolved into an interchannel location. Some 2 to 3 m above the base of the thin-bedded sequence a 2 to 3 m sandstone, well sorted, cross-laminated, and with an uneven surface, is present. Overlying this the siltstone interbeds are thin, though commonly disrupted by loading.

One kilometre due west, in Craig y Fedw [7907 8999], some 30 m of massive sandstones are overlain abruptly by the same thin-bedded sequence, in which the thinner siltstone interbeds are mostly evenly bedded but the thicker layers (about 5 cm) are disrupted, knobbly and gnarled, presumably due to loading. Bottom structures include groove-casts and flute-casts (currents from 085°), while small-scale cross-lamination indicates currents from the south-east.

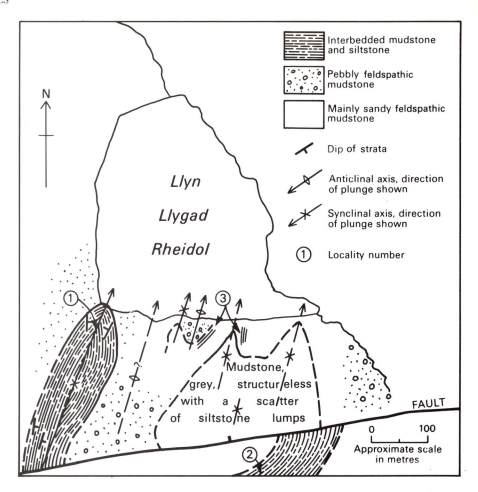

Figure 12 Location diagram for outcrops on the south side of Llyn Llygad Rheidol

On the north side of Afon Gwarin, in Craig yr Eglwys [8035 8946], the massive sandstones are very well exposed in a syncline (Figure 13). The total thickness appears undiminished, but has not been assessed accurately. The outcrop extends to an E–W normal fault which throws down north. Individual beds are up to 4 m thick, mainly of grey medium-grained feldspathic sandstone. Ten metres above the base is a lenticular bed up to 1.5 m thick consisting of a mix of feldspathic greywacke and mudstone in contorted wisps and clasts. Such a bed may reflect collapse of a channel margin and subsequent debris flow, or a local channel avulsion upstream. The overlying thin-bedded sequence is exposed in a syncline along the banks of Afon Hengwm [799 895] (Figure 13, Loc.4) 250 m due east of Lluest-newydd.

Farther north, the massive sandstones reach 50 m thick where they cross Afon Hengwm [8043 9060] (Figure 13). The beds are individually 1 to 2 m thick, and of usually massive, coarse-grained, rather poorly sorted greywacke, commonly showing slight grading with bedding lamination visible in parts which is convoluted near the tops of some beds. Bases of some beds show unevenness due to loading, while the base of the member is uneven due to the impersistence of beds of poorly sorted argillaceous greywacke with mudstone clasts in places [8043 9040]. Surfaces of strongly asymmetrical current ripples are present with strikes of 070° [8003 9070] and 059° [8003 9071] near the base, and 023° near the top [8007 9060]. Lee sides face WNW to NNW.

In the second outcrop area the member is well exposed on Plynlimon [e.g. 792 872; 799 863] where the basal sandstone sequence consists of two massively bedded, medium- to coarse-grained, high matrix feldspathic sandstones, each about 10.5 m thick and separated by about 6 m of unbedded splintery mudstone

similar to that of the main body of the formation (Figure 5). West of Drybedd the splintery mudstone between the sandstones thickens to about 32 m at Hirnant (Figure 5) and 37 m in the Peithnant [7575 8480]. The upper sandstone thins to 4.5 m, while the lower sandstone, which appears to be split by 6.5 m of interbedded mudstone and siltstone to the east [755 837] of Hirnant, cannot be traced north of Lle'r-neuaddau. Along the western side of the inlier the upper sandstone is less massive and commonly includes thin mudstone beds. In places [e.g. 7497 8532], it is contorted and partly disrupted into slump rolls or ball-and-pillow, and it is locally absent adjacent to Nant-y-môch Reservoir and on Drosgol.

The overlying thinly bedded facies is only about 12 m thick on Plynlimon. Its full thickness is seen, apart from approximately a metre at the top and bottom, in a line of crags [7895 8605 – 7895 8577] and on Dyll Faen [7785 8469 – 7786 8492], this latter case showing the transition to the Bryn-glâs Formation above. As in the type area, it is a thinning- and fining-upward sequence with increasing disruption of the bedding within the mudstones. The only other exposure of the upper boundary of the member is near Lle'r-neuaddau (p.36), where it is sharp and not transitional. At Hirnant the beds have thickened to 23.5 m, with a local 2.5 m sandstone in the upper part, and this thickening continues north-westward to 57 m along the upper part of a well exposed steep slope [749 846 – 749 854] west of the Rheidol. In this thick sequence there is a lenticle of unbedded mudstone [7488 8476 – 7468 8583] up to 12 m thick, overlain in its northern part by a sandstone up to 5 m thick which extends northwards to Nant-y-môch Reservoir [7485 8665]. The best section in this part of the outcrop [7497 8532] shows 8.5 m of sandstone at the base of the member overlain by approximately 35 m of the thinly bedded facies.

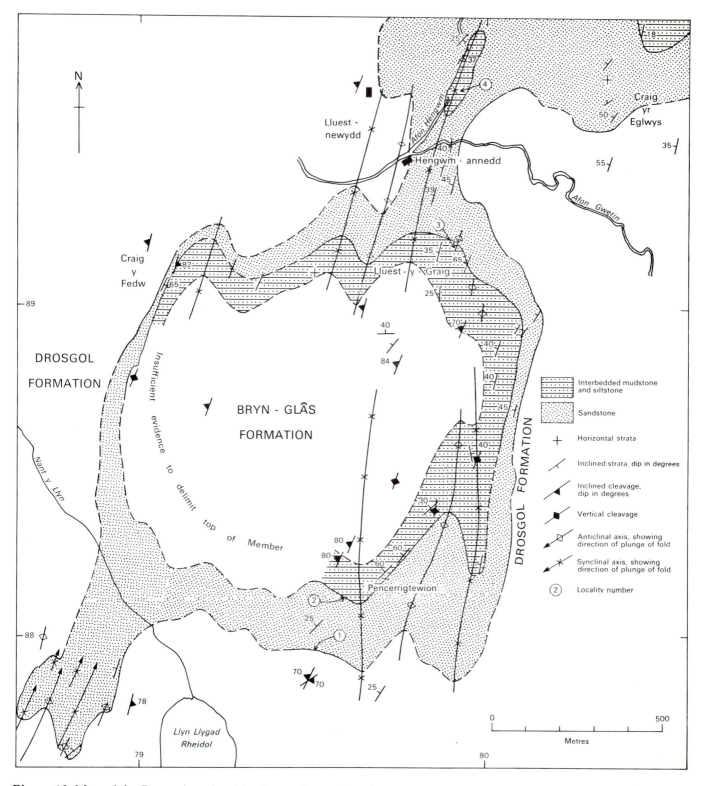

Figure 13 Map of the Pencerrigtewion Member at Pencerrigtewion

A thin massive sandstone is present on the southern slopes of Drosgol in the core of a small anticline [7540 8755] (Figure 10, Loc. 1), where 2 m are visible at its base. Some 4 m of thinly bedded strata overlie this sandstone; within this thickness there are sandstones up to 23 cm thick. North-eastwards towards the summit the massive sandstones become diminutive, and the sequence includes interbeds of sandstones up to 0.5 m thick at the base of a broadly fining and thinning upward sequence many metres thick. It is possible that the basal beds, some of which contain epiclasts, represent the edge of the channel sandstones to the north.

On the north side of Drosgol the massive sandstones seem to be absent, as they are in the southern part of Banc Llechwedd-mawr, but north-westwards they reappear and are 12 m thick at the summit of Banc Llechwedd-mawr (Figure 14, Loc.1), mirroring the situation on Drosgol. Some 3 m above their base a layer of dark grey feldspathic structureless mudstones of unknown thickness divides them. The underlying rock is tough splintery feldspathic mudstone with balled inclusions of feldspathic sandstone [7722 8982] (Figure 14, Loc.2). Such inclusions are taken to be the result of disruption of density contrasting channelised arenite and mud substrate; the sequence as a whole suggests a channel margin with possible slumping along it.

The overlying thinly bedded strata are also well exposed and again display a thinning upwards sequence. Within the basal 2 m, arenite beds are up to 45 cm thick; those above are much thinner and commonly discontinuous due to loading distortion, while some layers appear to be trains of isolated sand-starved ripples. To the SSW of the summit these thinly bedded strata appear to maintain a thickness of about 22 m with bedding most clearly evident in the lowest 15 m. North-west-facing cross-laminae were noted towards the southern end of Banc Llechwedd-mawr.

Carn Owen Inlier

Pencerrigtewion Member

Exposures of the member are good in this periclinal inlier (Figure 15); sections across the short axis of the pericline on the northern side of the Hafan Fault display massive sandstones which are of three distinct lithofacies: well bedded, massive quartzose sandstones; similar sandstones, contorted and disrupted within massive mudstone; and high matrix, crudely layered sandstone showing pillow-form structures and rafts of bedded sandstone.

The first lithofacies is exposed in a quarry (Quarry A, Figure 15; Plate 8) [7337 8809] in the eastern limb of the fold, which contains 17 m of medium grey, fine-grained, mainly quartzose, feldspar-bearing sandstone, in even beds up to 2 m thick (E44702, E44704–8). They contain a few intraclasts of mudstone, and commonly display bedding lamination. Their most striking characteristic is the concentration of convolute-lamination and loading deformities. These were examined by Mr G. E. Strong who noted the presence of pseudonodules (Plate 9). He also observed a set of flute-casts indicating currents to the SW. Dish-structure is also evident in these sandstones some 200 to 300 m to the north. Still farther north in the inlier rippled surfaces occur on which ripple crests trend 115°.

Westward of the quarry, a further 33 m of more massive siliceous and feldspathic sandstones are visible containing more abundant clasts of mudstone; lower still in the sequence, towards the core of the pericline, the sandstones are more massive, less well sorted, with pillow structure and with bedding poorly displayed. Their thickness may not be great, for dips appear to be low. The whole sequence of bedded sandstones in this eastern limb is taken to represent a channel-fill which, over all, is fining- and thinning-upwards. The top of the sequence is truncated, however, and sharply overlain by structureless mudstone of the Bryn-glâs Formation. Within the basal 3 or 4 m of this latter there are a few lenticular sandstones which represent final bursts of activity before the rather abrupt demise of the channel. The section in Figure 15 suggests that these channel sandstones are at least 90 m thick, greater than known elsewhere in the district.

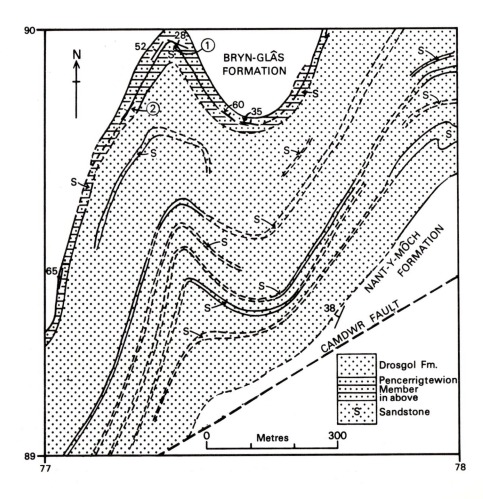

Figure 14 Drosgol Formation on Banc Llechwedd-mawr

Figure 15a. Map of Carn Owen area

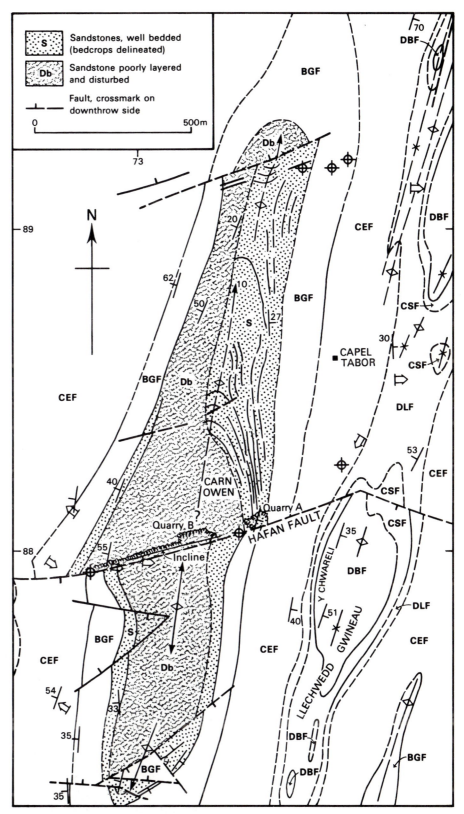

These bedded sandstones can be traced bed by bed as they climb the eastern flank of Carn Owen towards the fold axis. Many cross the axis, but they do not descend far down its western flank. Here the bedding disappears, and the linear strike features it has created are replaced by subcircular features caused by isolated masses of resistant sandstone in a complexly contorted condition within a mudstone surround.

The second lithofacies can be studied in a large quarry (Quarry B, Figure 15) [7312 8805] on or near the anticlinal axis. It has a face many tens of metres high and wide, in which are disposed large rolls and contorted rafts of well bedded massive fine- and medium-grained sandstones within dark grey structureless mudstones, the contortions distinguishing it from the first lithofacies and clearly a result of large-scale, very early, soft-sediment deformation.

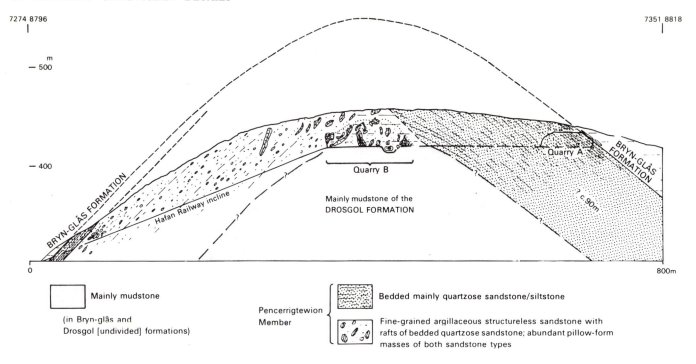

Figure 15b. Diagrammatic section through Carn Owen on a traverse c.70 m north of the Hafan Fault

Immediately west of this central zone, the same horizons are exposed in the western limb of the anticline, on the north side of the Hafan incline. The third variety of greywacke sandstones is here present, but there is scant sign of any primary bedding fabric. The mass of the sandstone is fine grained and argillaceous. In parts of it there are pillow-form bodies (Plate 10), many of which rest in contact and show plastic deformation of their margins. Some are of quartzose sandstone, in which original bedding is preserved; others are more argillaceous and homogeneous. Several possess thin envelopes of mudstone. Isolated rafts of the quartzose sandstones retain their original bedding. One such example on the north side of the incline towards the top, is tens of metres long and many metres thick. It is terminated abruptly, and dips west at about 80°. A vague mega-bedding can be discerned, suggestive of there being several thick sheets of rather structureless rock.

There seems little doubt that the three types of sandstones are all broad contemporaries, but those in the east show no major post-depositional disturbance, while those in the west show intensive

Plate 8 Carn Owen. Quarry A in evenly bedded massive sandstones of the Pencerrigtewion Member

disturbance: these two extreme types interface in the variety that crops approximately along the arch of the fold, where discrete masses of contorted quartzose sandstone are supported within dark grey silty mudstone. It is believed that these disturbances took place in soft, wet sediment and were not an end-Silurian tectonic event. Debate centres upon the precise processes of soft sediment failure; alternatives are subaqueous slumping and post-burial sediment failure.

Subaqueous slumping is a gravity-driven transportation of sediment *en masse*, requiring a bathymetric slope to induce the motion. At the time of their deposition the deposits in Carn Owen were located well away from the basin margin slope, so a basin-floor unevenness would be required to initiate slumping. The most likely unevennesses would have been levéed channel margins. Sediment failure on these might have produced disturbance of the type seen in Quarry B. The rocks on the western limb could be interpreted as channel-fill mass-flow deposits; if the sandstone body had been a very thick, rapidly loaded, channel sandstone emplaced as a result of a late avulsion of the main channel to the NE or east, then down-channel slumping and mass-flowage that incorporated foundered debris from the SE margin might have been a major process in the formation of these rocks. The loading of such a rapidly accumulated sand body on a wet mud substrate would effect stress, and this might have caused the upward injection of mud on the line of unloading along the top of the slump as seen in Quarry B. Basin-floor unevenness could have been caused by either a basement fault devolving upwards into a monoclinal warp of the basin sediments or by thin-skin folding of the basin sediments. There is no evidence for basement faulting having produced the unevenness but early fold movements are credible, for large folds were present in the basin margin sediments by the end of the Ordovician or early in the Silurian (Roberts, 1929, p.669). However, the lack of symmetrical arrangement of the disturbance around the axis of the Carn Owen Pericline implies that this structure, at least, was not active at this time, unless it was then a west-facing monocline.

An alternative explanation is that there was post-burial sediment failure due to fluidisation of parts of the sand body under an aquaseal of the silty mud of the Bryn-glâs Formation. The Carn Owen sand body is the thickest in the district, and a body of such size, if rapidly deposited, would trap a great quantity of water within its pore spaces. Dish structure, pseudonodules, and other wet-sediment loading effects preserved in the primary beds of Quarry A attest to this. The parts containing a high porportion of lithic detritus would compact considerably and, once capped by mud, this could have led to high pore water pressures; early diagenesis might have added to this excess water. Were an early anticlinal warp to have been imposed on this stratigraphy the tensional area of the hinge would have presented a weakness and formed a route of escape for the excess pore water. The same weakness might have facilitated the production of a diapir or 'flame' of mud from below. The release of water through the anticlinal arch could have led to the fluidisation of the sediments in the western limb of the fold, but pillows and mega-layering are unlikely products of such a process: again, the lack of symmetry of sandstone types about the fold axis tells against the theory, which seems less likely than that involving complex mass-movement of a rapidly loaded thick channel sandstone.

Banc Lletty-Evan-hen Inlier

PENCERRIGTEWION MEMBER

The sequence in this inlier closely resembles that at Carn Owen, and only the Pencerrigtewion Member appears at outcrop. Bedded sandstones and intervening splintery mudstones make up the upper two-thirds (about 80 m) of the exposed beds on the eastern limb of

Plate 9 Pseudonodules on a bedding-plane surface of an inverted fallen block. Quarry A, Carn Owen

the anticline, and pass laterally into disturbed beds on the western limb. The lowest third of the sequence, in the core of the fold, is disturbed throughout.

The bedded sandstone at the top of the member is best seen in two sections [7260 8601; 7265 8604] at the eastern end of Llyn Craigypistyll, which total some 24 m and are separated by a fault. A variety of bed forms is seen in these sections including parallel lamination, ball-and-pillow structure, graded bedding and layers of mudstone clasts. The ball-and-pillow structure occurs in several clearly defined beds lying between beds of massive medium-grained sandstones with some parallel laminations. A lower sandstone at least 18 m thick is present [7182 8534] about 250 m south of the Craigypistyll dam, with splintery mudstone above and below. These beds pass laterally southwards into disturbed beds [7175 8516].

The disturbed beds, which occupy the core of the fold and its western limb, are closely comparable with those at Carn Owen, and the sandstone component of the beds is well displayed in the crags [7153 8473 – 715 855] along the western side of Banc Lletty-Evan-hen. Rafts of undisturbed layered sandstone and mudstone are common [e.g. 7156 8540; 7170 8503]; at the latter locality there is a lateral passage into disrupted and contorted beds.

Plate 10 Drosgol Formation. Pillow-form sandstone at foot of the incline. [7298 8792] Carn Owen. A12605

Machynlleth Inlier

STRATA BELOW THE PENCERRIGTEWION MEMBER

These have two separate linear NNE-trending outcrops within the inlier, one on either side of the Brwyno Overthrust. The western outcrop (Figures 16 and 17) includes the full thickness of these beds, about 260 m. North of the Ogof Fault [694 923] dips vary between 45° and 60° to the WNW. Most of the strata is structureless splintery mudstone, divided by several sandstone units, six in the south (Figure 16) and seven in the north (Figure 17). These sandstone units are mostly massive and well jointed, contrasting with the structureless mudstone, and their outcrops form distinct topographical trenches. Exposure in the trenches is imperfect; most reveal massive sandstone but other thinly bedded lithologies are visible and may well be more common than is apparent.

The best exposures are at the northern end of the outcrop in Craig Caerhedyn (Figure 18), where the lowest of the seven sandstone units marks the base of the formation. The main body of structureless mudstone contains a scatter of small lumps of distorted synbasinal arenite. In the lower parts of the formation there is also a scatter of coarse sand grains of feldspar probably of extra-basinal origin, and towards the base exotic pebbles are locally common, though these are generally smaller than those seen near Plynlimon. Between the lowest two sandstone units is a sequence some 30 m thick of contrasting lithologies. Parts consist of an evenly bedded multilayered sequence of arenites, from laminae a few mm thick up to 4 cm, and argillites also up to 4 cm thick; other parts show

uneven layering in which the arenites are contorted or disrupted into load balls (pseudonodules). In other places this lithofacies is replaced laterally by a different one consisting of unlayered mudstone, pebbly and feldspathic in parts, a lithofacies which is very common in the Drosgol Formation but absent from the Nant-y-Môch Formation. The contact between the two lithofacies is highly transgressive and seemingly erosional [e.g. 7087 9659] (Figure 17, Loc.1); the interface is not, however, particularly sharp—at least as seen in exposure—because the layering is contorted in proximity to the contact. These relationships result from channelling of the multilayer sequence with partial collapse of the channel walls, the unbedded pebbly mudstone representing the channel fill (Figure 19). The channel in this section is bridged by a set of thin and even interbeds of siltstone and mudstone which, a few metres to the north [7093 9667] (Figure 17, Loc.2), themselves appear to have been channelled. Structureless sandy argillite pockets discordantly into such beds just below the second sandstone, which is here 23 m above the basal sandstone. The sequence between the two sandstones in Craig Caerhedyn probably represents wandering submarine channels with channel-fill and overbank sediments not far from the base of a submarine slope. It is arguable that the base of the Drosgol Formation should be taken locally at the bases of such channels, but it has proved more convenient to take the junction at the base of the lower of the two sandstones.

The eastern outcrop extends from the upper reaches of Cwm Einion [715 930] to the Llyfnant Fault [726 975]. It lies in the core of the main anticline of the inlier and exposes some 150 m of strata below the Pencerrigtewion Member. The continuous units of sandstone which divide the succession west of the overthrust and in the Plynlimon Inlier were not discerned here. The bulk of the beds are argillaceous and largely structureless, but they are very pebbly and sandy, feldspar being a common constituent. Bodies of conglomerate and sandstone are numerous, though their relationships one to another are unclear, especially when disposed on opposite sides of the anticline.

Two of the best areas of exposure are where the main anticline crosses the Dynyn valley [7177 9524] and the Brwyno valley [7207 9617]. A large body of conglomerate is exposed [7163 9538] (Figure 20, Loc.1) WNW of Moel Hyrddod:

	Thickness m
5. Conglomerate, as below, visible in places overlying the main section (Beds 1–4): possibly passing up into sandstone	?1.90
4. Conglomerate, as Bed 2. Contains large mudstone clasts, in places up to 25 cm in diameter, some of which possess silt laminae	3.33
3. Sandstone, grey to fawn, coarse. Some parallel lamination, also gritty streaks up to 2.5 cm thick. In form of an uneven lenticle	0 to 0.33
2. Conglomerate, grey, coarse and ill-sorted. Largely well rounded white quartz pebbles with a few white quartzite and igneous fragments; mudstone rip-clasts near middle up to 20 cm diameter	1.57
1. Sandstone, greywacke, grey-green, coarse	0.60

Although many pebbles in the conglomerate are flat they are not imbricate and hence give no indication of transportation direction. The mudstone clasts are much larger than the other pebbles, suggesting that they were less dense, as would be expected of mud. The wispy, cuspate outlines of these clasts, the contortion of the internal laminae, and the penetration of clasts by smaller pebbles attest to their softness when transported. Such a coarse body of sediment was probably quite proximal to its point of input, and is

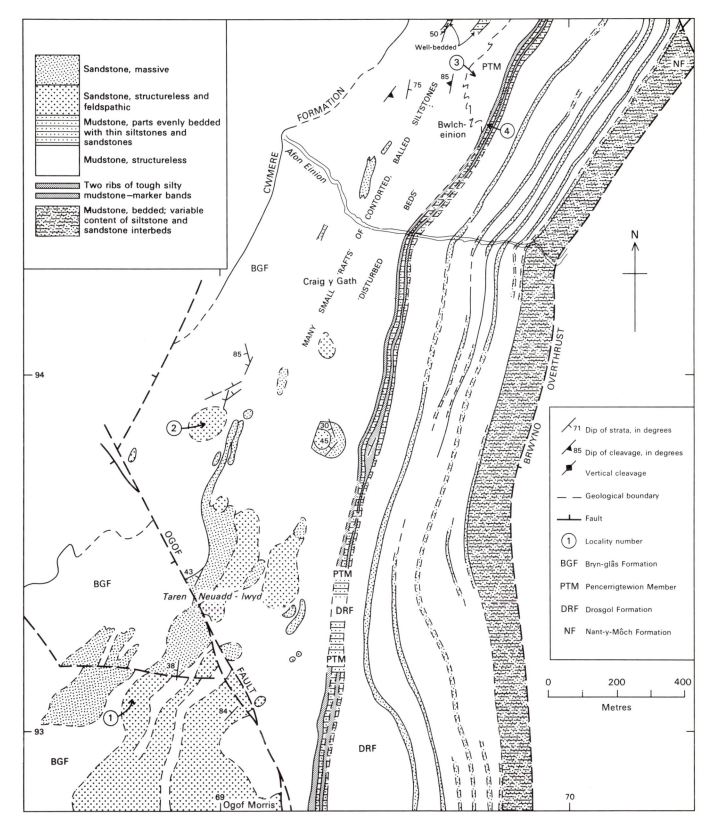

Figure 16 The Drosgol and Bryn-glâs formations between Ogof Morris and Bwlcheinion

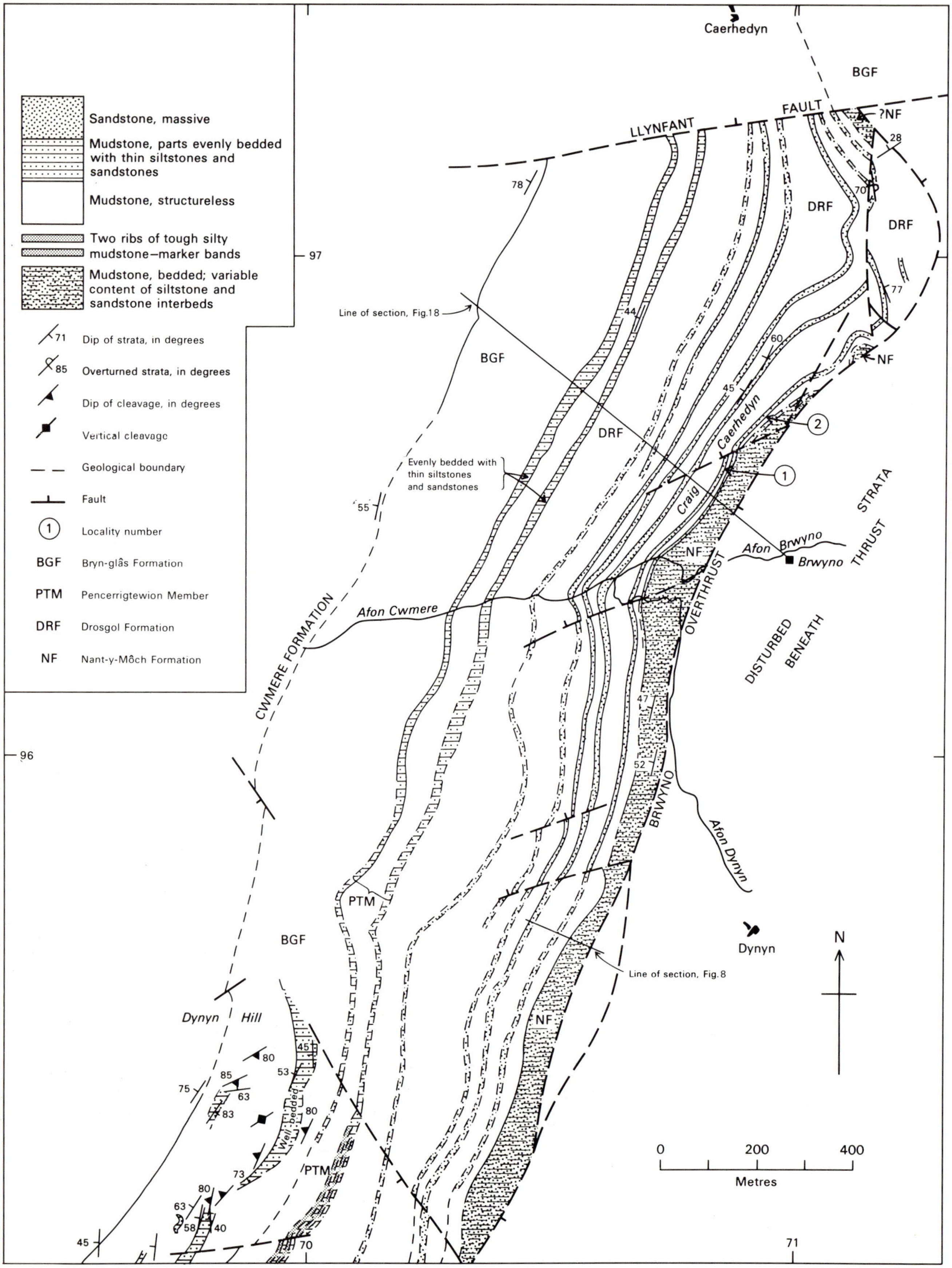

Figure 17 The Drosgol and Bryn-glâs formations between Bwlcheinion and the Llyfnant Fault

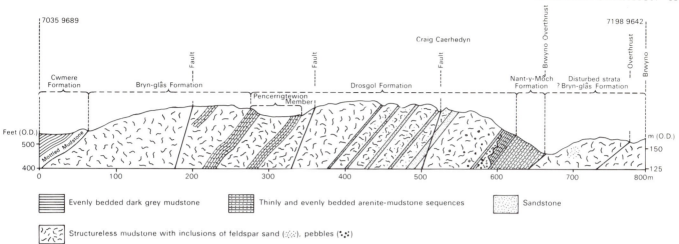

Figure 18 Section across Craig Caerhedyn

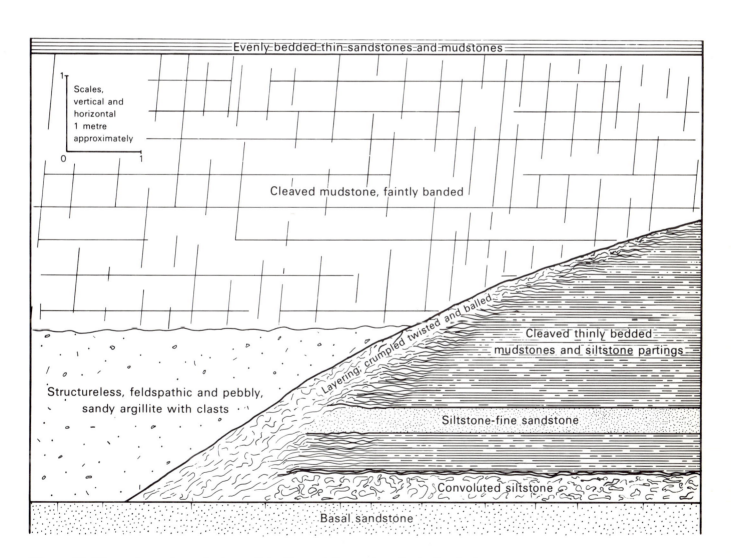

Figure 19 Channel-margin at the base of the Drosgol Formation, Craig Caerhedyn

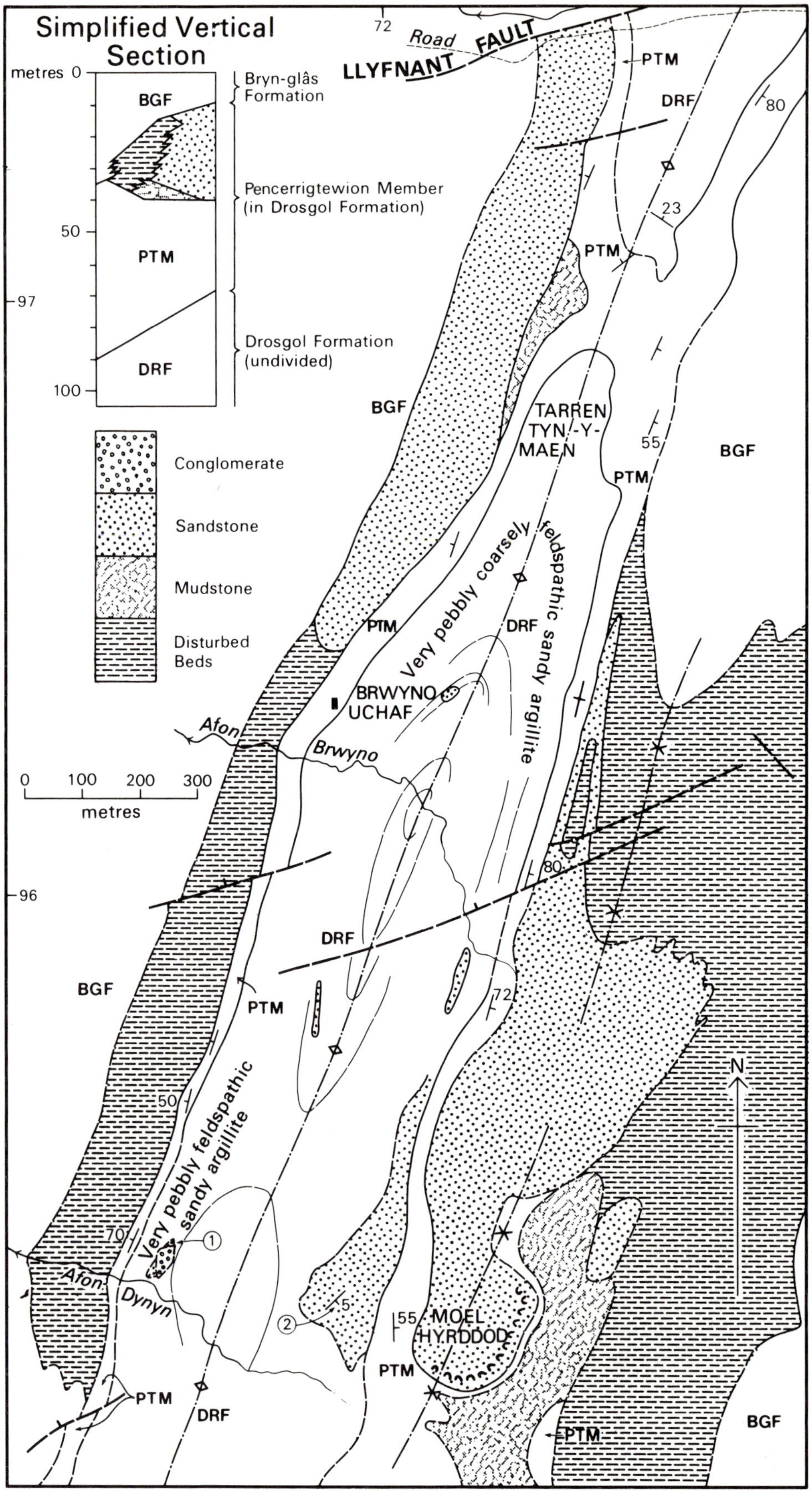

Figure 20a. The Drosgol Formation between Moel Hyrddod and the Llyfnant Valley

considered to be an example of a mass-flow of exotic detritus brought into the basin along a submarine channel with sufficient velocity to have eroded the local muddy and silty substrate of the channel floor or walls. Along the strike adjacent to this section, the arenaceous/rudaceous body thins abruptly, and its feather-edges are of non-pebbly sandstone overlain and underlain by mudstone.

Some 30 m away (Figure 20, Loc.2), on the eastward dipping limb of the anticline, there is a much larger arenite body, also lenticular in N–S section, and consisting of fairly well sorted, medium-grained, grey-green, massive sandstone. It contains small mudstone clasts (mud-flakes) but is not otherwise conglomeratic. It is up to 20 m thick and its base is discordant upon mudstone. This sandstone appears to be at a higher horizon than the conglomerate on the western limb, for it lies close beneath the thin-bedded unit of the Pencerrigtewion Member. Nevertheless it may represent part of the fill of the same channel. Rather than being more distal, which would contradict other nearby current evidence, its finer grain might be explained as a more marginal expression.

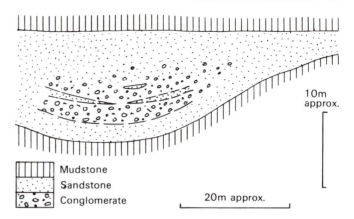

Mudstone
Sandstone
Conglomerate   20m approx.

10m approx.

Figure 20b Section across channel-fill [7163 9538] near Dynyn. Tectonic dip c.10° towards 280°. (Loc.1, Fig.20a)

Pencerrigtewion Member

This has three main outcrops within the inlier, those to the west and east of the Brwyno Overthrust and a third north of the Llyfnant Fault. The distribution of facies within the member is shown in Figure 4. Its thickness is very variable, ranging from nil to 130 m.

The western outcrop extends from Cwm Ceulan [688 900] northwards to the Llyfnant Fault [708 972]. In the southern part of this outcrop the member is estimated to be 60 to 90 m thick and is dominated by massive arenites which form a broad ridge extending south-westwards from Garn Wen [691 928] to Afon Clettwr and also the high ground of Moel y Garn. Its upper part is best displayed in Afon Clettwr [6773 9170 – 6788 9173] where c.33.2 m of bluish grey fine-grained, massive sandstone, largely homogeneous subgreywacke, with some poorly developed ripple cross-lamination, underlie interbedded mudstones and sandstones of the Bryn-glâs Formation. The massive sandstones extend southwards from Afon Clettwr but are less well exposed than to the north. They are seen [6909 9118; 6944 9110] on Moel y Garn on either side of the Moel y Garn Anticline, but southwards they are replaced by disturbed beds [6905 9088; 6904 9098], and dark grey irregularly cleaved mudstones with discrete sandstone rolls and partially disrupted beds of sandstone are exposed. The disturbed beds are also well exposed in an old roadside quarry [6881 9013] and in crags [6875 9005 – 6890 9005] in Cwm Ceulan where mudstones and sandstones have been involved in large-scale disruption as at Carn Owen (p.26).

North of Afon Clettwr massive sandstones form a ridge towards Garn Wen. The beds are well exposed, as at a point [6805 9197] where massive, mainly structureless, pale grey fine-grained sandstone is seen in a 10 m joint-controlled face dipping at 70° towards 240°, and at another [6873 9259] where abundant mudstone flakes lie in an otherwise structureless high-matrix sandstone. At the top of the member up to 15 m of feldspathic sandy mudstones are present [6786 9183 – 6843 9257], and these pass northwards into structureless sandstone. In many places, especially in the southern part of its outcrop [6794 9194], the sandy mudstone encloses discrete masses of sandstone and, less commonly, feldspathic mudstone, up to 2 m across. The sandstone masses are largely structureless, but a few show intense internal contortion indicating a west-to-NW translation of sediment.

Within 1 km to the SW of the Ogof Fault, the Pencerrigtewion Member shows the first signs of the disruption which becomes almost total on the northern side of the fault. Mudstone beds develop within the sandstones, and the sandstones become layered and develop ball-and-pillow structure and slump rolls, particularly in the upper part of the sequence. One section [6878 9312] (Figure 16, Loc.1), 1.2 km east of Cefn-gweirog, shows 12 to 15 m of sandstone with interbedded siltstone and mudstone, individual bedforms ranging from parallel laminated and ripple cross-laminated types to contorted and pillowed. Similarly, at Ogof Morris [6899 9284] (Figure 16), the continuity of the sandstone fails, so that it forms separate masses surrounded by mudstone. Some of these on Taren Neuadd-lŵyd [6909 9342] are tens of metres across. Northwards they become progressively smaller, more rounded, and isolated. In one outcrop [6898 9387] (Figure 16, Loc.2), a sandstone mass dips approximately with the inclination of the hillside, and shows boat-shaped curved bedding surfaces and arcuate scars within it. These features present an impression of arcuate sediment slips directed towards about 350° (or possibly about 170°). This evidence is believed to indicate a channelised sand body, with disruption of its margins by slumping and differential loading. Northward, by Craig y Gâth [6943 9425], the outcrop is mainly of mudstone closely similar to the overlying mudstones of the Bryn-glâs Formation, but clearly at a lower stratigraphic horizon. Rafts of thin-bedded siltstones appear in the mudstones and are presumed to be foundered interchannel deposits.

Some 200 m north of Bwlcheinion [6963 9470] the outcrop abruptly enters a sharp topographical 'trench' some 100 m wide (Figure 16, Loc.3; Figure 17), which continues northwards to the Llyfnant Fault. Within it are some 45 to 50 m of structureless mudstone with packets of thinly interbedded siltstones and mudstones, the latter being most persistent at the bottom and top. The siltstones are parallel and cross-laminated, and the deposits are interpreted as interchannel equivalents of the sandstones of Garn Wen and Moel-y-Llyn. The marker beds used to prove this equivalence are two thick beds of almost structureless silty mudstone or siltstone which mark the eastern side of the 'trench' just north of Bwlcheinion (Figure 16, Loc.4). These were readily mappable southwards to the east of Craig y Gâth, Taren Neuadd-lŵyd and Ogof Morris, where they underlie massive sandstone.

The outcrop east of the Brwyno Overthrust commences in the south at Moel-y-Llyn [7128 9169], which is a large body of massive, dark grey, fine-grained, greywacke sandstone and siltstone with many exposures around its slopes and on its summit. The sandstone is mostly fine grained (E46546, E44720), of subangular to subrounded quartz grains with rock fragments and albite in a chloritised matrix. Mudstone rip-up clasts are common in parts. Thick bedding is evident on the SE flank of the hill where sheets of structureless khaki and grey sandstones occur beneath the Bryn-glâs Formation. The outcrop is terminated on the southern flank of the hill largely due to the southward plunge of the Moel-y-Llyn Anticline, but it is clear that the sand body is thinning southward. Likewise it diminishes rapidly northward on Cerig Blaen-Clettwr-fawr [708 928] beyond which it cannot be traced. The underlying thinly bedded siltstones and sandstones are exposed in the core of the anticline which crosses the depression northwards from Waen Badell [7088 9187]. These beds are continuous throughout this eastern outcrop.

Farther north the member is well exposed in the valley side south and SE of Pemprys [7162 9419]. Some 30 m of dark grey, massive, fine-grained feldspathic sandstone recur here [720 939], underlying the Bryn-glâs Formation. The sandstone passes rather abruptly westward into evenly and well bedded siltstones, sandstones and mudstones, some 38 m thick [7154 9391]. This sequence includes layers up to 3 m thick in which the arenite is balled and contorted. The thin siltstones are cross-laminated, some showing indication of transportation currents from the south or SSW. There is at least one thin layer of structureless high matrix fine sandstone. The sequences in this slope appear to represent a small channelised sandstone and thinly bedded marginal deposits. On the northern side of the valley the massive channel sandstone of the upper part of the member passes for a short distance into a thin-bedded sequence [7229 9413] before becoming disturbed into balls and contortions still farther north.

Farther north along the eastern limb of the Pemprys–Tarren Tyn-y-maen Anticline, the disturbed beds revert to massive, structureless, high-matrix fine-grained sandstone on Moel Hyrddod (E46554) [7212 9525] (Figure 20). The massive sandstone of Moel Hyrddod contains an abundance of mudstone rip-up clasts and, together with the adjacent disturbed beds, is taken to be another channelised body, probably in continuity with a similar sandstone in the western limb of the anticline at Tarren Tyn-y-maen (Figures 4 and 20): if so, a SE–NW channel alignment is indicated.

The outcrop in the western limb of the Pemprys–Tarren Tyn-y-maen Anticline shows a different sequence, in that massive sandstone is absent south of Afon Brwyno. Some contorted sandstone balls are contained within the mudstone which replaces it, but they are not as abundant as they are in beds mapped as 'disturbed'.

On the north face of Tarren-Tyn-y-maen, the massive sandstones are missing from the succession displayed in the eastern limb of the anticline. The whole member here consists of thinly interbedded arenites and mudstones, as it does just south of Pemprys. Some 37 m are exposed, the upper parts probably representing levée and interchannel beds equivalent to the sandstone seen in the western limb. The arenites are normally less than 2.5 cm thick, but some near the middle of the member are as much as 12 cm with exceptions up to 30 cm.

Outcrops north of the Llyfnant Valley around Cefnmaesmawr [7227 9810] belong to the third area of outcrop, north of the Llyfnant Fault. The succession resembles that in the eastern part of the inlier between Moel-y-Llyn and Tarren Tyn-y-maen. An interval of structureless mudstone, which passes under Cefnmaesmawr farm, separates the bedded rocks from balled-up sandstones set in mudstone (Disturbed Beds), which are well displayed immediately SW and due north of the farm. Some of the sandstone balls are up to 2 m across, and coarse primary foliation in these disturbed beds is suggestive of soft sediment folds several metres across. At Cefnmaesmawr the disturbed beds pass into bodies of largely structureless feldspathic fine sandstone and siltstone, and a 6 m thick lens of this nature occurs in an outcrop 320 m SSE of the farm. The outcrop widens NE to Mynydd Cae-du [729 987] where it passes into a large mass of structureless fine sandstone, probably about 20 m thick [7269 9872]. Northwards, the sandstone thins and is again replaced by disturbed beds, though up to 2 m of sandstones and mudstones, equivalent to the basal part of the sandstone, are present in a transition zone [7277 9887]. These are interpreted as a discrete channel on Mynydd Cae-du, with associated channel margin deposits (Figure 4).

BRYN-GLÂS FORMATION

Plynlimon Inlier

The thickness of the Bryn-glâs Formation increases southwards from about 130 m north of Eisteddfa Gurig to some 180 m in the type area, decreasing again northwards to around 100 m at the Nant-y-môch Reservoir; it is at least 100 m thick at Hyddgen and possibly considerably more. The beds are predominantly unlayered mudstone, apart from a sandstone up to 12 m thick east of Bwlchystyllen.

The obvious lithological change from the Pencerrigtewion Member to the unlayered mudstone of the Bryn-glâs Formation is exposed only in a narrow gorge [7564 8459] SSW of Lle'r-neuaddau and on Dyll Faen (p.24). The gorge is aligned along the strike, with its eastern side a dip slope (60° towards 284°) within the Pencerrigtewion Member. In its western side interbedded mudstones and sandstones of that member are overlain by the unlayered mudstones of the Bryn-glâs Formation. The junction is sharp, but does not appear to be erosional. There is a continuous stream section in the Bryn-glâs Formation from the southern end of the gorge to Afon Rheidol [7536 8439]. Most of the exposed sequence is unlayered, but there are some intercalations, up to 0.6 m thick, of multi-layered mudstones and sandstones resembling those of the Pencerrigtewion Member.

The main part of the formation is well exposed over most of its outcrop. In the northern part of a section [7952 8477 – 7948 8522] along Afon Tarenig the sediments are layered, the layering marked by colour-banding or by thin siltstone beds and laminae. The layered rocks appear to be rafts (p.15), and in places [7950 8513] pass through disrupted beds into typical unlayered, phacoidally cleaved, mudstones. Similar bedding is developed in parts of other sections along eastward flowing tributaries [7917 8559 – 7946 8536; 7879 8504 – 7946 8497] of the Tarenig. Unbedded mudstone, with small wisps and rounded masses of arenite and a few discrete sandstone 'rolls', is well exposed in the type area on Bryn-glâs hill [767 823] and in gorges up to 10 m deep along Nant Fuches-wen [7674 8101–7688 8095], Afon Castell [7592 8149 – 7660 8119] and Nant Ceiro [7544 8227 – 7560 8220; 7582 8231 – 7620 8234].

There is an interesting development of arenite within the upper part of the Bryn-glâs Formation at Pen Cerig, east of Bwlchystyllen (Figure 21). The greater part of the deposit is a massive fine to medium-grained sandstone (Facies C_1 of James, 1972, p.295), with a maximum observed thickness of 12 m (Figure 21, Loc.1), but to the south (Figure 21, Loc.2) there are minor channels within the sandstone filled with an arenite/shale-clast melange (Facies C_2 of James, 1972, p.296 and fig. 3) showing a NW or NNW direction of transport. The sandstone thins eastwards, and at several places (Figure 21, Loc. 3, 4, 5) is replaced laterally by disordered mudstones with sandstone pillows. This suggests that the sandstone may originally have had a greater lateral extent before undergoing partial penecontemporaneous disruption. In the west (Figure 21, Loc.6) the sandstone passes laterally, by interdigitation, into mudstone; together with its thin development just to the north (Figure 21, Loc.7), this indicates that it is dying out to the NW and is probably the fill of a small local channel. This is contrary to the interpretation (James, 1972) of the deposit as a channel-fill at the easterly limit of a fan (his member 6), which includes the sandstones of Carn Owen and extends westwards across much of the Machynlleth Inlier (where it falls within our Drosgol Formation).

North of the Camdwr Fault, on Cefnyresgair, there are exposures of structureless, medium to dark grey, splintery mudstone with a scatter of arenite lumps including crumpled thin silty

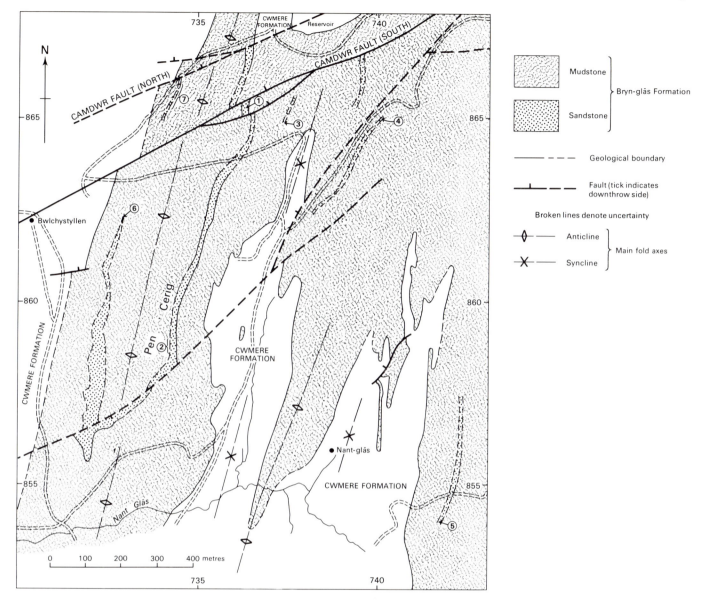

Figure 21 Distribution of sandstone within the Bryn-glâs Formation near Bwlchystyllen

laminae [e.g. 7437 8805]. There are more regularly bedded thin siltstones in places [7470 8819; 7473 8831], and a distinct mega-stratification can be discerned in the broad features of the topography. These features may reflect lateral continuity of packets of fairly evenly bedded siltstone laminae within structureless mudstone; such packets represent periods of distal or inter-channel turbidity currents. In one exposure [7532 8955] granular feldspar is present in the mudstone, an occurrence normally exclusive to the underlying Drosgol Formation.

Similar topographical features to those of Cefnyresgair also occur on the outcrops at Banc Llechwedd-mawr, Esgair [787 911] and Ochr Lygnant [795 920]; immediately NE of Crip y Frân [754 883] they outline two folds. A small quarry [7825 9126], NE of Hydd-gen, exposes several metres of structureless medium-dark grey mudstone.

The topmost beds of the Bryn-glâs Formation and their relation-ship to the overlying Cwmere Formation are clearly displayed in a number of sections. Jones (1909, pp.475-482) defined the upper

limit of his Bryn-glâs Group by reference to sections near Eisteddfa Gurig and in Nant Fuches-gau [7666 8087], both of which are still accessible. About 200 m WNW of Eisteddfa Gurig the topmost Bryn-glâs Formation is exposed along the south side of a track in the core of a southerly plunging anticline. On its western limb [7951 8409], grey to bluish grey irregularly cleaved splintery mudstone with local poorly developed bedding (Bryn-glâs Formation) is overlain by bedded and regularly cleaved grey mudstone (Cwmere Formation). The junction is not sharp but can be located to within a few centimetres. Abundant *Glyptograptus persculptus* occur in a 15 cm bed of dark grey mudstone 1.2 m above the junction. In Nant Fuches-gau there is a similar sequence, but the bed with *G. persculptus* (F.7 of Jones, 1909, p.481, and fig. 5) is now obscured.

A dip slope at the top of the formation is particularly well displayed on the western side of Cribyresgair [7969 8424 – 7991 8508] north of Eisteddfa Gurig, and in the intricately folded rocks east of Bwlchystyllen (Figure 21).

Carn Owen Inlier

Here the formation comprises massive structureless mudstone no more than 40 m thick, much less than around Plynlimon. Its base is exposed in Quarry A of Figure 15 [734 881], where it unevenly overlies beds of massive greywacke sandstone probably related to the sandstones of Pencerrigtewion. Impersistent and thin layers of similar sandstone are present in the basal metre or so. Other sections [7305 8724; 7290 8723], near its base also display interbedded sandstone and mudstone. Some of these sandstone layers display ball-and-pillow structure, and they become thinner upwards. All these sections indicate that the change in the conditions of deposition at the base of the Bryn-glâs Formation was not instantaneous.

The topmost beds of the formation are seen in a line of low crags [7292 8671 – 7328 8750] exposing dark grey splintery mudstone with a few discrete sandstone balls.

Banc Lletty-Evan-hen Inlier

The main outcrop in this inlier is terminated northwards by the Camdwr fault-belt; the topmost beds occur in a further small outcrop on the side of the Leri gorge (Figure 29, Loc. 1) immediately north of the fault belt. The formation has a thickness of 45 to 55 m. There are numerous small sections in grey splintery mudstone, in places with sandstone 'balls' which vary in size from a few centimetres to a metre or more in diameter. An old quarry [7200 8536], south of Llyn Craigypistyll, displays dark grey sheared silty mudstone on its north side, faulted against grey very silty mudstone with rounded 'rolls' of argillaceous sandstone to the south.

The upper part of the formation, and its junction with the Silurian, are seen in two sections in the Leri gorge [713 855]. The first of these (Figure 29, Loc.2) shows:

	Thickness m
Cwmere Formation	—
Bryn-glâs Formation	
Mudstone, grey, splintery; phacoidal cleavage; mainly unbedded with some irregular arenite wisps, but layered in a few places	c.6.0
Mudstone, dark grey, shaly, weathering to rusty brown	? 9.0
Mudstone, grey, splintery (as highest mudstone, but poorly developed bedding more widespread)	seen 16.8

The second section (Figure 29, Loc.3) shows:

	Thickness m
Cwmere Formation	—
Bryn-glâs Formation	
Mudstone, grey, splintery; phacoidal cleavage	c.12.0
Mudstone, dark grey, shaly; weathering to rusty brown; fine-grained sandstone beds to 30 cm in basal part	c.5.5
Mudstone, grey, splintery; phacoidal cleavage	seen ? 6.0

The dark grey shaly mudstones in these sections are an unusual component of the formation. They are softer than the splintery mudstones above and below them, and are not exposed outside the gorge sections.

Bwlch-glas and Alltgochymynydd inliers

These small inliers lie in the core of the Coed Dipws Anticline. The southern inlier, near Bwlch-glas, is only 150 m long by a maximum of 20 m wide, and lies to the west of a NNE-trending fault. There is one exposure [7012 8739] of dark grey unbedded splintery mudstone.

In the second inlier, just west of Alltgochymynydd, the top 73.5 m of the formation is well exposed in Afon Cyneiniog [7040 8822 to 7049 8818]. From west to east the western limb of the anticline reveals:

	Thickness m
Cwmere Formation	—
Bryn-glâs Formation	
Mudstone, massive, medium to dark grey with inclusions of pale grey arenite, up to 1.2 m thick and down to small wisps and knots. The basal 1 m contains subparallel thin interbeds of pale silt, uneven, and clearly disturbed penecontemporaneously	5 to 5.5
Sandstone and mudstones in layers up to 12 cm thick. Basal 30 cm in thin layers up to 1.5 cm thick, lenticular cross-laminated (starved ripples)	2
Mudstone, medium to dark grey, containing parallel and subparallel thin streaks and crumpled layers of pale grey siltstone up to 0.3 cm thick	6
Mudstone with isolated, twisted inclusions of arenite	45
Siltstone, thinly bedded, pale grey, up to 1.5 cm thick, strongly cross-laminated. A few layers of darker grey colour are unlaminated and reach 8 cm thick. Some arenite load-balls also occur within interbedded mudstones. Dips of 75°W are steeper than normal suggesting penecontemporaneous mass disturbance	14
Mudstone, massive, non-feldspathic, micaceous	seen 1

The eastern limb of the fold reveals a sequence which differs in detail and seemingly is thinner; a sedimentational explanation seems the most likely for these differences. These southerly exposures of the Bryn-glâs Formation show the strong influence of distal or marginal turbidity currents.

Cwmere Inliers

These two inliers (Figure 22) lie in the core of the Moel y Garn Anticline. In the small northern inlier only massive mudstone is exposed [685 891]. In the southern and larger inlier, some 40 m of beds crop out [682 885]. The rocks are divisible into two: a lower, bedded sequence some 22 to 25 m thick, base not exposed; and an upper one, about 15 m thick, of massive unlayered mudstone with dispersed inclusions of balled-up and gnarled sandstone and siltstone.

The lower sequence is well exposed, and 5 to 6 m of the lowest beds are visible near the core of the anticline [6817 8853] including sandstones up to 22 cm thick. Near the top of the sequence, a strike-ridge [6819 8845 – 6821 8853] is formed from about 4 m of beds, mainly well bedded, medium to pale grey and fairly well sorted sandstones, in layers up to 30 cm thick. The thicker layers are uneven due to penecontemporaneous disturbance and show convolute bedding and pseudonodules. In places the layers lose their individuality and the 4-m sequence becomes a single 'balled-up' sandstone. The top part of this sequence is well exposed in a N – S joint face [6821 8860], which reveals some 5 m of thinly interbedded siltstones/sandstones and mudstone partings. In the lower part of this section, the arenite layers are up to 30 cm thick, but they thin upwards to about 7 cm, this indicating a waning of the turbidity impulses in response either to the abandonment of the channel or to a general regressive phase.

The wider significance of these beds is debatable in the absence of precise correlation with other sequences of the Bryn-glâs Formation. With only 15 m of massive mudstone overlying it, this bedded sequence is clearly higher in the succession than the thick turbidite

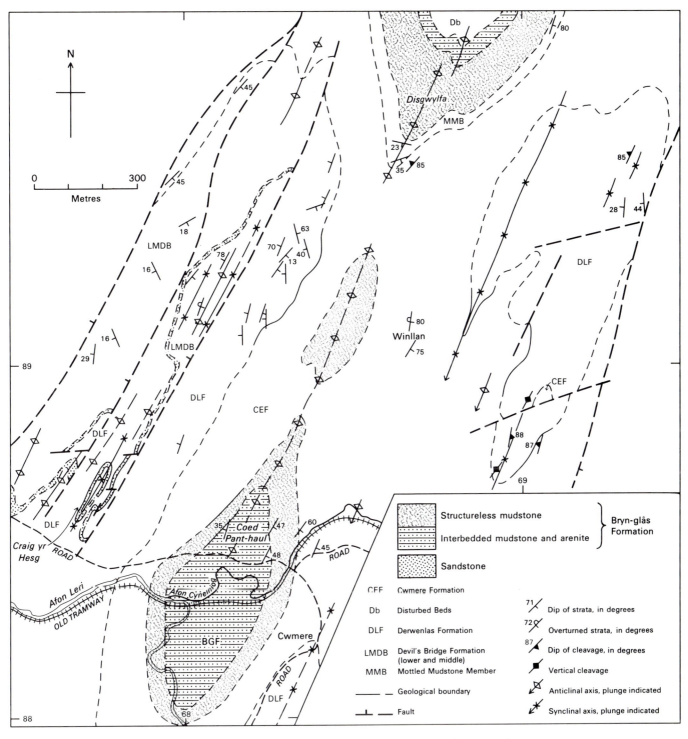

Figure 22 The geology of the southern extremity of the Machynlleth Inlier, 2.5 km east of Talybont

sandstones of the Pencerrigtewion Member on Plynlimon; it is probably higher even than the turbidites on the eastern side of Carn Owen, also placed in the Pencerrigtewion Member. It is conceivable that these sequences indicate the westward progradation of turbidites into the basin. If so, they could be fitted into a submarine-fan model (e.g Mutti, 1977), with the inner fan existing somewhere to the east or NE of Plynlimon, but such a model may be inapposite to a basin the size of this one. Were the model to be

sustained, however, then these beds at Cwmere would be better placed in the Pencerrigtewion Member.

The overlying sequence of massive mudstone is visible in natural exposures [6829 8873; 6824 8866], where some 6 m overlie a near-horizontal bedded sequence which includes sandstones up to 15 cm thick. In the northern inlier [685 891] only these massive mudstones are exposed.

Machynlleth Inlier: west of the Brwyno Overthrust

The two-fold division of the formation near Cwmere (p.38) is also discernable in places at the southern end of this inlier. At Disgwylfa [687 898] the lower bedded rocks are thinner, and the individual beds are also thinner, giving a more distal (or marginal) aspect to the sequence. Northwards [6912 9024], the bedded rocks are 35 m thick, and individual sandstones are up to 1 m thick with well developed bottom structures (especially flute-casts) indicating a SE derivation of sediment. The overlying massive mudstones are about 25 m thick. North of this locality the bedded regime becomes disrupted, and passes laterally into massive mudstone with small arenite masses. On the opposite side of the anticline, east of Fronlas, bedded rocks form the basal quarter of the sequence, passing laterally both northwards [6885 9064] and southwards [6874 9032] into massive mudstones.

Northwards the entire formation thins. In Afon Clettwr [6759 9172 – 6773 9170] it is about 30 m thick, the thinnest known in the district. Here a complete section through the formation, which shows the greater part of it to be layered, is as follows:

	Thickness m

Cwmere Formation

| Mudstone, grey, colour-banded. *Glyptograptus persculptus* common, especially about 2 m above base. Horizon of cone-in-cone concretions at base (approximately 13 m SE of the bridge) | — |

Bryn-glâs Formation

Mudstone, grey, silty; mainly unlayered, but with indistinct bedding in top 60 to 90 cm. Rounded structureless mass of grey fine-grained sandstone (60 × 30 × 30 cm) embedded in mudstone 30 m from bridge	6 to 9
Mudstone, grey, layered, with thin (up to 3 cm) silt laminae. Contorted adjacent to 90-cm-wide crush zone opposite old adit	c.3.7 to 4.6
Mudstone, grey, with beds of fine-grained sandstone up to 30 cm thick. The mudstone contains sandstone laminae, commonly corrugated; the sandstone ranges from evenly bedded through internally contorted to well developed ball-and-pillow structure	12.5
Mudstone, grey, with sandstone beds gradually becoming predominant towards base. Internal contortion of sandstones, but no development of ball- and-pillow structure	6.4

Drosgol Formation

| Sandstone, massive | — |

Northwards from Afon Clettwr the formation passes within 500 m from this predominantly layered sequence into more typical unlayered mudstones enclosing rounded lumps and wisps of arenite. The formation thickens to about 70 m at the Ogof Fault. Rafts of layered rocks occur within these mudstones, most notably about 1 km east [686 931] of Cefn-gweirog where, over an area of about 300 m by 150 m, there are numerous sections in massive sandstone and in interbedded mudstones and sandstones.

From the Ogof Fault (Figure 16) northwards to Bwlcheinion, the Bryn-glâs Formation is largely inseparable from the underlying Pencerrigtewion Member. Around Bwlcheinion many exposures reveal the Bryn-glâs Formation to be a body of structureless mudstone enclosing random masses of sandstone [6958 9479] and thinly bedded sequences of mudstones and siltstones [6954 9480]. Some portions of thinly multilayer sequences are not internally contorted, though they commonly possess anomalously steep dips [e.g. 6913 9435]. Here some 5 m of siltstones and mudstones in layers 2 cm to 4 cm thick are interbedded in medium to dark grey structureless mudstone. The siltstones are ripple-bedded, and some layers consist merely of a train of isolated starved-ripples. Elsewhere, as at Bwlcheinion farmyard [6968 9469] the ripple bedding was loaded, and this has produced small bladders of arenite along the bases of the layers. Such exposures of thinly multilayer sequences may represent portions of interchannel turbiditic deposits which were subjected to traction reworking of their arenitic bases. The channel may have been the one which is partially filled by the sandstones at Ogof Morris to the south (Figure 16), and which developed after the bulk of the Drosgol Formation had been laid down.

Some multilayer sequences persist intact along the strike for tens of metres [690 940; 697 950; 700 954]. In each of these three localities the outcrop is slightly curvilinear. When the regional westerly dip is removed these rocks are left with E–W strikes and with residual dips that can be accounted for by rotational slip of wet sediment. If the sandstones of Ogof Morris partially infill a channel, it is possible to view these postulated slips, and at least part of the overall disruption of the Bryn-glâs Formation hereabouts, as collapses of the northern side of this channel during the decline of its activity. Fracturing displayed in the smooth surface of the road [6966 9468] through Bwlcheinion farmyard is compatible with such a view, for when the regional westerly dip is removed from the rocks the fractures become low angle southerly dipping normal faults.

Machynlleth Inlier: east of the Brwyno Overthrust

On Llechwedd Llŵyd [711 910] and the eastern slope of Moel-y-Llyn the Bryn-glâs Formation consists of massive structureless mudstone which, in at least one place, is feldspathic [7148 9165]. It is very thin, possibly even thinner than the known minimum of 30 m in Afon Clettwr (p.40), and overlies a large body of massive, feldspathic sandstone assigned to the Pencerrigtewion Member. This thinness can be explained as depositional attenuation over a bathymetric 'high' but it is also possible (p.14) that, in this western area, the 'Pencerrigtewion event' arrived late.

On the NW side of Moel-y-Llyn, the outcrop of Esgair Foel-ddu has a much higher content of thinly bedded siltstone–mudstone turbidites. Since these beds are folded, the total thickness of the formation is difficult to assess. Some westward dipping beds [7031 9262] are affected by minor folds which trend at 040°.

Still farther northwards the formation maintains its predominant lithology of massive mudstone with small arenite inclusions.

Penrhyngerwyn Inlier

The strata are typical splintery unbedded mudstones, in places with sandstone masses (ball-and-pillow structure). The best sections are in Afon Ddu [6680 9397 – 6683 9370]. North of the road bridge, a 6-m gorge displays mudstone with well developed ball-and-pillow structures. This mudstone is separated from unbedded mudstone by a crush zone, dipping at 35° towards 283°, which forms the eastern side of the gorge. The crush zone appears to be continued in the road cutting NE of the bridge (Jones and Pugh, 1935b, p.413; James, 1972, p.301). A massive bed of fine- to very fine-grained sandstone, at least 3 m thick, with local poorly developed parallel lamination, crops out in the centre of the pericline.

CHAPTER 4

Silurian—general account

The Ordovician Bryn-glâs Formation is overlain by a thick succession of well bedded strata which, apart possibly from the basal 1 m, are of Silurian (Llandovery Series) age (Figure 23), and, throughout the district, appear to be perfectly conformable with the underlying rocks. The distinct even multilayering and the parallel cleavage of these Silurian beds contrast sharply with the splintery, phacoidal cleavage and massive nature of the mudstones in the Bryn-glâs Formation. The passage between the two is quite abrupt (Figure 30 p.66). About 1 m above the junction lies an ubiquitous and easily identifiable thin layer of graptolitic mudstone containing *Glyptograptus persculptus*, indicative of the basal graptolite zone of the Llandovery Series.

Silurian rocks crop out over by far the larger part of the

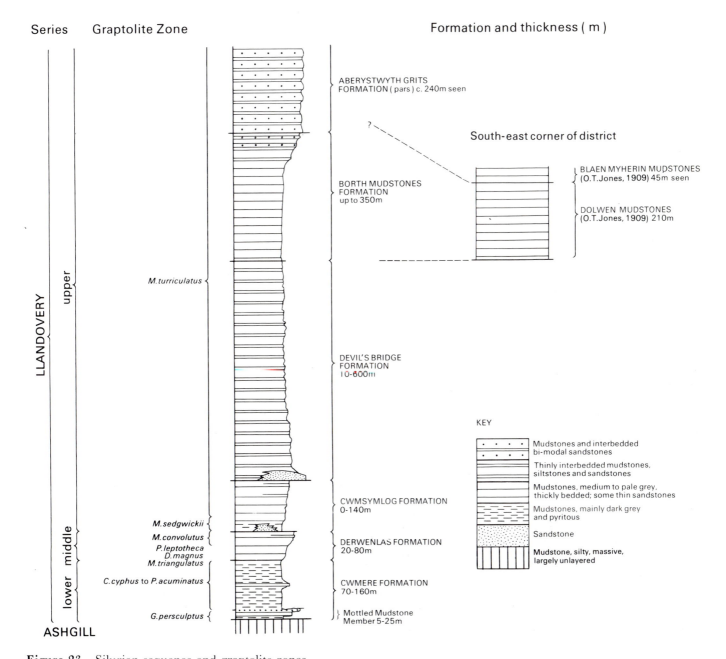

Figure 23 Silurian sequence and graptolite zones

district, extending from the Ordovician inliers across the remainder of the land area, so that the higher parts of the succession lie towards the four corners of the district (Figure 1). Jones (1909) was the first to describe these rocks in detail, and later his initial classification was modified and extended northwards by him and Pugh (1916; 1935a); the lithostratigraphy of the current work is based largely on their scheme, with only minor adaptations and additions. Complimentary descriptions of local geology include those by Hendricks (1926) and Davies (1933) to the south, Jehu (1926) and Pugh (1929) to the north, and W. D. V. Jones (1945) to the east. The biostratigraphy used is based on graptolites and, while the presence of some biozones is not confirmed, there is reason to believe that the Llandovery sequence is complete up to and including much of the *Monograptus turriculatus* Zone. The name '*Monograptus' leptotheca* Zone (Jones and Pugh, 1916) has been used widely in Welsh stratigraphy and is used in preference to the equivalent '*Monograptus' argenteus* Zone (Marr and Nicholson, 1888).

The basic component of the sequence is the sedimentary rhythm illustrated in Figure 2b. The formations are largely natural groupings of like variants of this rhythm. These Silurian formations comprise a sequence which, in general, coarsens upwards, though upward fining is evident over several restricted portions. Sedimentary differences in the sequence are largely the products of changes in the density of the inflowing turbidity currents and in the balance of their products as against pelagic fall-out. Some of these differences are abrupt, others diffuse. Each abrupt difference commonly occurs at the same horizon over the whole mid-Wales basin, and such differences are viewed as stemming from whole-basin, or extrabasinal, influences. Formations based on these differences are thus approximately chronostratigraphic. The more gradational differences in the sequence commonly reveal lateral variation from one subfacies to another, and these probably arose through intra-basinal sedimentation evolution. Formations with such parameters are markedly diachronous, and particularly characterise the upper Llandovery.

CWMERE FORMATION

The dominant rock type of this formation is mudstone; arenaceous detritus is a minor constituent apart from certain restricted intervals. Except for the colour-banded pale and medium grey mudstone of the Mottled Mudstone, the general appearance of the rock is dark grey, and a high pyrite content gives rise to rusty hues on weathered surfaces and in debris.

The name derives directly from the Cwmere[1] Group of Jones and Pugh (1916, p.348) with a minor adjustment at

---

[1] The name Cwmere is that of a farmhouse [6990 9605]. It is named on modern 1:10 560 maps as Cymmerau, but is not named on the 1:50 000 map. A nearby house, Cymerau [6961 9629], is named on the 1:50 000 map but not on 1:10 560 maps. It is this latter house which is used as a reference point in this account. Another locality called Cwmere [683 882], 3 km ESE of Talybont, is also used as a reference point in this account.

the top; the basal strata have been separated as the Mottled Mudstone Member, a name adapted from the 'Mottled Beds' of Jones and Pugh (1916, p.346). The outcrop of the formation circumscribes that of the Bryn-glâs Formation. Because of intricate folding it is widely distributed across the district.

The formation is nowhere exposed in its entirety; indeed even good partial exposures are uncommon. Its variations in thickness are determinable, therefore, only from calculations based on outcrop width. Inevitably these are imprecise. The effects of folding, including repetition in numerous minor folds, and the possible bulk incompetence of the sediments in strongly folded belts, are unknown. Estimates of thickness range from about 145 m in the SE around Ponterwyd, to between 70 m and 90 m in the centre, NE of Bontgoch, and over 100 m in the NW near Tywyn.

The Cwmere Formation includes the *Glyptograptus persculptus* Zone at the base, but the top is slightly diachronous, approximating to the top of the *Coronograptus cyphus* Zone.

Mottled Mudstone Member

The basal 5 to 25 m of the Cwmere Formation are characterised by mudstone which is colour-banded medium grey and pale grey to fawn. The medium grey mudstone is homogeneous, in layers some 4 cm to about 30 cm thick, whilst the paler layers are usually 2 to 6 cm thick though they can be as much as 20 cm. These are unlaminated, but usually liberally flecked with dark spots and streaks which are burrow infillings considered to be *Chondrites*. At the interfaces between pale and darker mudstone, dark grey lenticular phosphate concretions are common (Plate 1a). In some layers bioturbation has extended from the overlying pale mud to disrupt and distort the concretions (Plate 1b). The phosphate was therefore a very early diagenetic accretion initially in a soft condition. Arenite is a negligible component except at the very top of the member. Where present it occurs as thin siltstone partings beneath the medium grey mudstone layers.

Near the top of the member the sedimentary rhythm loses the pale, mottled (bioturbated) mudstone layer, together with the underlying phosphate concretions; these are replaced by dark grey, laminar, graptolitic mudstone. The general appearance of the rock is darker, and weathered surfaces are prone to rusty staining from the high pyrite content. The pyrite is visible as blebs and stringers lying preferentially along some bedding planes, and is commonly associated with thin silty or sandy partings. This 'anoxic' phase of the member occupies the top 3 m or so, and is capped in many places by a 'bundle', up to 2 m thick, of sandstones and siltstones in mudstones. Where these prominent arenite layers are absent, the top 3 m or so are difficult to distinguish from the overlying similar beds, despite the greater thickness of the homogeneous mudstone layers in the former. This arenite 'bundle' has two distinct lithological variants. In some areas it comprises either a single bed of ill-sorted, high-matrix feldspathic sandstone or two sandstones up to 1 m apart. These sandstones are very like those common in the Drosgol Formation, and are of a type which does not recur in the overlying succession of the district; they probably equate with the Cerig Gwinion Grits near Rhayader (Kelling and Woollands, 1969). They indicate a brief recurrence of sedimentation more typical of the late-Ordovician. Beds of this type are most common in the north of the district, immediately SW and west of Machynlleth, but occurrences are also known near Bwlch-glas (p.65) and in the SE near Ponterwyd (p.64). Elsewhere the arenite 'bundle' is of finer grain, and commonly the layers are much thinner than 0.3 m. However, particularly in the SE of the district, NE of Ponterwyd, there are sandstones up to 1 m thick, suggestive of a source to the SE or east of the district. The arenites have not been found north of the Dovey estuary, or on the SW flanks of the Machynlleth Inlier. Near Penrhygerwyn, their position is occupied by two layers of sandstone nodules from 2.5 to 15 cm across.

Strata above the Mottled Mudstone Member

In the main body of the formation the mudstone is of two sorts (Plate 11).

a. *Medium grey mudstone* In some layers this mudstone shows normal grading and in places possesses vague laminae; otherwise it is homogeneous. Pyrite is present near the top of some layers, as pinhead size globules. It is turbiditic.

b. *Dark grey mudstone* (Cummins, 1959; Rickards, 1964, p.436). This mudstone is of the same type as that of parts of the Nant-y-Môch Formation; it is a finely stratified compress of very thin discontinuous films or layers of ovoid aggregates of pale grey silt particles, in a dark grey finer mudstone containing abundant black films and small clasts. Most of this black matter is thought to be graptolitic and algal debris, and thus carbonaceous. Pyrite is a common constituent, again as small globules. It is hemipelagic.

Arenaceous detritus is present mainly as siltstone laminae occurring only at the bases of some of the homogeneous mudstone layers. It commonly contains lenticular aggregates of crystalline pyrite. The bases of these laminae are very sharp, and some are minutely irregular showing very small erosive discordance with the laminae of the dark grey

Plate 11 Cwmere Formation. Layers of hemipelagic and turbiditic mudstone [7371 9012] Nant Rhyddlan. Polished section perpendicular to bedding

mudstone below. Where siltstone layers are absent the base of the homogeneous mudstone is similarly sharp. Interfaces between the top of homogeneous mudstone layers and darker graptolitic mudstone are also sharp. The thicknesses of mudstone layers are usually within the range 1 to 5 cm, while the silt laminae are normally no more than 2 mm thick, though thicker layers occur, even up to 20 cm, mainly in the SE of the district.

A distinct and widespread 'bundle' of sandstones and siltstones occurs at about the middle of the formation, seemingly near the top of the *acuminatus* Zone (Jones, 1909, p.484). Adams (1963, p.373) assigned similar 'grits' near Ponterwyd to the higher '*Monograptus rheidolensis* Zone' (= *Lagarograptus acinaces* Zone), but he provided no graptolitic evidence to support this assignation.

Characteristically, these arenites are rather dark grey, fine-grained sandstones and micaceous siltstones. They are either non-laminated or show parallel planar lamination. Cross-lamination is uncommon, occurring within some arenite layers as thin paler grey laminae, sporadically rippled and, in places, displaying loading structures. The general dark colour is imparted by a liberal scatter of black grains which are probably fragments of organic matter, mainly algal and graptolitic. The paler layers lack this organic matter, though pyrite has developed in them. The difference between the two types of layers is attributed to the winnowing of bed-load by intermittent stronger currents, perhaps contour currents, in an otherwise low energy transport regime. The arenites are responsible for a fairly persistent low ridge, which in many places has been 'blurred' by intensive folding. In the SE, around Ponterwyd, where the total 'bundle' is about 3 m thick, there are individual sandstones up to 1 m thick. Farther north, this topographic feature is well expressed along the western side of the Machnylleth Inlier though the strata are not well exposed. Athough the package there is probably still about 3 m thick, it consists of sandstones and siltstones usually no more than 6 cm thick. The feature is still evident north of the Dovey estuary, but there are no exposures and the individual sandstones are probably even thinner. Nevertheless, the 'bundle' has a very wide extent, Pugh (1929, p.274) recognising it some 35 km to the NE around Llanymawddwy.

Conditions of deposition

The depositional regime which produced the Bryn-glâs Formation ceased abruptly. The small thickness of grey mudstone below the *Glyptograptus persculptus* Band is considered to have been deposited under anoxic conditions, and to represent a brief period of adjustment to the cessation of late-Ordovician regressions before the onset of the transgressive pulses that characterise the Llandovery. In view of the dark colour and pyritous nature of the beds it is assumed that the sediments were inhospitable to an infauna. Disturbance of the bedding may, therefore, be more physical than biological, for instance due to dewatering and the sinking of iron sulphide (now pyrite) through wet mud.

The first of the Llandovery transgressive impulses may be reflected in the dark grey pyritous and abundantly graptolitic mudstone of the *persculptus* Band; bottom conditions were certainly anoxic at the time. The main part of the Mottled Mudstone, however, reflects the arrival into the basin of flushes of turbidite mud and the prevalence of oxic bottom waters which allowed an active burrowing benthos to colonise the tops of turbidites and the ensuing hemipelagic muds immediately after their deposition. The unusually large thicknesses of the bioturbated parts of the rhythms does not necessarily imply that the pauses between density flows were long; it is more likely that the oxicity of the underbottom sediments permitted deep penetration by the infaunas. Persistent phosphatic layers occur in these beds, emphasising the link between the oxic conditions of Division 3ii and the subjacent apatite enrichment (Figure 2b). Sea-water phosphorus was probably 'fixed' by marine organisms, and cycled via their remains into the sediments, where it became mobile in pore-waters (Notholt and Highley, 1979, p.11). This association suggests that the phosphorus was again fixed as apatite at a critical pH value in the pore-waters that became established between turbidity flows; the thickness of the bioturbated mud suggests that this occurred at depths of 1 – 10 cm below the sea-floor.

The transgression resumed probably when the oxicity of the bottom waters of the early *persculptus* Zone (Mottled Mudstone) was replaced quite suddenly by anoxic conditions just below the top of the zone, and this condition persisted throughout the deposition of the rest of the Cwmere Formation. During this time, low density (extremely 'distal') turbidity currents introduced mainly mud grade sediment, but the dominant sediment was hemipelagic mud. Turbidite silt or fine sand occasionally extended to the district, and for a short time in the middle of the Cwmere 'episode' was dominant. The turbidites are thickest in the SE and wane to the north and NW, indicating a source to the east or SE: cross-lamination observations and a few sole markings support this view. The source might have been the shelf to the east of Rhayader (Kelling and Woollands, 1969, pp.276–280). Within the hemipelagic sediments the only current indicators are graptolite stipes, and in a few exposures a NNE–SSW alignment has been noted, suggesting the bottom waters were flowing either to the NNE or to the SSW, parallel with the basin slope.

DERWENLAS FORMATION

The formation consists dominantly of a thinly multilayer pile of turbidite mudstones and fine arenites. Colours range from dark grey near the base, to pale grey and greenish grey near the middle, and back to a darker grey towards the top.

There is a lithological passage from the Cwmere Formation up into the Derwenlas Formation. Dark grey rusty-weathering mudstones give way upwards to less dark, more blocky, and slightly coarser mudstones. These latter are resistant to weathering and commonly form a bold scarp with the formational boundary approximately at its foot. The lithological change roughly coincides with the top of the *cyphus* Zone but can lie either slightly above or slightly below this horizon. Apart from the inclusion of the *Monograptus triangulatus* Zone, the formation is the lower part of the Derwen Group of Jones and Pugh (1916), and their type section is retained (Figure 38). There is no obvious stratigraphic break within the formation, and the zones

Monograptus triangulatus to *M. convolutus* appear to be fully represented.

The Derwenlas Formation scarp is best displayed on Moel Golomen [700 876] and north of Moel-fferm [708 891], and between Melin-y-cwm [691 949] in Cwm Einion and Derwenlas [719 991]. Other important outcrops are those between Brynbras [748 797] and Disgwylfa Fâch [737 838], east of Cwmsymlog, where sandstone occurs in the middle of the formation, on the eastern flank of Cerrigyrhafan, and in the strike-faulted ground of the forests around and south of Esgair Fraith. North of the River Dovey, the railway cuttings south and east [637 964 – 647 968] of Coed y Gofer afford good sections, and the outcrop is well displayed northward to Tyddyn-y-briddel [640 984]. One of the best complete sections through the formation is alongside the old coach road at Derwenlas [719 991], just north of the district (Jones and Pugh, 1916, pp.354 – 356) (Figure 38).

The Derwenlas Formation possesses greater variety across the district than does the Cwmere Formation. This diversity is accounted for by the genesis of the dominant deposits. They are mainly turbidites, with strongly preferred directional inputs. The formation is thickest (100 m) in the south around Goginan, and thinnest (20 to 35 m) to the NE around Cerrigyrhafan and near Rhiw-gam [797 949]. These thickness variations reflect lithological differences, for the proportion of arenite varies directly with the thickness. The highest proportion of arenite thus coincides with the area of greatest thickness near Goginan, and the thin sequences are dominantly mudstone.

The thickest individual sandstones, up to 30 cm, and the thickest exposed arenite dominated sequence (37 m) are in Afon Stewy (p.75), though a borehole [7302 7833] in the Rheidol valley about 3 km SSW of Ponterwyd proved an even greater thickness (45 m) of sandstone. Near Ponterwyd, arenites are thinner, up to 22 cm, and sparser, dominating only up to 18 m of the sequence. Northward of Cwmsymlog the arenites are individually thin and of finer grain, though for several kilometres they dominate many metres of the formation, being thinly interbedded with grey mudstone and commonly uneven to lenticular and cross-laminated. These arenites occur within the *convolutus* Zone, diagnostic graptolites having been found both below and above them. This position is different from that of similar arenites in the Ystrad Meurig area (Cave, 1979, p.522). Although barely 10 km to the south, the arenites there are lower in the sequence, lying within the *Diplograptus magnus* or *Pribylograptus leptotheca* zones (Jones, 1938, p.lxxxiii).

Within the formation there are several intervals, usually under 2 m thick, of dark grey graptolitic 'shale'. It is from these graptolitic bands that most of the graptolites have been collected. Jones (1909, p.490) recorded similar bands in the Rheidol Gorge [759 797], giving to each the name of its most characteristic graptolite. Similar bands were observed later (Jones and Pugh, 1916) in the Derwenlas section just north of the district, and it was considered that the bands of each section were identical. This is probably generally true, but some sections contain more graptolitic shale bands than others, so not all these bands are continuous across the whole district. It is possible that some graptolite bands split laterally, while others are restricted in extent.

Benthic fossils, mainly brachiopods, have also been recorded (Appendix 3, locs. 1–6, ?7–8; also Challinor, 1928b), particularly in lenticular concentrations at the bases of turbidites.

Conditions of deposition

At the base of the Derwenlas Formation, turbidity currents were already affecting deposition, for the mudstones of the *triangulatus* Zone are noticeably paler and coarser than those of the *cyphus* Zone. Concurrent with the increasing dominance of turbiditic sedimentation was a change to more oxygenated bottom conditions (Cave, 1979, pp.75–76). The sequence can be interpreted as an outer-fan, or a lobe, prograding into the basin.

The Derwenlas Formation in this district has a high arenite content in the *convolutus* Zone, though farther south, at Ystrad Meurig, this contrast appears earlier (p.45). The locus (or depocentre) of thick arenaceous deposition thus shifted northwards with time (Figure 24a), and in the *convolutus* Zone seems to extend from east of Aberystwyth, where the arenites are thickest and coarsest, north-westwards towards Aberdyfi, though both thickness and coarseness wane distally in that direction. This indicates that sediment was transported towards the west or NW. Sediment input to the basin, therefore, seems to have been in the region of the slope near Rhayader where the contemporary submarine canyon described by Kelling and Woollands (1969) was located.

In the Aberdyfi area the formation, although largely argillaceous, is still fairly thick (c.50 m); thicker for instance than at Derwenlas to the east (c.37 m), and in the Afon Lluestgota area to the SE (c.20 m). Such a pattern of thickness can be explained if the main flow of turbidites from

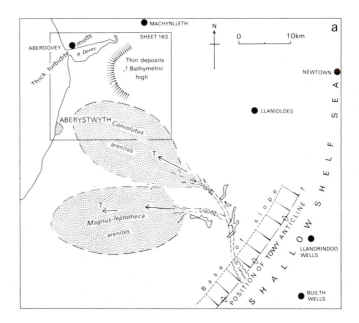

Figure 24a. Bodies of arenaceous turbidites, Derwenlas Formation

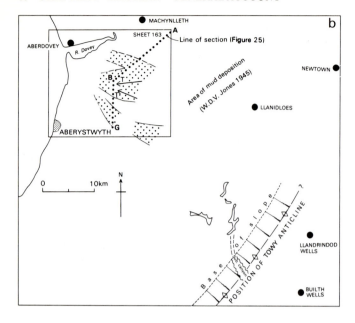

Figure 24b. Channel sandstones at the base of the Devil's Bridge Formation

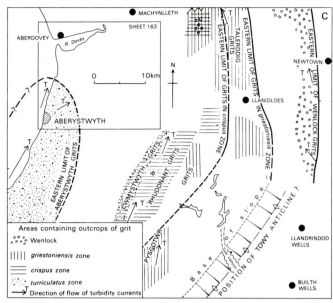

Figure 24c. The encroachment eastwards during the mid-Silurian of northward delivered turbidite sands

the SE entered the district along its southern margin, say near Goginan or Ponterwyd, passed south of the Lluestgota dropping most of its sand fraction there, but carried a large body of mud farther to the NW (Figure 24a). Clearly the energy of these turbidity currents expired between Aberdyfi and the Lleyn peninsula, for little detritus reached this latter area (Baker, 1981).

In many places the topmost beds of the formation are more argillaceous, more pyritous, and darker in colour. These indicate that less oxic conditions of deposition recurred, and that the area had become either more distal or, more probably, that the volume of input had waned and so its depositional 'lobe' or 'fan' had contracted. The changes are probably basinal indicators of the marine transgression that was active on the shelf at about this time.

CWMSYMLOG FORMATION

This formation is again varied in composition, but essentially consists of very thinly colour-banded mud-dominated thin turbidite units, the arenites commonly being less than 5 mm thick. Lateral facies changes mean that the formation locally fails. It is named from the hamlet [698 838], some 12 km inland from Aberystwyth, in an area where it has its maximum development of about 140 m. There are two major forestry road sections in this type area which display most of the lithological variants within the formation. The type-section [6845 8408 – 6868 8415], NE of Pen-bont Rhydybeddau, displays all but the basal few metres (p.84); the second section [7023 8167 – 7100 8183], at Pen y Graig-ddu, shows a complete sequence (p.83).

The sequence is composed essentially of fine-grained mudstone; arenite is important only locally. Apart from the basal few metres, which in most areas are very dark grey graptolitic mudstones, (the '*Monograptus- sedgwicki* shales' of

Jones and Pugh, 1916, p.359), it is largely colour-banded pale grey and medium grey to bluish grey. Colour variants are locally significant. Over most of the district the base of the formation is clearly defined, with a marked contrast between the basal very dark grey mudstones and the underlying paler mudstones of the Derwenlas Formation. Where the dark mudstones are absent, as for example in the central part of the district between Nantperfedd [707 868] and Blaen-Ceulan [707 902], the base of the formation is more difficult to discern. In many places, as near Ponterwyd in the SE and around the Dovey Estuary in the NW, the Cwmsymlog Formation is between 20 m and 30 m thick. In the centre of the district, as for instance near Cwmere [6835 8820], there are places where the formation is absent; yet close by, to the SE at Cwmsymlog and to the west in the Glan-fred Borehole, it is at least 140 m and 105 m thick respectively. In the NE, the formation is moderately thick, up to 65 m, but here the top of the formation is demonstrably diachronous (Figure 25).

The binary colour banding which is common, and is probably the most distinctive feature, is an arenite-muted representative of the tripartite sedimentary unit common in the district (Figure 2b). Individual units are commonly under 2 cm thick and, though thin siltstones and sandstones are present in the thick southern sequences, many are no more than 3 mm thick. Evidence of contemporary disturbance is common and mudstones of different colours are mixed in places (Cave, 1971, plate 1); flame structures protrude from one layer up into the next; concordant, and sub-concordant to curving, glide planes duplicate or remove bands, and there are mainly small or 'mini' normal step faults together with low angle rotational faults. Listric fracture surfaces are very common, and there is a more penetrative even cleavage than in the underlying Derwenlas Formation. However, major disruptions of the bedding, such as slumps, have not been observed. As far as can be seen, and excluding post-depositional movement, the

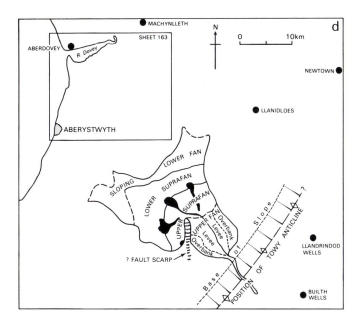

Figure 24d. Plan of the Navy Fan and Navy Channel superimposed on a map of mid-Wales

banding is like that of other formations, of even thickness and extensive. The presence of thin arenites characterises the top part of the formation, particularly in the south, and they are commonly about 1 cm thick, with as many as 15 per 0.3 m of strata. The main lithological variants (informal 'members') within the formation are colour variants of the mudstones; there is also an arenite facies in the central part of the district. In most places the base of the formation approximates to the base of the *Monograptus sedgwickii* Zone but the upper parts belong to the *Rastrites maximus* Subzone of the *M. turriculatus* Zone.

Dark grey mudstone 'member': ('*Monograptus-sedgwicki* shales' of Jones and Pugh, 1916, p.359)

The basal few metres of the formation is, in most places, very dark grey graptolitic mudstone. This contains much pyrite as very finely disseminated globules and as large accretions, 1 to 2 cm long, commonly aligned preferentially on bedding planes. The sedimentary units are normally arenite free and dominated by graptolitic hemipelagic mudstone.

The mudstone acts as a good stratigraphical marker. By reason of the topographic hollow which generally coincides with its outcrop, it can be traced easily, and the base of the formation in such places is thus readily mappable and well defined. Its thickness varies in a similar way to that of the formation as a whole. Thickest near Cwmsymlog in the south centre of the district, at about 25 m, it is more commonly between 2 m and 5 m thick, as for example in the north and in the SE.

Green mudstone 'member'

Green mudstone occurs at or near the top of the formation in many places but is best developed in the type-area. Most of it exhibits all the banding characteristics of the normal grey or bluish grey mudstone. The same thin silty laminae and

minor disturbances of the bedding are present, the only difference being that the bands are dark and pale green. There is, however, a bed of dull green mudstone which in the south occurs within the 'member' but in places in the north is its sole representative. Within this bed banding is absent or obscure, so the mudstone is tough and rather massive, producing a prominent positive topographical feature in places. While a cleavage is well developed in most of the formation, in this mudstone it is poorly developed and uneven or anastomosing, not unlike that in the Bryn-glâs Formation. The combination of this characteristic and the lack of banding suggests penecontemporaneous disturbance (mass movement or bioturbation).

The thickness of the mudstone is not constant across the district. It is uncommonly thick (80 m) at Pen y Graig-ddu and probably reaches 100 m around Cwymsymlog, but thicknesses in the order of 5 m are more usual and there are many places where it seems to be absent. Jones and Pugh (1916, pp.360–361) noted the presence of the unit, likening it to the 'green flags' seen in the lower (middle Llandovery) part of their Derwen Group.

This 'member', or more specifically the massive mudstone within it, makes another useful marker bed, particularly in the NE, whereby the diachronous nature of the top of the formation can be demonstrated. Near Esgair Fraith [745 915] green mudstone, and in particular the 'massive' bed up to 3 m thick, occurs just below the Devil's Bridge Formation, but traced to Rhiw-gam [7973 9488], where it is up to 10 m thick, up to 35 m of grey mudstones progressively appear above it. These beds are not an intercalated wedge but are considered to represent a lateral passage from the Devil's Bridge Formation.

Red mudstone 'member'

This 'member' is of very limited extent, occurring in the type section and about 500 m to the WSW. In these areas it occurs about 4 m and 11 m respectively below the top of the formation, but apart from its dull red to purple colour it has the same characterstics as the green or grey banded mudstones. In the type section it is 7.3 m thick.

Arenite facies

Between Craigypistyll and Cwm Ceulan the Cwmsymlog Formation is thin or absent, probably by reason of facies change rather than hiatus. The equivalent strata are dominated by thin greywacke interbeds—mainly siltstones and fine sandstones with some medium- to coarse-grained sandstones. If the vertical limits of these beds could be fixed and mapped with certainty it would be possible to consider this facies as a discrete body and as another 'member' of the formation. In practice the beds are too thin and too closely similar to both the topmost Derwenlas strata and the lowermost Devil's Bridge Formation for such an approach to be practicable. They can, however, be identified by both the coarseness and ill-sorted nature of some of the arenites and by their fossil content and have been reported on by Strong (1979). Even so, in places they have been described together with the Derwenlas Formation because the two cannot be clearly demarcated (p.76). In some places the base of these

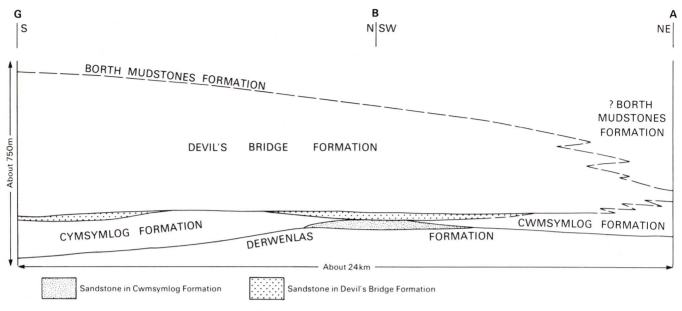

Figure 25 The Cwmsymlog Formation with its sandstones and the relationship to the overlying Devil's Bridge Formation and its basal sandstone (G, B and A are shown in Figure 24b)

beds is marked by thicker, parallel-laminated sandstones which have yielded *M. convolutus*, *M. lobiferus* and *M. sedgwickii*. These fossils indicate that these sandstones are contemporaries of the dark grey mudstone 'member', and possibly also the top beds of the Derwenlas Formation.

It was found (Strong, 1979) that *Tc-e* and *Td-e* of the Bouma sequence (Figure 2a) predominated in these arenaceous beds, and by Walker's (1967, p.30) standard this indicates a 'very distal' depositional environment. Consequently, it might be expected that the facies would exhibit wide uniformity but this is not so. The facies is restricted to a SE–NW tract, and Strong showed that the incidence of current-ripple lamination is higher at Bryn Brith [705 875], Moel Golomen, and Bwlch yr Adwy [719 869] than it is farther north at Cwm-byr or south at Craigypistyll [715 857]. There was thus a higher flow regime in the centre of the arenaceous tract than there was nearer its margin, leading Strong to the conclusion that there was a shallow depositional trough through the area in which the general current flow was longitudinal from the ESE. In detail his evidence records a slight splay in the current directions, and this he interpreted as the result of lateral spillage from the high energy tract of the 'channel' on to lower energy areas.

Conditions of deposition

The rocks at the base of the formation (*sedgwickii* Zone) record a change in the overall pattern of sedimentation, as a phase of 'black' mud deposition widely affected the basin. This produced dark grey graptolitic mudstone of much greater thickness than the graptolite bands of the Derwenlas Formation, and is thought to represent a major resurgence in the Llandovery marine transgression of the shelf discussed by Ziegler and others (1968, p.774) and Bridges (1975, pp.84–92). The consequences of this resurgence were a sudden waning of the input of detritus from the shelf and anoxicity in the bottom waters of the basin. The basin sediments were thus largely laminated hemipelagic muds, in which carbonaceous debris was preserved and ferric compounds were reduced to sulphides. However, even these sediments possessed rhythmic banding, which stemmed from repetitive intercalations of non-laminar homogeneous mud, thus indicating the continued influence, albeit weak, of turbidity currents. The Rhayader canyon of Kelling and Woollands (1969) was still active at this time, and so it is logical to interpret the arenaceous facies found east of Talybont as a distal component of the fan of detritus spewed forth by this canyon. Strong's assessment (1979) of this facies as a 'distal' channel-fill would accord with this thesis.

Above the 'dark grey mudstone member', and thus near the base of the *turriculatus* Zone, the paler banded mudstones are also thought to be largely turbiditic, but deposited under more oxic conditions. During the marine transgression of the shelf, the head of the Rhayader canyon had been severed from its catchment of coarse detritus (Kelling and Woollands, 1969, p.280), but it may not have been abandoned as a conduit for sediments of finer grain. The source of the mud turbidites of the Cwmsymlog Formation is not certain, though their abrupt increase of thickness in the southern areas opposite the mouth of the Rhayader canyon is suggestive. In general, however, this facies is comparable with the arenite starved 'Facies D' of Mutti and Ricci Lucchi (1972). It is widely represented at this and later *turriculatus* times e.g. as the Rhayader Pale Shales in the east, which may be slope or base-of-slope deposits, and in the Cwmsymlog Formation which was deposited farther into the basin. Without identifiable and clearly channelised inner, middle and outer fan morphologies at this time it is difficult to categorise these deposits as channel levée or interchannel. If, therefore, a fan model is at all applicable, then the highly mud-dominated sediment input and the low density of the current system produced one different from that described by Mutti and Ricci Lucchi. In comparison with many modern submarine fans, the distance of these 'distal' Facies

D turbidites from their possible source canyon at or near Rhayader is small, though modern, yet small, submarine fans are known e.g. the Navy Submarine Fan (Normark and others, 1979).

Colour variations in this formation are due to the state of oxidation of the contained iron, and this state is considered to be more a result of the conditions under which the sediment was deposited than the state of oxidation of the detritus at source (Cave, 1979, p.524).

DEVIL'S BRIDGE FORMATION

The rocks of this formation are turbidites of the Bouma *Ta-e* pattern (Figure 2a), though mostly base-missing. They produce a multilayer arenite/mudstone rock in proportions varying considerably between about 2:1 and 1:6 in the south, and with even higher proportions of mudstone in the north (Plate 12). In many places there is a division, up to 18 m thick, of thick arenites (individual beds up to 30 cm) with mudstone interbeds at the base of the formation. These are the 'massive grit-bands' of Jones and Pugh (1916, pp.360, 365; 1935a, p.278). Where these are present there is a closely defined base to the formation, but elsewhere the main criteria for distinguishing it from the underlying Cwmsymlog Formation are the thinness and close spacing of the arenites in the latter. Confusion can arise when these Cwmsymlog Formation arenites are thicker than normal, especially where, as in the Glan-fred Borehole [6305 8812], there is an absence of lateral control. No pattern in the differences of arenite inter-bed thickness has been observed within the main part of the formation other than several weakly defined 'upward-thickening' cycles, possibly progradational, in the Nant Rhŷs section (p.91). In the NE the thicker arenites are largely concentrated into bundles, while other parts of the formation are largely mudstone. It appears that the Devil's Bridge Formation not only passes upwards into the Borth Mudstones Formation (Dolwen Mudstones in the SE), but also laterally into them in the NW of the district.

The formation is an adaptation of O.T. Jones' Devil's Bridge Group, which he described as 'The group of mudstones with regular alternations of thin grits' (1909, p.516) succeeded by the Myherin Group, 'a group of mudstones almost devoid of grits'. It was first defined in three traverses, one of which follows the Rhayader road from Devil's Bridge, where mudstones with thin sandstones crop out between Devil's Bridge [742 770] and just east of The Arch [766 755]. This could properly be termed the type section; it is south of the district and has not been re-examined. Nearby good sections within the Aberystwyth district are those in Nant Rhŷs [7874 8158 – 7920 8148] (p.91), about 1.1 km SSE of Llysarthur, and in new road cuttings [7146 8103 – 7187 8128] (p.92) about 3 km east of Goginan.

The most extensive outcrops lie towards the south of the district between Talybont and Ponterwyd, and in the southeastern extremity. A long, structurally complex, outcrop through Nantperfedd [707 868] and Bryn Mawr [723 904] links this southern area with a further large area of outcrop in the extreme NE. Other outcrops occur on the NW flank of the Machynlleth Inlier and north of the Dovey.

The formation is thickest at about 600 m around Goginan,

in which area an informal tripartite division has been recognised (p.49). In the SE it thins to some 460 m, and in the NE to 300 – 340 m, though between these areas there is a more radical diminution to some 160 m towards the Plynlimon Inlier. In the NW there is marked thinning which, coupled with a lateral passage into the Borth Mudstones Formation, reduces it to about 10 m at the northern margin of the district (Figure 25).

The arenite interbeds are generally medium or pale grey greywackes, mostly fine sandstones, very fine sandstones and siltstones. The great majority are current-ripple-laminated *Tc* and parallel-laminated Bouma *Td* arenites (Figure 2a). Graded *Ta* and *Tb* intervals are confined to the basal beds of the formation. The dominant constituent is quartz, with minor feldspars, micas and angular to subangular rock fragments. The thickness and separation of the arenite layers remain very constant across even the largest of exposures. There is no evidence of channelling or channel-fill sequences except perhaps locally at the base of the formation. Scour-casts are uncommon, except again in the sandstones at the base, but there is an abundance of trace fossils in the form of arenite casts on undersides. Many of these were either trails on the mudstone substrate when it formed the boundary layer, or burrows along the mud/silt or sand interface. Forms such as *Dictyodora* and *Nereites* are common.

The mudstone interbeds are non-laminar, homogeneous, pale to medium grey and, in places in the lower part of the formation, green. Mudstones in the higher parts of the formation tend to be darker grey: commonly a pale grey or fawn rim is present at the top of each individual mudstone. These mudstone tops are non-laminar, but some possess a scatter of darker grey blebs, probably the effect of bioturbation e.g. *Chondrites*. For the most part, therefore, the sedimentary unit of the formation is a Division 3ii mudstone (Figure 2b) in which there are concentrations of phosphate at the 2/3ii interface. This phosphate is commonly evident as dense and hard black concretions of apatite in thin lenticles. On well weathered surfaces these acquire an ash-grey to white skin.

Between Goginan and Cwmerfyn an ill-defined tripartite division of the formation has been recognised. The green colour, so well displayed at the top of the underlying Cwmsymlog Formation of this area, persists into the Devil's Bridge Formation and characterises its lowest subdivision here. The packet of thick arenites, the base of which defines the base of the formation, is normally present at the bottom of this subdivision, which has a maximum thickness of about 190 m. In the middle subdivision, the mudstones are darker grey and greenish grey; in comparison with the underlying mudstones they are better cleaved, and have weathered surfaces stained characteristically rusty-brown. The arenites are slightly thinner, 1 to 3 cm thick, and contrast in colour with the mudstones, so appearing pale grey. The subdivision is some 100 m thick at maximum, but its outcrop is even more restricted than those of the subdivisions above and below. As a consequence there is a binary subdivision (in which the lower and middle subdivisions are combined but the upper subdivision remains separate) in areas on an arc from Ponterwyd through Craigypistyll [712 857] to the Leri valley [662 889] east of Talybont. The uppermost subdivision is up

Plate 12 Devil's Bridge Formation, mainly mud turbidites with arenites subordinate [7639 9842]. Coed Rhiw-lwyfan

to 310 m thick, comprising dark grey, well cleaved mudstone, with arenite interbeds which are also rather dark grey. The latter are slightly finer grained than those below, being mainly siltstones which show parallel laminae and which probably represent the *Td* interval.

Fauna

The Devil's Bridge Formation has yielded graptolites widely but generally sparsely, while brachiopods and other shells have been obtained only from a few exposures in the south (Appendix 3, locs. 12–18).

The graptolites indicate the Zone of *M. turriculatus*, the bottom part of which, the *maximus* Subzone, is evident in the underlying Cwmsymlog Formation. The lower parts of the Devil's Bridge Formation yield collections usually without *Rastrites maximus*, thus indicating (Rickards, 1976, p.165) the upper part of the *turriculatus* Zone. However, *R. maximus* has been collected from several places high in the formation. It seems that *R. maximus* is absent where sandstones are most common and where they host the graptolite remains. That happens to be the lower part of the formation; the higher parts contain layers or films of dark grey, non-turbiditic

mudstone which are graptolitic, and it is from these that *R. maximus* has been obtained. It seems that, in the Devil's Bridge Formation, the absence of *R. maximus* from a collection of graptolites is an untrustworthy indication of position within the *turriculatus* Zone. *R. maximus* had a delicate stipe and very slender thecae; it might not have survived transportation in a turbulent density flow, and almost certainly would have been hydrodynamically dissimilar from most of the contemporary species. Though selective transportation may thus explain the absence of *R. maximus* in the turbidite sandstones of the lower part of the formation, another factor may be unrepresentative collections. The delicate nature of the skeleton of *R. maximus* resulted in its being preserved only as fragments; in the relatively coarse lithology of a turbidite sandstone the isolated thecae of *R. maximus* can be very difficult to discern.

The brachiopods are generally small, poorly preserved, and present in turbidite sandstones. They are, therefore, presumably exotic. Other animal remains are present with them but are very sparse; they include crinoid and trilobite fragments. The stratigraphical implications of the brachiopod fauna are imprecise, but compatible with the *turriculatus* Zone.

Conditions of deposition

Deposition of the Devil's Bridge Formation appears to have involved little modification of the basin inherited from the Cwmsymlog Formation. The turbidites are coarser and thicker than those beneath, reflecting an increased input of detritus. It appears likely that the marine transgressive pulse which commenced at the start of the upper Llandovery was diminished, or even reversed initially, and that the supply of shelf detritus to the basin as a whole thus became greater.

BORTH MUDSTONES FORMATION

The formation is dominantly turbidite mudstone, commonly in thick beds up to 40 cm thick and normally with very subordinate thin arenite layers. In many places the mudstone layers possess pale grey upper rims as seen commonly in the Cwmsymlog Formation.

Keeping (1881, p.145) considered that the Aberystwyth Grits show a complete passage into the (subjacent) 'Metalliferous-slate group' (mainly equivalent to the Devil's Bridge Formation) from rocks with thick and abundant sandstones into rocks with thin and few sandstones. Jones (1909, p.520) recorded a comparable observation, but most previous authors have treated the mudstones exposed in the foreshore and cliffs west of Borth as part of the Aberystwyth Grits (Jones and Pugh, 1935a, p.280; Price, 1962, p.543; Crimes and Crossley, 1980, p.822).

The name Borth Mudstones was first used to describe the mudstones overlying the Devil's Bridge Formation in the Glan-fred Borehole [6305 8812]; their base was defined at a depth of 31.39 m (Cave, 1975a, p.7). The top of the formation is defined by the conformable base of the Aberystwyth Grits Formation on the foreshore [5961 8769] at Craig y Delyn (Harp Rock) (Cave, 1976, p.24). The formation thus consists of those rocks in the west of the district, mainly mudstones, which succeed the Devil's Bridge Formation but are not Aberystwyth Grits. Its thickness is difficult to determine because of poor exposure, but it appears to thicken northwards, probably partly due to lateral passage from the Devil's Bridge Formation, and has a maximum thickness of at least 350 m.

The outcrop of the Borth Mudstones provides a 2.5 km border on the eastern side of the Aberystwyth Grits outcrop from Blaen-geuffordd [648 805] northward; it broadens to cover most of the area under Cors Fochno (Borth Bog), then reappears north of the Dovey in a 2 to 3 km wide belt around Aberdyfi. The formation is typified in the foreshore between Upper Borth [6073 8888] and Harp Rock.

The mudstones are medium to dark grey, homogeneous and almost wholly turbiditic, belonging to the Bouma *Te* interval. Traces of silty parallel laminae are present in the lower parts of some beds which may belong to *Td*. The mudstone layers are of differing thicknesses, mostly between about 8 and 40 cm, with the thinner bedding occurring in the lower parts of the formation, as in the Glan-fred Borehole. Most are of Division 3ii (Figure 2b), but there are bundles where Division 3i is present and these are taken to represent brief periods when the basin waters were anoxic and the boundary layer unsupportive of benthic fauna. Graptolites have been obtained from these 3i layers. Cone-in-cone concretions commonly occur in the mudstones, usually at about the middle of the mudstone layer; they rigidly follow one particular level within any one layer. In shape they are ellipsoidal, with the a-axis and the b-axis lying along the plane of bedding and with the cones aligned with the short c-axis. In content they consist largely of mudstone with an enrichment of silica and ferruginous carbonate (Denaeyer, 1948, pp.382–388). Like those of the Aberystwyth Grits the mudstones of the Borth Mudstones are soft. When wet their surfaces can be rubbed into a muddy smear, which is not true for mudstones from lower formations. This difference may be stratigraphically based: equally it can be argued that it is geographical and related to the possibly lower metamorphic grade in the western part of the district.

The subordinate arenites are mainly fine-grained sandstones and siltstones developed as base-missing Bouma sequences *Tcd*. They are medium to pale grey, commonly parallel-laminated and cross-laminated. Convolute lamination is common in the thicker sandstones. Thicknesses are usually between 0.3 cm and 2.5 cm, and rarely more than 8 cm. Some sandstone layers in the Glan-fred Borehole are 'loaded' into the underlying mudstones, and some of the thinner layers have been completely disrupted into 'loadballs'. The regular, small-scale puckering of sand layers that is common in the arenitic layers in other thinly bedded formations of the district also affects some of the arenites of the Borth Mudstones.

Several prominent beds of a quite different nature from the rest of the formation lie towards the top of the sequence, westwards from near Careg Milfran [600 885]. They are comparatively thick (up to 30 cm) fine-grained silty mudstones, almost homogeneous, rather dark grey, and not well differentiated into a Bouma sequence (Figure 2a) though a basal layer of parallel-laminated siltstone or fine sandstone (*Td*) is present in some. In some the homogeneous portion contains small flakes of mudstone, presumably rip-ped from the substrate and buoyed by their low density into the upper parts of the flow, and paler grey, ill-defined blebs or clasts of siltstone. These beds are the earliest of the Harp Rock Type beds (p.55) which are more numerous in the basal parts of the Aberystwyth Grits and form the three prominent beds of Harp Rock (Plate 13). The 'clasts' within them become more obvious higher in the sequence as their numbers and sizes increase.

Fauna

In contrast to the graptolites from the Devil's Bridge Formation those from the Borth Mudstones were not obtained from the arenaceous bases of turbidites. Differences between the modes of transportation in the two formations may, therefore, be responsible for differences in the faunas not inherent in the living communities (see p.50).

Two major collections of graptolites have been made from the Borth Mudstones. One came from an outcrop near Careg Milfran (p.95), the other from the Glan-fred Borehole (Table 1). Both indicate the *turriculatus* Zone, but *R. maximus* is present only in the borehole, and there only in the bottom 2.05 m of the formation. The absence of *R. maximus* might be taken to indicate that most of the Borth Mudstones lies above the *maximus* Subzone. Nevertheless, since *M. halli*, *M.* cf. *sedgwickii* and *R. maximus* occur in the overlying Aberystwyth Grits (Table 2) there is clearly a strong doubt about the validity of this conclusion, and it seems more likely that the whole of the Borth Mudstones belongs to the *maximus* Subzone.

Stratigraphical relationships and sedimentation

While there can be no doubt of the stratiform superposition of the Borth Mudstones above the Devil's Bridge Formation in the Glan-fred Borehole, nor of the Aberystwyth Grits above the Borth Mudstones on the foreshore, the survey has shown that there is strong diachroneity within the Borth Mudstones. They appear to be the basinal mud-rocks distal to the arenaceous detritus of both the Devil's Bridge Formation, with an input from an easterly quarter, and part of the Aberystwyth Grits, with their input from the south (Figures 26 and 27). Unfortunately the current knowledge about the graptolite faunas of all these rocks is inadequate, and refined correlation is impracticable. For instance, the Borth Mudstones and the Grogal Sandstones (Anketell and Lovell, 1976, p.102) are the formations immediately below the Aberystwyth Grits at their distal and proximal ends respectively. The former belongs to the *maximus* Subzone; the latter has yielded no fauna but lies between beds yielding *M. sedgwickii* and *M. turriculatus*. There is thus no faunal substance for the view that the base of the Aberystwyth Grits is younger in the north than in the south (Jones *in* Wood and Smith, 1959, p.191). Nevertheless, it is quite possible that the lobe of arenaceous detritus which is the Aberystwyth Grits failed to extend or prograde northward much beyond the northern limit of the outcrops near Borth; if so, the Borth Mudstones farther north may be partly contemporaneous with the lobe (Figure 27).

This interpretation involves the concept of a bimodal source for the Borth Mudstones. Much of the mudstone and the thin arenites comparable with those in the Devil's Bridge Formation probably had their origins in the basin margin to the east and SE. There was, however, also the growing influence of the basin margin to the south and SW, and this was contributing mud and, towards the top, arenaceous layers comparable with those at Harp Rock in the Aberystwyth Grits. There is convincing evidence of palaeocurrent alignments in the orientation of the trace fossils *Palaeodictyon* and *Squamodictyon*. Their parent organisms inhabited the sub-surface layers of sea-bottom mud, so current influence upon their orientation must have been from inter- or end-turbidity currents immediately prior to the succeeding turbidity flow: however their disposition can indicate only current alignment and not direction. The recorded west to east current directions from mudstones between Wallog and Borth (Crimes and Crossley, 1980) draw upon the historic belief (Jones, 1938)

that the Ystrad Meurig Grits represent detritus fed from an Irish Sea landmass to the west. Palaeocurrent indicators in the Ystrad Meurig Grits do not, however, underpin this belief (Cave, 1979, p.552), so there seems to be no *a priori* reason for interpreting an E–W alignment of trace fossils as indicating currents from the west; it is at least equally plausible to look upon this alignment as a result of currents flowing westwards from the eastern margin of the basin. There is, however, a single record of flute casts indicating flow to the east in the Borth Mudstones (p.95). It is possible that a lobe of turbidite sandstones preceded the Aberystwyth Grits and extended to the west; the flute casts may have been the product of a turbidity current diverted off the eastern flank of that lobe. Further observational data are needed, however, before any broad interpretation can be put forward with confidence.

Table 1   Graptolites from the Glan-fred Borehole

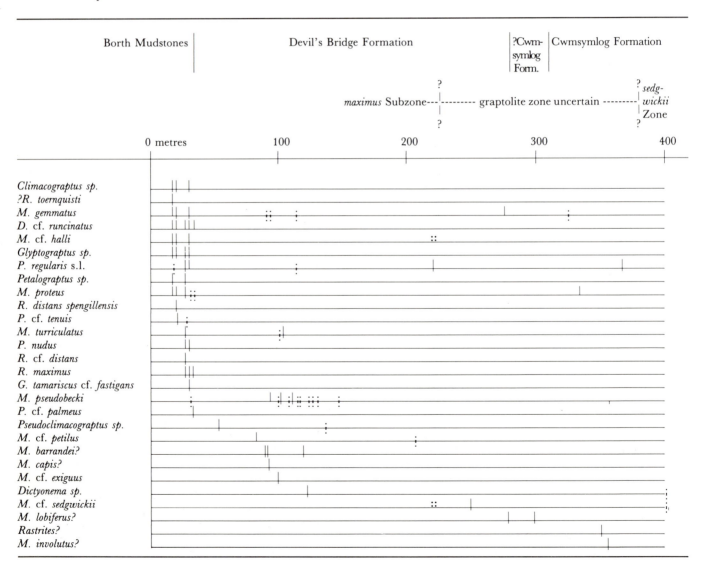

___|___ = present   __|__ = cf. species   ___ = aff. species   __¦__ = ?identification   __::__ = ?*M.* cf. *halli* or *M.* cf. *sedgwickii*

Table 2   Graptolite faunas in the Aberystwyth Grits

Localities	1	2	3	4	5	6	7	8	9	10	11	12	13	14	15	16	17
Climacograptus sp.										?							P
?Diversograptus ramosus										?				?			
D. runcinatus		X												P			P
Glyptograptus sp.		P								P							
?Lagarograptus tenuis	?																
?Monoclimacis galaensis												?					
?Monograptus acus																?	
M. barrandei	P		cf.														
M. cf. *communis*									cf.								
M. exiguus											P		P		P	P	
M. cf. *gemmatus*														cf.			
M. halli		X	cf.	X			?	?	P	P		cf.			X		cf.
M. knockensis														P			
M. lobiferus?								X									
M. marri		X								P		P	P	P	P		P
M. proteus	?	X	P				?	P		cf.				P	P	X	P
M. cf. *pseudobecki*			cf.			?								?			
?M. rickardsi																?	
M. cf. *sedgwickii*	cf.	cf.	X						?						X	?	X
M. turriculatus										cf.	cf.	X		P			?
Petalograptus palmeus			P														P
P. sp.		P													?		
Pristiograptus cf. *nudus*																	cf.
P. regularis s.l.	P		P				?			P					P		P
P. regularis regularis	P		cf.				?						P				
P. cf. *regularis solidus*	X		cf.	cf.		?											
Rastrites cf. *distans*															cf.		
R. linnaei	P							P	cf.								
R. maximus											P						
R. sp.	?																

P present   cf. qualified specific identification   ? generic and specific identification doubtful   X specific identification uncertain

Coastal localities
1 Fallen block on foreshore [5950 8716]; *maximus* Subzone
2 Fallen block on foreshore [5924 8648]; might be *sedgwickii* Zone, but preservation difficult
3 Fallen block on foreshore [5914 8629]; low *turriculatus* Zone, probably *maximus* Subzone
4 Cliff section [5887 0542]; zone probably as 3
5 Cliff section [5886 8538]; zone probably as 3
6 Cliff section [5877 8518]; zone probably as 3
7 Cliff section [5874 8510]; zone probably as 3
8 Cliff section [5878 8506]; zone probably as 3
9 Cliff section [5868 8492]; *maximus* Subzone or higher in *turriculatus* Zone

10 Cliff section [5857 8425]; probably upper *maximus* Subzone or slightly higher in *turriculatus* Zone
11 Road section [5810 8094]; maximus Subzone
12 Road section [5804 8078]; probably *maximus* Subzone
13 Cliff section [5804 8078]; probably *maximus* Subzone
14 Allt-wen cliff [5768 7960]; zone as 10

Inland localities
15 Old quarry [5982 8343 – 5975 8349], Clarach valley; zone as 9
16 Old quarry [6090 8185], Cefnhendre Farm; zone as 9
17 Old quarry [6354 8285], near Peithyll; zone as 9

ABERYSTWYTH GRITS FORMATION

The formation is the most widely known of any in the district and its rhythmic alternations of turbidite mudstone with comparatively thick turbidite sandstones makes it also the most distinctive of the district. The formation was first defined (Keeping, 1881), as dark, regularly alternating greywackes and shales, typified in the sea cliffs at Aberystwyth, and extending in a crescent shaped outcrop (as depicted by yellow dots on the Geological survey Old Series one-inch map-sheets 57 and 59 SE) from Borth in the north to south of Llangranog in the south and extending some 15 km inland. Keeping mistakenly regarded them as the oldest rocks in the area, but they were later (Jones, 1909)

shown to be the youngest and to be restricted to the west of the Plynlimon dome. The base of the formation has been defined at the base of the lowest of three prominent turbidites in Harp Rock (Plate 13) where it conformably overlies the Borth Mudstones Formation. Nowhere is its top seen. It is estimated that about 240 m, possibly 300 m, of beds are exposed in the district.

Although Jones and Pugh (1935a, p.280) commented on the greywacke sole structures, it was Rich (1950) and Kuenen (1953) who realised that the greywackes were the deposits of density currents. The detailed sedimentology of the coastal sections was described in a now classic paper (Wood and Smith, 1959). Subsequent work on the coast has involved the examination of various parameters to evaluate

Figure 26 Facies formations in part of the upper Llandovery

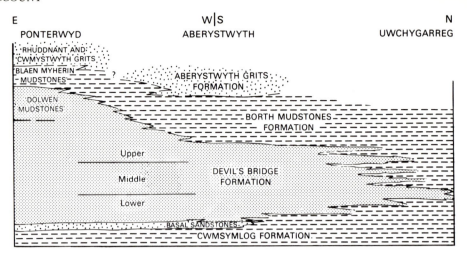

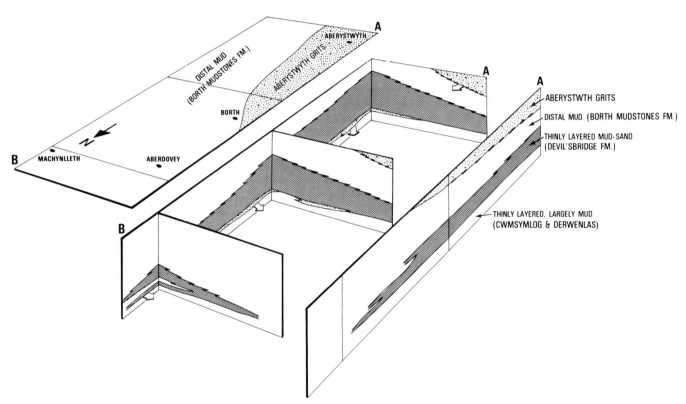

Figure 27 A sedimentary model explaining the disposition and relationships of the Devil's Bridge, Borth Mudstones and Aberystwyth Grits formations

turbidite proximality (Lovell, 1970), and the possible identification of contourites near the base of the formation in the Newquay area to the south (Anketell and Lovell, 1976).

Apart from attracting attention as a well exposed, classic sequence, the formation has raised many problems for both structural geologists and sedimentologists in that the establishment of unequivocal criteria for differentiating between soft sediment deformation and tectonic structures within it is extremely difficult. Opinions vary between the extremes of regarding the deformation as wholly tectonic (Price, 1962), and regarding it as the expression of soft sediment, intrabasinal, down-slope creep (Davies and Cave, 1976). The present chapter deals only with those structures

that are wholly confined to an individual bed; the remainder are dealt with in Chapter 6.

Lithologies

The formation consists of interbedded turbidite mudstones and sandstones, with a few thin beds, described here of Harp Rock Type, in the lower part of the sequence. One thin volcanic ash bed is also known, whilst siliceous-sideritic cone-in-cone nodules are locally present. Sandstone to mudstone ratios vary from 1:5 at the base of the formation, to 1:1 in the south. The sandstones rarely exceed 0.3 m in thickness.

Plate 13   Aberystwyth Grits. Harp Rock (Craig y Delyn) [5955 8755], 1.6 km SSW of Upper Borth. The lowest of the three prominent turbidites marks the base of the Aberystwyth Grits Formation. The rupture on the left (inset) has allowed upward injection of sediment in which the cleavage is disorientated. A12955

TURBIDITE SANDSTONES AND SILTSTONES

These mostly range in thickness from a centimetre or so to 0.2 m, though some are just over 0.3 m thick. They conform with the Bouma *Ta-e* classification of turbidites (Figure 2a). Sole markings are abundant; flute casts, groove casts and bounce casts are spectacularly displayed on the sea cliffs. Most sandstones lack the *Ta* interval, and the thin ribs usually lack both *Ta* and *Tb* intervals. Convolute lamination (Kuenen, 1953) is widespread in units over 25 mm thick,

and has been variously explained (Rich, 1950; Kuenen 1953; Leppard, 1978). Most of the sandstones are fine-grained. The chief constituent minerals are quartz (75–80 per cent), feldspar (commonly oligoclase-albite), muscovite and carbonate minerals (Wood and Smith, 1959).

Certain distinctive turbidites in the lower half of the exposed succession are here termed the 'Harp Rock Type'. They comprise clasts of mudstone and siltstone in a silty matrix.

They were first described (Wood and Smith, 1959) as 'fragmented beds' and 'slurried beds', both of which were regarded as forming a continuum: in this account the former variety is called Type 1, and the latter variety Type 2.

Type 1 is seen only in the coastal sections from immediately south of Clarach Bay [5862 8352] to Wallog. The individual beds are thicker than the normal turbidite sandstones, being between 0.24 and 0.38 m thick. They are everywhere underlain and overlain by normal turbiditic sandstone. The basal turbiditic sandstone varies from 2 to 20 cm thick. It has flutes and worm track casts on its sole, and most examples appear to only have the Bouma *Tb* and *Tc* intervals, though a single one has been noted with an *Ta* interval. Most are fine grained, but coarse sand is present in the *Ta* interval. The junction with the overlying Type 1 bed appears to be gradational. The rock consists of homogeneous muddy siltstone, full of intraformational clasts of mudstone and fine-grained laminated sandstone. The mudstone clasts occur either as mudflakes up to 5 cm or, less commonly, as thin rafts up to 0.6 m in length with local recumbent folds. The sandstone 'clasts' vary between ellipsoidal and spherical, but many have been modified by immediate post-depositional loading. The boundaries of many of the sandstone 'clasts' are diffuse. The overlying turbidite sandstones rest sharply on the Type 1 deposits and most appear to have erosive bases. They vary in thickness from 5 to 7 cm, and are either *Ta*- or *Tab*-absent Bouma units. In a few cases the base of the turbidite is load casted. In such cases many load-balls and pillows, some still attached by umbilical cords to the parent bed above, are scattered throughout the underlying Type 1 deposit. Many of these sandstone ball-and-pillows are difficult to distinguish from intraformational sandstone 'clasts' within the deposit : indeed they may be of

similar origin, the difference lying in the degree of incorporation. This origin may stem from nearly synchronous arrivals of two turbidity flows, one immediately behind but above the other, creating a density inversion which facilitated vertical mixing of the upper sediment load into the lower.

Type 2 (Plate 14) is different from Type 1 in having no overlying bed of turbidite sandstone, whilst the clasts in the deposit are well sorted; only in a few cases do they exceed 10 mm and the large rafts of mudstone are absent. Like Type 1, it is ubiquitously underlain by a turbidite sandstone. This is 50 to 100 mm thick and, either a *Ta*- or *Tab*-absent Bouma unit that, in some cases, is convoluted. The junction with the overlying deposit is either sharp or slightly erosional. Type 2 consists of muddy siltstone with abundant scattered, diffuse, fine-grained sandstone wisps and balls and mudstone flakes; none of these is more than 10 mm in diameter. The layers are 0.1 m to 0.17 m thick, and concentrations of sandstone clasts are common in their basal parts. The tops of the deposits are sharp and overlain by turbidite mudstones.

TURBIDITE MUDSTONES

These form the main bulk of the mudstones and represent the Bouma *Te* interval (Division 2, Figure 2b). They are blue-grey and fairly homogeneous except that they are usually slightly graded. They are affected by a fracture cleavage. In the lower part of the succession a paler blue-grey mudstone has been noted, especially in the sea cliffs north of Wallog. It occurs above very thin (up to a centimetre) siltstones, or above inter-turbidity deposits such as black graptolitic mudstones. These paler mudstones are obviously turbiditic; their differences from the main bulk of the turbidite mudstones may well reflect a different source area.

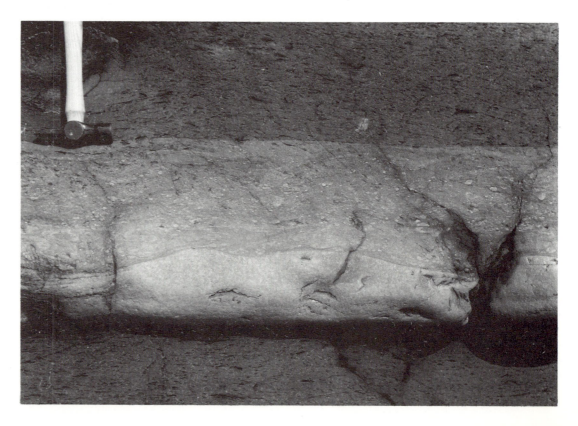

Plate 14
Aberystwyth Grits, Harp Rock. Type 2 Turbidite [5920 8640] 600 m north of Wallog. A12953

INTER-TURBIDITE DEPOSITS

These have been noted only in the northern part of the coastal outcrop, north of Clarach Bay. They are represented by the two types of division 3 (Figure 2b). The first consists of black, fine-grained pyritous and carbonaceous beds, generally only a few millimetres thick, and rich in graptolites; the second is of similar thickness, but pale grey to fawn with little pyrite. Both types have fairly sharp tops and bottoms.

Siliceous-sideritic nodules exhibiting cone-in-cone structures occur sparingly in layers within the turbidite mudstones in the lower part of the succession. They are generally ovoid in shape and 0.1 m to 0.15 m in diameter. Many, however, coalesce. In such cases the composite nodule is either linear, curvilinear or branching, and up to 0.5 m long.

One very thin 25 mm volcanic ash bed has been noted within the succession on the southern side of Clarach Bay [5862 8532]. It occurs at the top of a turbidite mudstone, and is very fine-grained but crudely graded. X-ray diffraction analysis by Mr R. Merriman reveals it to contain abundant chlorite and clay-grade mica, probably a little feldspar, but only a small amount of quartz (film X7507).

Distribution of lithologies

The structural style of the formation and the fact that the coast provides only a strike section make it impossible to measure a vertical sequence through the formation. Although the Borth Mudstones clearly underlie the Aberystwyth Grits at the defined base of the formation on the coast at Harp Rock, there is a possibility that they may be in part the lateral equivalent of beds that elsewhere form part of the Aberystwyth Grits (Waters, 1974). The evidence for this is twofold. First the Aberystwyth Grits show marked changes from south to north, reflecting their increasing distance from their southerly source (Wood and Smith, 1959; Lovell, 1970). It can be argued, therefore, that the Borth Mudstones may be the distal part of a turbidite lobe represented by the Aberystwyth Grits. Second, although the graptolite faunas of these formations are enigmatic they are so closely comparable as to support a lateral equivalence.

If it is assumed that the outcrop of the formation is broadly synclinal with the lowest beds occurring in the north near Borth and in the south near Llangranog (Wood and Smith, 1959), then a traverse from south of Borth to Aberystwyth should provide a vertical section through the lower part of the formation; by reference later in this section to 'low' and 'high' in the local succession, we mean relative to such a traverse. A traverse in an eastward direction from the coast at Aberystwyth should give the same result, but exposure is poor.

The base of the formation has been defined (Cave, 1976) at Harp Rock [5958 8761], at the base of the lowest of three prominent Type 2 beds interbedded in mudstones (Plate 13). In the New Quay area, at the presumed proximal end of the outcrop, Anketell and Lovell (1976) have defined the lower limit at the base of the lowest Aberystwyth Grit turbidite facies sandstone. The junction within the district as a whole is gradational and arbitrary. Thus thicker turbidite sandstones of Aberystwyth Grits type appear, together with a few

Type 2 deposits, in the uppermost part of the Borth Mudstones. Above the base of the Aberystwyth Grits thicker turbidite sandstones become more common but, because of the folding and the gradational base of the formation, the mapped base inland is no more than the lower boundary of a crude envelope, and fold cores of one formation may occur within the other in the boundary zone.

In a north to south traverse across the coastal exposures there is an increase in the amount of sandstone as the succession is ascended. Sandstone/mudstone ratios are as follows at various points along the coast: Harp Rock 1:5; Wallog 1:2.8; Clarach 1:1.8; Aberystwyth 1:1.1; Allt-wen 1:0.9. Along this traverse, and indeed throughout the district, the dominant lithology is interbedded turbidite sandstone and mudstone. Harp Rock Type turbidites and cone-in-cone concretions appear only in the lower half of the local succession, and so are only seen from Clarach northwards in the coastal traverse. Type 2 appears earlier in any one succession and disappears first: on the coast Type 1 is seen only between Clarach and Wallog.

Faunas

Graptolites occur mainly in the very thin interturbidite dark mudstones but, south of Aberystwyth, they occur also in the basal part of thin turbiditic sandstones as a derived fauna. Shelly fossils are extremely uncommon though a derived fauna of the brachiopod *Eocoelia* has been reported from the base of one turbidite sandstone on the coast [5826 8270] at Aberystwyth (Bates, 1982).

The ichnofauna of the formation is ubiquitous, Crimes and Crossley (1980) reporting some twenty genera. The most common trace fossils are meandering parallel-sided ridges some 1 to 5 mm across, preserved as convex half-reliefs on the soles of turbidite sandstones. The most striking trace fossils are *Palaeodictyon* and its subgenus *Squamodictyon*, which occur respectively as hexagonal and elliptical nets as half-reliefs and full-reliefs on the soles of turbidite sandstones or within the underlying mudstone (Crimes and Crossley, 1980). The nets may be up to 0.5 m in diameter and represent an open-burrow system filled from above by the succeeding turbidite.

Although the Aberystwyth Grits have been said to represent both the *Monograptus turriculatus* and *M. crispus* zones, (Wood and Smith, 1959, p.163) collections during the present survey, have shown that, within the district, they fall wholly within the *turriculatus* Zone. The presence of *M. halli* and *M. cf. sedgwickii* in the lower part of the succession suggests horizons almost as low as the *sedgwickii* Zone but, assuming the succession to be structurally simple, most of the sequence appears to be of *maximus* Subzone age with the uppermost beds of the district just including the higher part of the *turriculatus* Zone. There is no evidence within the district of the presence of the *crispus* Zone.

Conditions of deposition

The Aberystwyth Grits represent a major change in the type of sediment reaching this part of the basin. Firstly, they are derived from the south to SW quadrant (Wood and Smith, 1959) as opposed to the eastward derivation of the underlying formations; secondly, the turbidites are coarser and thicker than those in the rest of the Llandovery. The reason for the sudden change in sedimentation style is almost certainly tectonic. The new style begins in the basin near the base of the *turriculatus* Zone at the arrival of the Aberystwyth Grits (Cave, 1979), which is the beginning of a prolonged break in sedimentation in parts of SW Dyfed (Zeigler and others, 1969) that only ended in either latest Llandovery or Wenlock times. The shelf area in SW Dyfed is known to have been active in the late Llandovery (Zeigler and others, 1969), and was an obvious source of detritus.

This interpretation is complicated by the suggestion (Anketell and Lovell, 1976; Crimes and Crossley, 1980) that contourites (deposits of geostrophic contour-following bottom currents involved in the deep thermohaline circulation of the world ocean) are present in the lowest part of the Aberystwyth Grits and in the underlying Grogal Sandstones of the New Quay area [380 600]. Their diagnosis rests on a comparison with modern contourites, and on a divergence in noted current vectors. Current vectors from the SW have been obtained from sole marks and cross-lamination in the lower parts of individual beds, and vectors from the SE from cross-lamination either in the upper parts of beds or throughout a particular bed. They concluded that contour currents were, completely or partly, reworking sediments brought in by turbidites. It would appear that their main evidence for recognising contourites lies in the divergence from the normal current vector, for the lithological distinction between contourites and very distal turbidites are equivocal and we have not recognised any on lithological criteria.

It has been postulated (Crimes and Crossley, 1980) that, for nutritional reasons, the alignment of *Palaeodictyon* and *Squamodictyon* burrow systems is parallel to bottom currents. Crimes and Crossley have noted that at Aberystwyth and Clarach the alignment of long axes and turbidite sole structures are parallel, but that between Wallog and Borth the burrow alignment in beds thought to belong to the Borth Mudstones is orientated west–east, suggesting that bottom currents here flowed at right angles to the main turbidity currents. In the Aberystwyth Grits, however, the palaeocurrent evidence is entirely consistent with that of the turbidite sole structures.

The present day outcrop represents part of a lobe of a turbidite fan derived almost entirely from the south to SW quadrant. Both Wood and Smith (1959) and Lovell (1970) have commented on the southward changes that reflect increasing proximality. In the district, the beds are relatively distal in nature, for washouts and imperfect grading are unknown. Lovell (1970) has shown that the sandstone/mudstone ratio (see p.57) is the best indicator of proximality.

The location of the eastern limit of the Aberystwyth Grits lobe is imprecisely known but the formation is absent on the eastern side of the Plynlimon dome, for the Dolwen and Blaen Myherin mudstones of *turriculatus* Zone age there succeed the Devils' Bridge Group. Above these two dominantly mudstone formations, but still within the *turriculatus* Zone, are the coarse southerly-derived turbidites of the Cwmystwyth Grits, and the slightly more distal Rhuddnant Grits of predominantly *crispus* Zone age (Cave, 1978a and 1979). It has been suggested (Cave, 1979) that the

Aberystwyth Grits and those other, slightly younger turbidites to the east are the expression of the northward growth of a turbidite fan, the deposition of successive lobes of this fan having taken place in the hollow between its precursors and the basin margin to the east.

DOLWEN MUDSTONES FORMATION

This formation was first defined as the Dolwen Mudstones (Jones, 1909, p.516), the lower component of the Myherin Group. It is about 210 m thick (Jones, 1909, fig. 13).

The outcrop is limited to the high ground [7865 7915–7975 7945] on the southern side of Afon Myherin, and thence northwards along the edge of the district [7995 8045]. Its base is transitional from the Devil's Bridge Formation, and its top transitional into the Blaen Myherin Mudstones Formation. It comprises a multilayer sequence of mudstones and fine-grained sandstones or siltstones, the rhythms generally being 5 to 10 cm thick. The arenite component is usually 5 to 10 mm thick with some reaching 25 mm; it is more commonly 25 to 50 mm thick in part of the upper portion of the sequence. The mudstones are grey to bluish grey, the sandstones fine grained (a few medium grained in the basal few mm) and graded, some with a laminated component above. In the lower part of the sequence the mudstones are homogeneous, thickly bedded and poorly cleaved. A few shells and crinoid columnals have been noted (Appendix 3, locs. 19, 20) but, though graptolites have been recorded, none was found during the survey.

BLAEN MYHERIN MUDSTONES FORMATION

This formation was first defined as the Blaen Myherin Mudstones (Jones, 1909, p.516), the upper component of the Myherin Group. It is present only in the south-eastern extremity of the district, and is not exposed. Sections and soil debris to the east of the district, show the beds to be dark grey and grey mudstones, commonly with vivid orange-brown weathering. The transition from the Dolwen Mudstones is marked by a deepening of the colour of the mudstone, a rapid diminution in the thickness and number of sandstone beds, and the development of orange-brown weathering.

BIOSTRATIGRAPHY

Shelly fossils occur at a few localities in the southern part of the district, in the Derwenlas and overlying formations (Appendix 3). However, graptolites are relatively common throughout the succession and all the Llandovery zones up to, and including, the *Monograptus turriculatus* Zone have been recognised. Indeed, several of these zones were first established in the present district, near Ponterwyd by Jones (1909), and close to the northern margin, south of Machynlleth, by Jones and Pugh (1916). Zonal distributions of individual species collected during the present survey are shown in Table 3, and the main local characteristics of the

Table 3 Zonal distribution of Silurian graptolites in BGS collections, Aberystwyth district

	1	2	3	4	5	6	7	8	9	10	11
?Climacograptus innotatus	1	.	.	.	.	.	.	.	.	.	.
C. medius	1	.	.	4	.	.	.	.	.	.	.
C. cf. *miserabilis*	1	.	.	.	.	.	.	.	.	.	.
C. normalis	1	.	.	4	5	.	.	.	.	.	.
?Diplograptus modestus	1	.	.	.	.	.	.	.	.	.	.
Glyptograptus persculptus	1	.	.	.	.	.	.	.	.	.	.
G. persculptus — thin variant	1	.	.	.	.	.	.	.	.	.	.
Diplograptus modestus parvulus	.	2	.	.	.	.	.	.	.	.	.
Glyptograptus tamariscus s.l.	.	2	.	4	5	6	7	8	.	.	.
Parakidograptus acuminatus	.	2	.	.	.	.	.	.	.	.	.
Climacograptus rectangularis	.	.	.	4	5	.	.	.	.	.	.
Cystograptus penna	.	.	.	4	.	.	.	.	.	.	.
C. vesiculosus	.	.	.	4	.	.	.	.	.	.	.
Glyptograptus sinuatus	.	.	.	?	cf.	6	7	.	.	.	.
Orthograptus aff. *mutabilis*	.	.	.	4	.	.	.	.	.	.	.
Pseudoclimacograptus (*Metaclimacograptus*) *hughesi*	.	.	.	4	5	6	7	8	.	.	.
Rhaphidograptus toernquisti	.	.	.	4	5	6	7	8	?	?	.
Atavograptus atavus	.	.	.	4	5	.	.	.	.	.	.
Coronograptus cyphus	.	.	.	4	.	.	.	.	.	.	.
C. gregarius	.	.	.	4	5	6	7	8	.	.	.
Lagarograptus acinaces	.	.	.	4	.	.	.	.	.	.	.
?Monograptus difformis	.	.	.	4	.	.	.	.	.	.	.
M. revolutus s.l.	.	.	.	4	5	6	.	.	.	.	.
Pribylograptus incommodus	.	.	.	4	cf.	.	.	.	.	.	.
P. sandersoni	.	.	.	4	.	.	.	.	.	.	.
Climacograptus aff. *alternis*	.	.	.	.	5	.	.	.	.	.	.
Orthograptus aff. *bellulus*	.	.	.	.	5	.	.	.	.	.	.
Petalograptus ovatoelongatus	.	.	.	.	cf.	6	7	8	.	.	.
Monograptus communis communis	.	.	.	.	5	6	.	.	.	.	.
M. communis s.l.	.	.	.	.	5	6	7	8	9	cf.	.
M. denticulatus	.	.	.	.	cf.	cf.	7	8	.	.	.
M. toernquisti brevis	.	.	.	.	5	.	.	.	.	.	.
M. triangulatus major?	.	.	.	.	5	6	.	.	.	.	.
M. triangulatus separatus	.	.	.	.	5	?	.	.	.	.	.
M. triangulatus triangulatus	.	.	.	.	5	cf.	.	.	.	.	.
Pribylograptus argutus	.	.	.	.	?	6	.	8	.	.	.
Rastrites longispinus	.	.	.	.	5	6	cf.	8	.	.	.
Diplograptus magnus	.	.	.	.	.	6	.	.	.	.	.
Glyptograptus (*Pseudoglyptograptus*) *vas*	.	.	.	.	.	6	.	.	.	.	.
Pseudoclimacograptus (*Metaclimacograptus*) *undulatus*	.	.	.	.	.	?	7	8	9	10	.
Monograptus austerus	.	.	.	.	.	6	.	.	.	.	.
M. capis	.	.	.	.	.	6	7	8	.	.	.
M. pseudoplanus	.	.	.	.	.	6	.	.	.	.	.
M. triangulatus fimbriatus	.	.	.	.	.	6	.	.	.	.	.
Pribylograptus leptotheca	.	.	.	.	.	6	7	8	.	.	.
Pristiograptus concinnus	.	.	.	.	.	?	.	8	?	.	.
P. fragilis	.	.	.	.	.	6	7	8	.	.	.
Petalograptus minor	.	.	.	.	.	.	?	8	.	.	.
Pseudoclimacograptus (*Clinoclimacograptus*) *retroversus*	.	.	.	.	.	.	7	8	9	10	.
Monograptus argenteus	.	.	.	.	.	.	7	.	.	.	.
?M. cerastus	.	.	.	.	.	.	7	.	.	.	.
M. involutus	.	.	.	.	.	.	cf.	cf.	9	cf.	.
M. lobiferus	.	.	.	.	.	.	7	8	9	.	.
M. millepeda	.	.	.	.	.	.	7	.	.	.	.
?M. nobilis	.	.	.	.	.	.	7	.	.	.	.
?M. triangulatus s.l.	.	.	.	.	.	.	7	.	.	.	.
Pristiograptus regularis regularis	.	.	.	.	.	.	?	.	.	10	11
P. regularis s.l.	.	.	.	.	.	.	7	8	9	10	11
Rastrites peregrinus	.	.	.	.	.	.	cf.	8	.	.	.
Cephalograptus cometa extrema	.	.	.	.	.	.	.	8	.	.	.
C. cometa cometa	.	.	.	.	.	.	.	8	?	.	.
Climacograptus scalaris s.l.	.	.	.	.	.	.	.	8	cf.	.	.

Table 3 continued

	1	2	3	4	5	6	7	8	9	10	11
Glyptograptus serratus s.l.	.	.	.	.	.	.	.	8	cf.	.	.
G. tamariscus linearis	.	.	.	.	.	.	.	8	.	.	.
Orthograptus cyperoides	.	.	.	.	.	.	.	8	.	.	.
O. insectiformis	.	.	.	.	.	.	.	8	.	.	.
Retiolites perlatus	.	.	.	.	.	.	.	8	.	.	.
Monoclimacis? crenularis	.	.	.	.	.	.	.	8	?	.	.
Monograptus clingani	.	.	.	.	.	.	.	8	.	.	.
M. communis rostratus?	.	.	.	.	.	.	.	8	.	.	.
M. convolutus	.	.	.	.	.	.	.	8	?	.	.
M. decipiens	.	.	.	.	.	.	.	8	9	.	.
M. delicatulus	.	.	.	.	.	.	.	8	9	.	.
M. gemmatus	.	.	.	.	.	.	.	cf.	.	10	11
M. limatulus	.	.	.	.	.	.	.	8	9	.	.
M. sedgwickii	.	.	.	.	.	.	.	?	9	10	.
M. undulatus	.	.	.	.	.	.	.	8	cf.	cf.	.
Pristiograptus jaculum	.	.	.	.	.	.	.	8	9	10	.
Rastrites ?approximatus geinitzi	.	.	.	.	.	.	.	8	.	.	.
R. hybridus	.	.	.	.	.	.	.	8	9	10	.
R. spina	.	.	.	.	.	.	.	8	.	.	.
Acanthograptus sp.	.	.	.	.	.	.	.	.	?	10	.
Dendrograptus sp.	.	.	.	.	.	.	.	.	9	10	.
Dictyonema sp.	.	.	.	.	.	.	.	.	9	10	.
Climacograptus simplex	.	.	.	.	.	.	.	.	9	10	.
Glyptograptus elegans	.	.	.	.	.	.	.	.	9	.	.
Petalograptus cf. kurcki	.	.	.	.	.	.	.	.	9	.	.
P. palmeus s.l.	.	.	.	.	.	.	.	.	9	10	11
P. tenuis	.	.	.	.	.	.	.	.	9	10	cf.
?Diversograptus ramosus	.	.	.	.	.	.	.	.	9	10	11
D. runcinatus	.	.	.	.	.	.	.	.	?	10	11
Lagarograptus tenuis	.	.	.	.	.	.	.	.	9	10	.
?Monograptus communis obtusus	.	.	.	.	.	.	.	.	9	.	.
M. cf. distans	.	.	.	.	.	.	.	.	?	10	.
M. elongatus?	.	.	.	.	.	.	.	.	9	.	.
M. intermedius?	.	.	.	.	.	.	.	.	9	.	.
M. knockensis	.	.	.	.	.	.	.	.	cf.	10	11
M. marri	.	.	.	.	.	.	.	.	?	10	11
M. planus	.	.	.	.	.	.	.	.	?	10	11
M. cf. pragensis ruzickai	.	.	.	.	.	.	.	.	9	10	.
M. proteus	.	.	.	.	.	.	.	.	9	10	11
Pristiograptus nudus	.	.	.	.	.	.	.	.	?	10	11
P. regularis solidus	.	.	.	.	.	.	.	.	?	cf.	11
Rastrites distans	.	.	.	.	.	.	.	.	9	cf.	cf.
R. fugax	.	.	.	.	.	.	.	.	9	10	.
R. linnaei	.	.	.	.	.	.	.	.	9	10	11
Glyptograptus incertus?	.	.	.	.	.	.	.	.	.	10	.
G. cf. tamariscus fastigans	.	.	.	.	.	.	.	.	.	10	aff.
Orthograptus sp.	.	.	.	.	.	.	.	.	.	10	.
Petalograptus cf. altissimus	.	.	.	.	.	.	.	.	.	10	.
P. ovatus	.	.	.	.	.	.	.	.	.	10	.
P. wilsoni	.	.	.	.	.	.	.	.	.	?	11
Pseudoplegmatograptus obesus	.	.	.	.	.	.	.	.	.	10	.
Diversograptus cf. rectus	.	.	.	.	.	.	.	.	.	10	.
?Monoclimacis galaensis	.	.	.	.	.	.	.	.	.	10	.
?Monograptus acus	.	.	.	.	.	.	.	.	.	10	11
M. barrandei	.	.	.	.	.	.	.	.	.	10	.
M. exiguus	.	.	.	.	.	.	.	.	.	10	11
M. halli	.	.	.	.	.	.	.	.	.	10	11
M. petilus	.	.	.	.	.	.	.	.	.	10	.
M. pseudobecki	.	.	.	.	.	.	.	.	.	10	11
M. rickardsi rickardsi	.	.	.	.	.	.	.	.	.	?	11
M. spiralis	.	.	.	.	.	.	.	.	.	10	11
M. cf. tullbergi	.	.	.	.	.	.	.	.	.	10	.

Table 3 continued

	1	2	3	4	5	6	7	8	9	10	11
M. turriculatus	.	.	.	.	.	.	.	.	.	10	11
?Pristiograptus variabilis	.	.	.	.	.	.	.	.	.	10	.
Rastrites distans spengillensis	.	.	.	.	.	.	.	.	.	10	.
R. maximus	.	.	.	.	.	.	.	.	.	10	.
Climacograptus nebula	.	.	.	.	.	.	.	.	.	.	11
Monograptus dextrorsus?	.	.	.	.	.	.	.	.	.	.	11
M. nodifer?	.	.	.	.	.	.	.	.	.	:	11
?M. veles	.	.	.	.	.	.	.	.	.	.	11

(key to zones: 1 *G. persculptus*; 2 *P. acuminatus*; 3 *A. atavus* and *L. acinaces*; 4 *C. cyphus*; 5 *M. triangulatus*; 6 *D. magnus*; 7 *P. leptotheca*; 8 *M. convolutus*; 9 *M. sedgwickii*; 10 *M. turriculatus* — *R. maximus* Subzone; 11 *M. turriculatus* — upper part.)

zones are summarised below. Several of the more common species are illustrated in Figure 28.

Glyptograptus persculptus Zone. *G. persculptus* is the only common species, the earliest records being in the *persculptus* Band, 1 m above the base of the Mottled Mudstone Member of the Cwmere Formation.

Parakidograptus acuminatus Zone. The fauna is dominated by *P. acuminatus*, *Climacograptus normalis* and *Orthograptus* cf. *truncatus* (Jones, 1909), and includes unusually early examples of *Glyptograptus tamariscus* s.l.

Atavograptus atavus and *Lagarograptus acinaces* zones. In the Ponterwyd area *A. atavus* and *Rhaphidograptus toernquisti* are well represented in the fauna of the *atavus* Zone, and both species are associated with *Monograptus rheidolensis* (= *L. acinaces*) in the overlying *acinaces* Zone, according to Jones (1909). Elsewhere in the district, neither zone has been unequivocally proved during the present survey.

Coronograptus cyphus Zone. Compared with the fauna of earlier zones, there is a marked increase in variety. In addition to *C. cyphus*, other commonly occurring species are *A. atavus*, *Monograptus revolutus* and *R. toernquisti*.

Monograptus triangulatus Zone. This zone is generally represented by two graptolite horizons, separated by 2 m of barren strata. *Monograptus toernquisti brevis*, *M. triangulatus triangulatus* and *M. triangulatus separatus* are probably confined to this zone. *M. revolutus*, usually present only in the *cyphus* Zone, is fairly well represented. *A. atavus* is recorded from the lower bed but has not been found in the upper one, where the earliest examples of *Monograptus communis communis* and of *Rastrites*, represented by *R. longispinus*, together with unusually early examples of *Monograptus* cf. *denticulatus* occur. In some sections this upper graptolite horizon coalesces with shale containing a *magnus* Zone fauna, but elsewhere there is a separation of up to 1.5 m of barren beds.

Diplograptus magnus Zone. There is a diagnostic association of *D. magnus* (common) with *Monograptus triangulatus fimbriatus*, *Glyptograptus* (*Pseudoglyptograptus*) *vas*, *Monograptus pseudoplanus* and *M. capis*. At some localities a thin mudstone

containing a *magnus* Zone fauna occurs above the main *magnus* mudstone.

Pribylograptus leptotheca Zone. Rickards (1976, p.163) observed that *P. leptotheca* seemed to be more common in Wales and *M. argenteus* more common in the Lake District. However, in the Aberystwyth district both are equally common and make their appearance at this level, except for a single record of *P. leptotheca* in the *magnus* Zone. *Monograptus millepeda* and possible examples of *Monograptus cerastus* are characteristically confined to this zone, and *P.* (*M.*) *hughesi* is also present, although elsewhere its upper limit is believed to be in the *magnus* Zone.

Monograptus convolutus Zone. There is an extensive variety of graptolites with *M. convolutus*, *M. decipiens*, *M. limatulus*, *M. lobiferus*, *Pristiograptus regularis* s.l. *Pseudoclimacograptus* (*Clinoclimacograptus*) *retroversus* and *P.* (*Metaclimacograptus*) *undulatus* characteristically well represented. *Monograptus clingani*, recorded from Wales for the first time, and *Monoclimacis? crenularis* are also present. A few examples of *Cephalograptus cometa cometa* and *C. cometa extrema* have been recorded in the upper part of the zone, but they are inadequate proof of the *cometa* Band of Jones and Pugh (1916).

Monograptus sedgwickii Zone. *M. sedgwickii* is characteristically abundant, associated with the diagnostic species *Climacograptus simplex*, *Lagarograptus tenuis*, *Monograptus* cf. *pragensis ruzickai* and *Petalograptus* cf. *kurcki*. Of the remainder of the rich fauna, *P. regularis* s.l. and *Pseudoclimacograptus* (*M.*) *undulatus* are very common, and also represented are *Petalograptus palmeus* s.l., *P. tenuis* and *Monograptus proteus*, the earliest examples of which usually occur in the overlying zone.

Monograptus turriculatus Zone, including the *Rastrites maximus* Subzone at the base. The assemblage immediately above the *sedgwickii* Zone contains forms occurring for the first time, such as *R. maximus*, *Monograptus spiralis* s.l., *Pseudoplegmatograptus obesus*, possible examples of *Monograptus exiguus* and *Rastrites linnaei*, together with *P. palmeus* s.l. and *M. proteus*, two species whose lower stratigraphical range is normally the *maximus* Subzone but which, as mentioned above, are

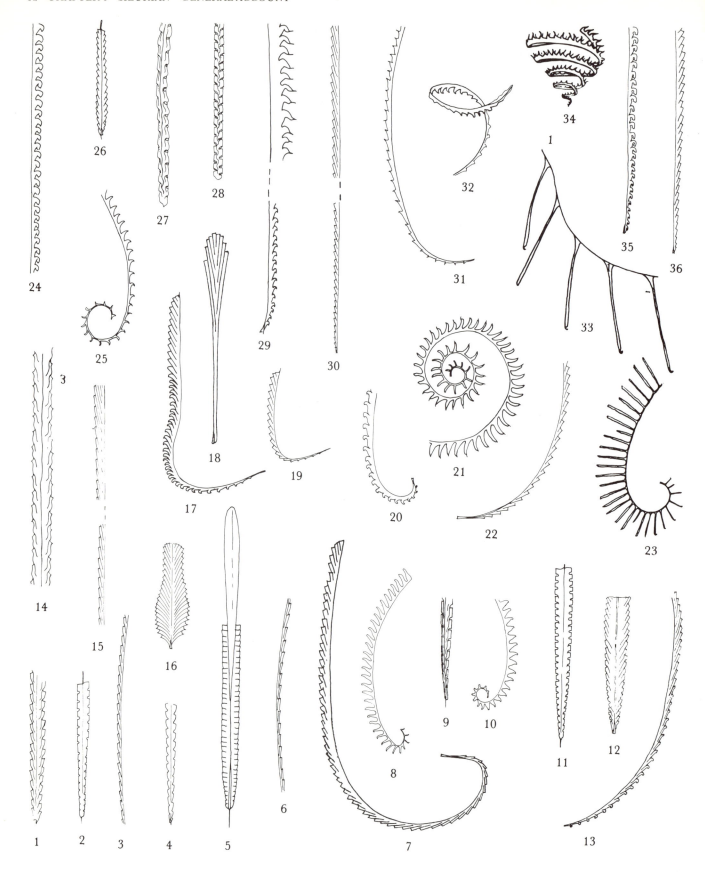

Figure 28   Some characteristic Silurian graptolites (for details see p.63)

Figure 28 Some characteristic Silurian graptolites.

The species are arranged approximately in ascending stratigraphical sequence from bottom left to top right. Their stratigraphical ranges are shown in Table 3. All figures are ×2, except 5 × 1.3; and 14, 27, 28 × 4.

1. *Glyptograptus persculptus* (Salter); 2. *Climacograptus normalis* (Lapworth); 3. *Atavograptus atavus* (Jones); 4. *Glyptograptus tamariscus* (Nicholson); 5. *Cystograptus vesiculosus* (Nicholson); 6. *Pribylograptus sandersoni* (Lapworth); 7. *Coronograptus cyphus* (Lapworth); 8. *Monograptus triangulatus triangulatus* (Harkness); 9. *Rhaphidograptus toernquisti* (Elles and Wood); 10. *Monograptus triangulatus fimbriatus* (Nicholson); 11. *Climacograptus rectangularis* (McCoy); 12. *Diplograptus magnus* H. Lapworth; 13. *Monograptus revolutus* Kurck; 14. *Glyptograptus* (*Pseudoglyptograptus*) *vas* Bulman and Rickards; 15. *Pribylograptus leptotheca* (Lapworth); 16. *Petalograptus ovatoelongatus* (Kurck); 17. *Monograptus argenteus* (Nicholson); 18. *Cephalograptus cometa cometa* (Geinitz); 19. *Monograptus limatulus* Törnquist; 20. *Monograptus clingani* (Carruthers); 21. *Monograptus convolutus* (Hisinger); 22. *Coronograptus gregarius* (Lapworth); 23. *Rastrites longispinus* Perner; 24. *Monograptus lobiferus* (McCoy); 25. *Monograptus decipiens* Törnquist; 26. *Petalograptus tenuis* (Barrande); 27. *Pseudoclimacograptus* (*Clinoclimacograptus*) *retroversus* Bulman and Rickards; 28. *Pseudoclimacograptus* (*Metaclimacograptus*) *undulatus* (Kurck); 29. *Monograptus sedgwickii* (Portlock); 30. *Pristiograptus regularis regularis* (Törnquist); 31. *Lagarograptus tenuis* (Portlock); 32. *Monograptus proteus* (Barrande); 33. *Rastrites maximus* Carruthers; 34. *Monograptus turriculatus* (Barrande); 35. *Monograptus marri* Perner; 36. *Pristiograptus nudus* (Lapworth).

present in the *sedgwickii* Zone in this district. This characteristic *maximus* Subzone assemblage is associated at this level with *M. sedgwickii*, which is often numerous. Other forms not normally extending above the *sedgwickii* Zone, namely *C. simplex, Glyptograptus incertus?, M. communis* s.l., *M. involutus*, possible examples of *M. pragensis ruzickai, P.* (*C.*) *retroversus* and *P.* (*M.*) *undulatus* also occur. Higher beds within the subzone contain a typically varied fauna, with *R. maximus* fairly common at some localities, but *M. turriculatus* is rare. *L. tenuis* and *P.* (*M.*) *undulatus* are also present, their previously recorded upper range being the *sedgwickii* Zone. Above the *maximus* Subzone, *M. turriculatus*, together with *M. marri* and *Pristiograptus nudus* occur more frequently. *M. proteus* continues to be well represented, but *M. halli* is much less common than in the *maximus* Subzone.

The blurring of the boundary between the *sedgwickii* and *turriculatus* zones and that at the top of the *maximus* Subzone in this region is attributed to the rapid deposition and consequent thick sequence above the upper part of the *sedgwickii* Zone.

64

CHAPTER 5

Silurian—details

CWMERE FORMATION

Eisteddfa Gurig – Ponterwyd – Craigypistyll

West of Eisteddfa Gurig the Mottled Mudstone Member is well displayed in a number of sections. The junction between it and the underlying Bryn-glâs Formation (p.37) is seen in a track [7951 8409] adjacent to Afon Tarenig, with a 15 cm bed of dark grey mudstone (the *persculptus* Band) with abundant *Glyptograptus persculptus* 1.2 m above the junction. Jones (1909, p.475) described a parallel section in Afon Tarenig, a few metres to the north. An eastward-dipping strike section in typical colour-banded medium and pale grey mudstone extends northwards along Afon Tarenig for about 300 m, with the *persculptus* Band again visible at its northern end [7949 8440]. The topmost beds of the member are seen in a small quarry [7940 8416] west of the river, where 1.8 m of fine-grained argillaceous sandstone in beds up to 0.6 m thick with inter-bedded thin mudstones is exposed in two small *en-échelon* anticlines with axes striking at 190°. Higher beds are exposed along the northern side of the A44 road [7868 8292 – 7936 8290], the beds being disposed in a number of broad southerly-plunging folds. They are dark to very dark grey, rusty brown weathering mudstones with some paler beds, and with fine-grained sandstones up to 30 cm thick in the western part [7878 8288] of the section. At the top of the Mottled Mudstone [7519 8119] south of Bryn-glâs, there is an unusual development (p.43) of 30 cm of medium-grained feldspathic sandstone.

The lower part of the formation (*persculptus* and *acuminatus* zones) is exposed in Nant Fuches-gau [7665 8086 – 7669 8055]. This section is essentially as described by Jones (1909, p.481 and fig.5) except that the *persculptus* Band (his loc.F7) is now obscured. The highest of three 'gritty mudstones' (fine-grained sandstones) described by Jones from this section can be traced along the hillside to the west, the best section [7640 8096] showing 5.5 m of dark grey mudstone with a few sandstone beds up to 7.5 cm thick, overlain by 0.6 m of fine- to medium-grained laminated sandstones, overlain by a further 3.0 m of mudstone with some thin sandstone beds. This sandstone is about 40 m above the top of the Mottled Mudstone, and probably correlates with the sandstone beds near the top of the *acuminatus* Zone in the Rheidol gorge (Jones, 1909, fig.11 and pl. xxv). The formation, from the upper part of the *acuminatus* Zone upwards, is displayed in the Rheidol gorge south of Ponterwyd (Jones, 1909, pp.483 – 488, 491 – 492 pl.xxv).

Between Ponterwyd and Dinas power station the higher part of the formation is seen in another gorge section [7454 8208 – 7469 8153] along Afon Rheidol immediately south of the Dinas dam. This shows at least 25 m of dark grey mudstones with many sandy laminae and fine-grained sandstone beds up to 5 cm thick. A packet of fine-grained sandstones (p.44), totalling 3 m in thickness with individual beds up to 0.9 m thick, is present near the top of the sequence and is repeated through the section by folding and faulting. Poorly developed flute marks and cross-lamination indicate a current flow from the SE. These sandstones are probably at the same horizon as 'grits' recorded by Adams (1963) from the Dinas tunnel portal [7454 8217] just to the north, though his records of sedimentary structures indicate a current flow from the SW.

South-east of Disgwylfa Fâch a long stream section [7419 8376 – 7402 8338] displays beds above the Mottled Mudstone; they are dark, grey rusty weathering mudstones with a 0.9 m bed of fine-grained grey sandstone at the northern end of the section. West of the hill, 4.6 m of dark grey mudstone with deep orange-brown

weathering and an overturned dip of 81° towards 115° is seen in the stream [7300 8393] at Gwenffrwd-uchaf. Jones (1909, p.492) described these beds as being close to the junction between the *acuminatus* and *atavus* zones. Some 2 km to the WNW there are a number of sections around the junction between the Cwmere and Derwenlas formations in and adjacent to a track [7114 8482-7084 8501] 0.7 km south of Craigypistyll. In one [7103 8495] dark grey to black mudstone with thin pale stripes has yielded *Atavograptus atavus*, *Climacograptus medius?*, *?Coronograptus cyphus* and *Rhaphidograptus toernquisti*, indicative of the *cyphus* Zone at the top of the Cwmere Formation. These beds are overlain by more closely striped beds with many triangulate monograptids at the base of the Derwenlas Formation.

The Mottled Mudstone Member has three separate outcrops in the Craigypistyll area (Figure 29). Immediately south of Craigypistyll (Figure 29, loc.14) the Bryn-glâs Formation is overlain by 4.6 m of grey, faintly reddish brown weathering mudstone, succeeded by 1.2 m of massive grey medium-grained sandstone with thin mudstone partings, in turn overlain by typical rusty weathering mudstones of the main part of the Cwmere Formation. The total thickness of 5.8 m is unusually small for the member. Between the two components of the Camdwr Fault these beds are rather thicker, and a continuous section (Figure 29, loc.15) on the north side of the gorge shows:

	Thickness m
Cwmere Formation (above Mottled Mudstone Member)	
Mudstone, dark grey, weathers to rusty brown	—
Cwmere Formation (Mottled Mudstone Member)	
Sandstone, grey, medium-grained, slightly feldspathic, well cemented	0.6
Mudstone, grey, colour-banded	c.7.6
Mudstone, grey, darker and weathering to red-brown in upper part, with *G. persculptus*	1.5
Bryn-glâs Formation	
Mudstones, dark grey, unbedded, splintery	—

Immediately north of the Camdwr Fault a small quarry (Figure 29, loc.16) shows 0.6 m of dark grey mudstone with abundant *G. persculptus* overlain by 5.5 m of paler grey mudstone.

There are excellent sections in beds above the Mottled Mudstone along the gorge at the western end of Llyn Craigypistyll. Dark grey, rusty brown weathering mudstones, with silty laminae and parallel laminated fine-grained sandstones up to 2.5 cm thick, are exposed between the two components of the Camdwr Fault (Figure 29, loc.4) and also north of the fault farther to the west (Figure 29, loc.5). Lithologically similar beds at the top of the formation are seen in the cores of two anticlines (Figure 29, locs.6 and 7) at the western end of the gorge. A nearby syncline in very dark grey mudstones (Figure 29, loc. 8) is confined by the main components of the Camdwr Fault.

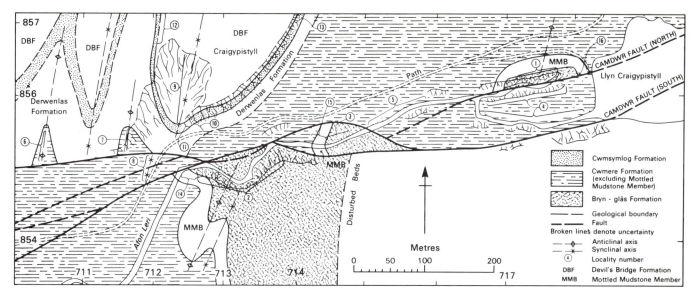

Figure 29 Sketch-map of the Craigypistyll area

Carn Owen – Nant Rhyddlan – Afon Lluestgota

Some 11 m of beds are exposed in a stream gully [7279 8815] on the west side of Carn Owen. They are considered to belong to the Mottled Mudstone, although only the basal 3 to 5 m of mudstone are medium grey and banded. The overlying 6 to 8 m are dark grey and rusty stained, and have been entered by a trial level [7273 8814]. The Mottled Mudstone is also exposed [7336 8793] on the east side of Carn Owen, where the *persculptus* Band, containing *G. persculptus* and consisting of dark grey, rusty weathering rubbly mudstone, is separated from the Bryn-glâs Formation by about 1 m of paler grey flaggy mudstone.

Beds belonging to the *cyphus* Zone are exposed in a roadside [7393 8825], where several metres of dark grey, rusty stained mudstones dip WNW at c.53° beneath the Derwenlas Formation. These beds are exposed again farther north [7405 8970] where a stream crosses the core of a tight anticline. Several metres of dark grey mudstone with rusty weather stains are overlain by some 11 m of rather dark grey, but non-rusty mudstones which probably mark the base of the Derwenlas Formation.

A sample of mudstone (RC1022; Plate 11) from Nant Rhyddlan illustrates banding typical of the formation; medium grey structureless mudstone considered to be turbiditic (Bouma *Te* interval, Figure 2a) alternates with dark grey laminar graptolitic mudstone considered to be largely pelagic.

In the strike-faulted area of Afon Lluestgota several forestry road sections expose the top part of the formation [e.g. 7463 9083 (Section I, p.74); 7412 9095; 7426 9115 (Section H, p.74)].

Bwlch-glas – Llechwedd Llûyd

Among several exposures in the Mottled Mudstone along the axis, and on the eastern side of the Coed Dipws Anticline, one [7025 8748], west of Bwlch-glas, yielded a graptolite fauna including *Climacograptus medius?*, *C.* cf. *miserabilis*, *C.* cf. *normalis* and *G. persculptus*. Feldspathic sandstone is present [7025 8756] at the top of the Mottled Mudstone (p.43) NW of Bwlch-glas, where dark grey pyritous and rusty weathered cleaved mudstones overlie 15 cm of feldspathic, non-laminated sandstone, on about 1.5 m of dark grey mudstone, on 34 cm of coarse, feldspathic, non-laminated sandstone, on grey banded and cleaved mudstone (in the roadway).

Similar beds are seen [7033 8761] where a small ridge is capped by c.30 cm of coarse, grading upwards to finer, feldspathic sandstone. Beneath are some 1.1m of dark grey mudstone containing more layers up to 4 cm thick of similar sandstone. Just below the road, on the eastern side of a strike-fault, the *persculptus* Band is exposed in an old excavation [7139 8987].

Blaeneinion – Cefn Coch

An extensive roadside exposure [7255 9275 – 7298 9278] shows dark grey, rusty coated, cleaved mudstones in the upper part of the formation repeated by several small folds. Slightly lower beds, with clearly displayed banding, are exposed in the stream [7286 9314], and this section continues eastward through some minor folding up to the top of the formation [7314 9303]. The top is also exposed in the roadside [7370 9450] where rusty stained dark grey cleaved mudstone crops out beneath the Derwenlas Formation. Similar mudstones near the middle of the formation are exposed in a gully [7392 9498], while nearby [7387 9480], and at much higher horizons, sandstones up to 7 cm thick are present.

Foel Goch Outlier

A small outlier of the Mottled Mudstone lies just beneath the Brwyno Overthrust east of Foel Goch. Well cleaved mudstones, dug for slates [7027 9265], are banded medium and pale grey, the latter being mottled by the darker grey spots of *Chondrites* that are typical of the Mottled Mudstone. Thin silty layers are common at the bases of the banded couplets. The *persculptus* Band was proved in a narrow hollow separating these rocks from the Bryn-glâs Formation [7024 9257].

Cwm Ceulan – Cwmere – Afon Clettwr – Cymerau

In Cwm Ceulan, 11 m of grey mudstone with pale bands, belonging to the Mottled Mudstone, are exposed upstream from Nant-y-nôd [7018 9046]. Farther upstream, and some 6 m higher in the succession, a gully exposes several metres of dark grey, rusty pyritous mudstone that contains thin (5 mm or less) siltstone partings which produce a small waterfall [7024 9061]. *?Diplograptus modestus* s.l. and

Climacograptus sp. are present at the foot of this waterfall. Faulting may cut out part of the middle of the section. South-west of Carregcadwgan an old quarry [6913 9005] shows 5.8 m of grey well cleaved mudstone with medium-grained sandstone beds up to 20 cm thick in the top 1.8 m. These sandstones are taken to mark the top of the Mottled Mudstone.

The basal 7.0 m of the Mottled Mudstone are seen in Afon Clettwr [6758 9173] at Pont Cwm-pandy. The strata comprise grey cleaved mudstones with some dark grey beds, and with many pale grey beds up to 2 cm thick from 1.2 to 2.7 m above the basal junction. Pyritised graptolites, including *G. persculptus*, are abundant, especially about 2 m above the base (p.40). Sections downstream [6728 9189 – 6756 9176] and in the roadsides immediately north and south of the river display most of the beds above the Mottled Mudstone, totalling at least 100 m. These sections demonstrate the monotonous nature of the strata (as described on p.43).

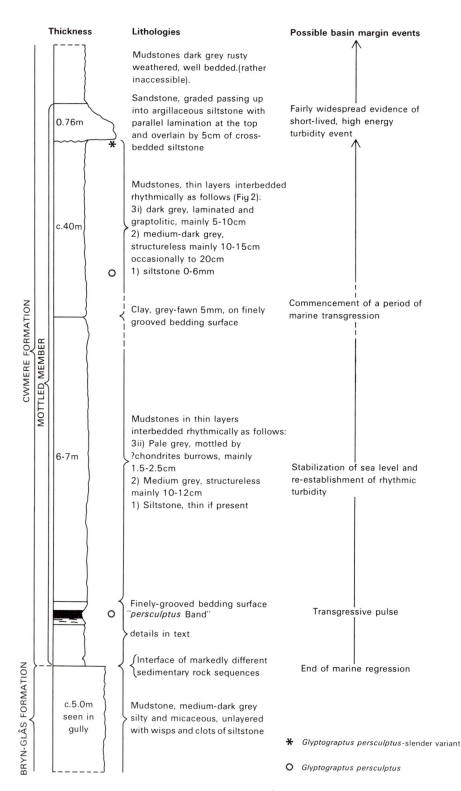

Thickness	Lithologies	Possible basin margin events
	Mudstones dark grey rusty weathered, well bedded.(rather inaccessible).	
0.76m	Sandstone, graded passing up into argillaceous siltstone with parallel lamination at the top and overlain by 5cm of cross-bedded siltstone	Fairly widespread evidence of short-lived, high energy turbidity event
c.40m	Mudstones, thin layers interbedded rhythmically as follows (Fig 2): 3i) dark grey, laminated and graptolitic, mainly 5-10cm 2) medium-dark grey, structureless mainly 10-15cm occasionally to 20cm 1) siltstone 0-6mm	
	Clay, grey-fawn 5mm, on finely grooved bedding surface	Commencement of a period of marine transgression
6-7m	Mudstones in thin layers interbedded rhythmically as follows: 3ii) Pale grey, mottled by ?chondrites burrows, mainly 1.5-2.5cm 2) Medium grey, structureless mainly 10-12cm 1) Siltstone, thin if present	Stabilization of sea level and re-establishment of rhythmic turbidity
	Finely-grooved bedding surface "*persculptus* Band" details in text	Transgressive pulse
	Interface of markedly different sedimentary rock sequences	End of marine regression
c.5.0m seen in gully	Mudstone, medium-dark grey silty and micaceous, unlayered with wisps and clots of siltstone	

CWMERE FORMATION

MOTTLED MEMBER

BRYN-GLÂS FORMATION

Figure 30 Cardiganshire Slate Quarry [6991 9595], near Cymerau. A section through the top of the Bryn-glâs Formation and the Mottled Mudstone Member of the Cwmere Formation

✳ *Glyptograptus persculptus*-slender variant

○ *Glyptograptus persculptus*

The Cardiganshire Slate Quarry [6991 9595] (Figure 30) SE of Cymerau, presents the best exposure in the district of the lower part of the formation. The Mottled Mudstone, including the *persculptus* Band, is completely exposed in the walls of the cascade at the southern end of the quarry. The basal 1.45 m are exposed half way down the cascade on its NW side and have been logged as follows:

	Thickness m
Mudstone, grey brown, tough, cleaved. Top surface grooved NW–SE by sliding of the superincumbent bed	0.220
'Persculptus Band': mudstone, grey-brown, soft, rubbly. Abundant small limonitic flecks. *G. persculptus*. Top of bed softer than lower part and laminated	0.305
Mudstone, grey, cleaved, tough	0.102
Mudstone, dark grey and shaly with limonitic specks and pale grey weathered spots	0.053
Mudstone, medium grey, tough, bedding lamination faintly visible on weathered surfaces together with layers of weathered pyrite particles	0.305
Shale, as below	0.083
Mudstone, medium grey, cleaved and slightly shaly	0.102
Shale, grey with limonitic (weathered) laminae, indeterminate graptolitic remains	0.102
Mudstone, medium grey, structureless, evenly cleaved	0.178
Bryn-glâs Formation	seen 0.610

Nearby, a quarry [6978 9601] exposes 7 m of cleaved mudstone, dark grey, soft and pyritous, with pyritised graptolites. Fracture surfaces are layered with hydrated oxides of iron. From 1 to 1.5 m below the top of the sequence *G. persculptus?*, *G. sp*, *Orthograptus sp.*, and possible examples of *Raphidograptus toernquisti* were identified and, from 0.5 m below the top, *Climacograptus sp.* and *G.* cf. *persculptus*. The quarry is up-dip of the main package of thin arenites in the formation, which suggests that these beds are not far below the *cyphus* Zone.

Llyfnant Valley

On the south side of the valley there are two synclinal outliers of beds low in the formation. In the eastern syncline, in quarries [7304 9734] NNE of Hiraeth, some 8 m of the lower part of the Mottled Mudstone are visible. They consist of medium grey and pale grey, banded, well cleaved (slabby) mudstones and pass up into a few metres of darker grey pyritous and rusty weathering mudstones that occupy the core of the syncline. These latter probably correlate with the dark mudstone below the sandstone noted in the Cardiganshire Slate Quarry (Figure 30). At the roadside [7302 9745] north of these quarries there is an adit in dark grey rusty mudstones. Some 40 m to the west the base of the Mottled Mudstone is exposed, and about 1 m above it are soft mudstones with limonitic speckles signifying the presence of the *persculptus* Band. Details on the north side of the valley are given in Figures 31a and b, see also Jones and Pugh, 1916, pp.353, 364).

Taliesin – Ynys-Hir

Almost the full thickness of the Mottled Mudstone is exposed [6684 9362 – 6684 9367] on the eastern side of Afon Ddu as follows:

	Thickness m
Cwmere Formation (Above Mottled Mudstone Member)	
Mudstone, grey to dark grey, rusty brown weathering, some silty laminae	seen 6.1
Cwmere Formation—Mottled Mudstone Member	
Discontinuous bed of weathered sandstone nodules (up to 30 cm diameter and 15 cm thick)	up to 0.15
Mudstone, dark grey, slightly rusty brown weathering	1.2
Discontinuous bed of weathered sandstone nodules (up to 30 cm diameter and 15 cm thick)	up to 0.15
Mudstone, grey to dark grey, slightly rusty brown weathering	6.7
Mudstone, grey, hard	c.4.9
Mudstone, very dark grey, rusty brown weathering, much pyrite	seen 1.2

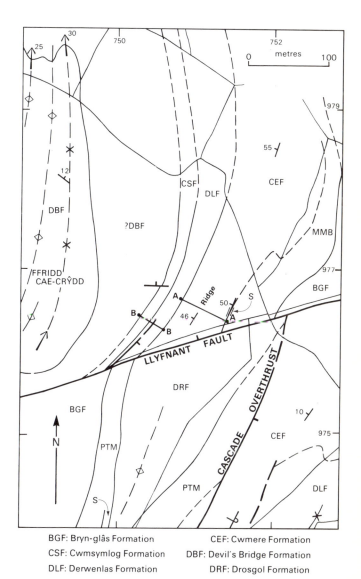

BGF: Bryn-glâs Formation
CSF: Cwmsymlog Formation
DLF: Derwenlas Formation
CEF: Cwmere Formation
DBF: Devil's Bridge Formation
DRF: Drosgol Formation

Figure 31a. Geological sketch of Fridd Cae-crŷdd

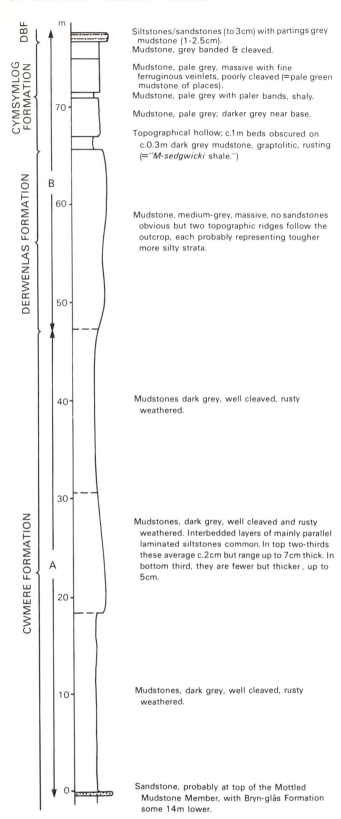

Figure 31b. Succession estimated from the outcrop near Fridd Cae-crŷdd

The beds of sandstone nodules may be equivalent to the sandstone beds recorded elsewhere (p.42) at the top of the member.

Dark grey ochreous weathering mudstones towards the top of the formation are discontinuously exposed in a forestry track [6702 9386–6698 9377] c.200 m to the NE. Graptolites from the northern, and stratigraphically lower, end of this section include *Atavograptus atavus*, *Climacograptus normalis*, ?*Coronograptus cyphus* and *Pseudoclimacograptus* (*Metaclimacograptus*) *hughesi* indicative of the *acinaces* Zone or *cyphus* Zone.

North of the Dovey Estuary

An exposure [6660 9898] to the NE of Mynydd y Llyn shows, in a downward sequence, c.15 cm of dark grey pyritous and shaly mudstone with *G. persculptus* (*persculptus* Band), c.15 cm of medium to pale grey cleaved mudstone, c.10 cm of dark, grey rusty striped mudstone with silt blebs, and c.18 cm of medium grey mudstone, overlying the Bryn-glâs Formation. The thickness of the Mottled Mudstone in this area is estimated at 13 m, little different from more southerly areas, but the thickness of the beds beneath the *persculptus* Band is clearly less than to the SE, and the most northerly record of this survey [6905 9988] confirms this trend. In a stream [6497 9994] to the west, dark grey graptolitic mudstone yielded ?*Climacograptus miserabilis*, *Diplograptus sp.* and *Orthograptus?* A nearby quarry [6027 9806] also displays the *persculptus* Band in its western face, apparently some 0.6 m above the Bryn-glâs Formation though the junction may be faulted.

DERWENLAS FORMATION

Eisteddfa Gurig – Ponterwyd – Syfydrin – Camdwr Fault

The major sections of Jones (1909) are still accessible and a composite section is shown in Figure 32. The base of the formation is taken at a bed of calcareous nodules (Jones, 1909, p.490; Sudbury, 1958, p.486) up to 30 cm in diameter, which approximately coincides with an upward change in lithology from dark grey, orange-brown weathering, thinly bedded, and commonly graptolitic mudstones to bluish grey, thickly bedded mudstones in which the dark grey graptolitic mudstones are restricted to specific 'bands'. Above the *leptotheca* Band the mudstones are paler (the Castell Group of Jones 1909) and contain numerous parallel-laminated siltstone and very fine-grained sandstone beds up to 5 mm thick. There are thicker fine-grained parallel-laminated sandstones, with individual beds up to 22 cm thick in places, in the middle part of the *convolutus* Zone, totalling about 3 m in Nant Meirch [7765 8125] and Nant Fuches-wen [7709 8079], and up to c.18 m [7504 7973] west of Tynffordd, where the section reads:

	Thickness m
Mudstone, pale grey, thin sandstone beds passing down into	—
Sandstone, pale grey, fine-grained, massive; thin mudstone beds	4.6
Mudstone, pale grey, interbedded with thin siltstones and sandstone beds	2.7
Interlayered pale grey mudstone and sandstone in beds up to 5 cm, sandstones thinner towards top	3.7
Interlayered pale grey sandstone and mudstone in 7.5 cm beds	7.3
Mudstone, pale grey to bluish grey, some thin sandstone beds in upper part	—

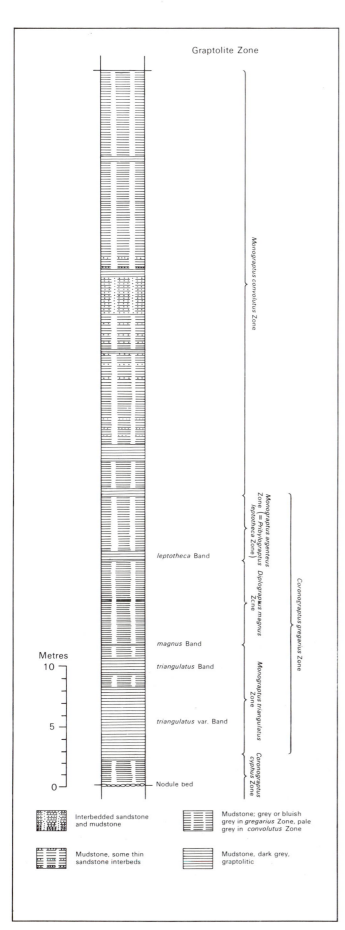

Graptolite Zone

Monograptus convolutus Zone

Monograptus argenteus Zone (= *Pribylograptus leptotheca* Zone)

Diplograptus magnus Zone

Coronograptus gregarius Zone

Monograptus triangulatus Zone

Coronograptus cyphus Zone

leptotheca Band

magnus Band

triangulatus Band

triangulatus var. Band

Nodule bed

Metres
10

5

0

Interbedded sandstone and mudstone

Mudstone, some thin sandstone interbeds

Mudstone; grey or bluish grey in *gregarius* Zone, pale grey in *convolutus* Zone

Mudstone, dark grey, graptolitic

The sandstones thin northwards to 3 m [7460 8024]. Pale grey to grey colour-banded mudstones, with thin siltstone laminae in places, are well exposed between this section and the Cwmere Formation [7485 8021] to the east. The total thickness of the formation here ranges from about 50 to 58 m; unlike the sequence around Goginan (p.45), it does not appear to be affected by the varying thickness of the arenaceous beds within it.

A stream section [7431 8189 – 7426 8183] south of Craignant-mawr shows:

	Thickness m
Mudstone, bluish grey, many fine-grained sandstone beds up to 2.5 cm thick in top 7 m; rare shell fragments (Appendix 3, loc.4)	c.11.0
Mudstone, dark bluish grey, rusty brown weathering; some very dark beds with graptolites including *Glyptograptus tamariscus linearis*, *Monograptus capis*, *M.convolutus*, *M.delicatulus*, *M.* cf.*gemmatus*, *M.* cf.*involutus*, *M. limatulus*, *M. lobiferus*, *Pribylograptus leptotheca*, *Pristiograptus regularis* and *Pseudoclimacograptus (Clinoclimacograptus) retroversus*. (indicative of the *convolutus* Zone)	4.6
Mudstone, bluish grey, discontinuous exposure; many thin sandstone beds, up to 2.5 cm thick in top 1.8 m; graptolite fauna c 5.5 m from base includes *M. capis*, *M.convolutus?* and *M.lobiferus*. (probably *convolutus* Zone)	10.4
Mudstone, pale bluish grey; abundant fine-grained sandstone beds up to 5 cm thick in top 5 m	10.1
Sandstone, pale grey, fine-grained	0.9
Mudstone, pale grey, many thin sandstone beds	1.2
Mudstone, dark grey, rusty brown weathering. Graptolites include *Monograptus* cf. *denticulatus*, *M. lobiferus*, *Petalograptus ovatoelongatus*, *?Pseudoclimacograptus (C.) retroversus* and *Rastrites* cf. *peregrinus*. (indicative of *convolutus* or *leptotheca* Zone). *Discinocaris* cf. *browniana* also present (Appendix 3, loc.3)	0.8
Mudstone, pale grey	4.3
Mudstone, dark grey, silty. Graptolites include *?Diplograptus magnus*, *Monograptus communis*, *M. revolutus* and *M. triangulatus* cf. *fimbriatus*. (*magnus* Zone)	4.6

The arenaceous beds within the *convolutus* Zone are locally well developed [7342 8305] (Figure 33, loc.1); at this locality 12.8 m of grey laminated sandstones (beds up to 22 cm thick in the lower 9 m, 5 to 10 cm thick above) are interlayered with mudstone beds up to 5 cm thick in the basal part of a long section. The sandstone beds die out northwards.

Laminated and convolute-laminated medium- to fine-grained sandstones in beds up to 60 cm thick are seen at Syfydrin [7270 8445], and interlayered laminated sandstones and mudstones form the basal 9 m of a cliff section [7217 8451] south of Llyn Syfydrin, the overlying beds being the usual pale bluish grey mudstones with thin sandstone beds. About 9 m below the sandstone packet at the top of the section, a forestry track [7188 8429] to the SW yielded a very extensive *convolutus* Zone fauna including *Cephalograptus cometa*, *M. capis?*, *M.* cf. *clingani*, *M.communis rostratus*, *M.convolutus*, *M.deci-piens*, *M.denticulatus*, *M.lobiferus*, *M.* cf.*undulatus*, *Orthograptus cyper-oides*, *Pristiograptus* cf. *concinnus*, *P.regularis*, *Pseudoclimacograptus (Meta-climacograptus) undulatus?* and '*Retiolites*' *sp.*

Figure 32 Composite section of the Derwenlas Formation in the Rheidol Gorge-Eisteddfa Gurig area

Figure 33 Folding in the Derwenlas and adjoining formations, south of Disgwylfa Fâch

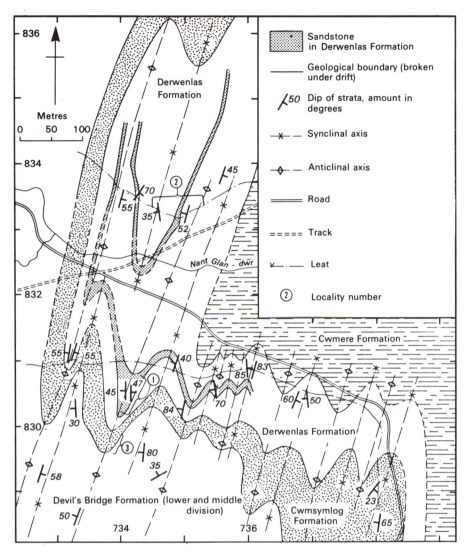

There is a small inlier in the topmost beds of the formation in the core of the Pen y Graig-ddu Anticline where it is cut by the gorge of Afon Melindwr; there are sections in pale grey silty mudstones with 3 to 6 mm sandstone beds every 12 to 25 mm in the forestry track (Figure 34) and the river. Northward [7111 8486], pale grey mudstones are underlain by interlayered pale grey and dark grey mudstones with a graptolite fauna including *Diplograptus magnus*, *Monograptus pseudoplanus?*, *M. triangulatus fimbriatus*, *Petalograptus* cf. *ovatoelongatus* and *R. toernquisti*, indicative of the *magnus* Zone. Higher beds, including a 6 m packet of grey laminated sandstones, are seen in sections [7068 8518 – 7054 8512] to the WNW.

Craigypistyll – Hafan Fault

The formation appears to be at least 35 m thick at Craigypistyll, thinning northwards to about 20 m at the Hafan Fault. Crags at Craigypistyll [7126 8555] (Figure 29, loc. 10) show:

	Thickness m
Cwmsymlog Formation	
Sandstone, fine- to coarse-grained, graded, with graptolites including *Monograptus sedgwickii*; some mudstone laminae	3.6
Derwenlas Formation	
Mudstone, bluish-grey, interbedded with fine-grained	

	Thickness m
laminated sandstones 2.5 cm thick at base, decreasing in thickness and abundance upwards; only thin silty laminae in top 3 m	7.3
Sandstone, grey, medium-grained, in 10 cm beds interbedded with mudstone	1.2
Mudstone, bluish grey, many 1 cm fine-grained sandstone beds in top 5.5 m	7.3
Mudstone, dark grey, rusty brown weathering, some paler stripes	3.7
Mudstone, bluish grey, darker in basal metre	5.8
Mudstone, dark grey, rusty brown weathering, with paler stripes. Abundant graptolites include *Coronograptus gregarius*, *M. triangulatus fimbriatus*, *M. t.* cf. *triangulatus* and *Rastrites longispinus?* (*magnus* or, less likely, uppermost *triangulatus* Zone)	—

The 1.2 m sandstone may be the equivalent of sandstones in the *convolutus* Zone in other areas.

Near the Hafan Fault, interbedded coarse- to medium-grained flaggy sandstones and dark mudstones of the Cwmsymlog Formation [7172 8720] are underlain by 3 m of pale to medium grey flaggy mudstones, and then by dark grey, rusty weathering mudstones with an extensive graptolite fauna including *M.* cf. *convolutus*, *M. limatulus*, *M. lobiferus* and *Pribylograptus argutus*.

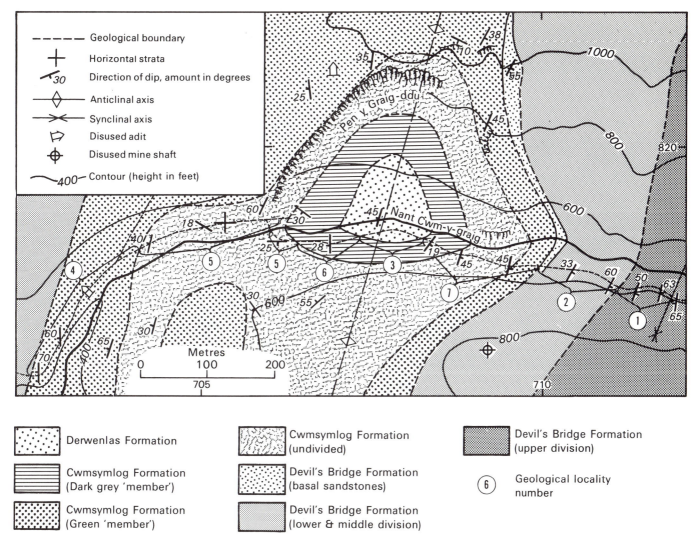

Figure 34 Geological sketch-map of the Pen y Graig-ddu area

Legend:
- - - - - Geological boundary
- Horizontal strata
- /30 Direction of dip, amount in degrees
- Anticlinal axis
- Synclinal axis
- Disused adit
- Disused mine shaft
- ~400~ Contour (height in feet)

Derwenlas Formation

Cwmsymlog Formation (Dark grey 'member')

Cwmsymlog Formation (Green 'member')

Cwmsymlog Formation (undivided)

Devil's Bridge Formation (basal sandstones)

Devil's Bridge Formation (lower & middle division)

Devil's Bridge Formation (upper division)

(6) Geological locality number

Llechwedd Gwineau–Afon Lluestgota; Moel Cyneiniog

Between Llechwedd Gwineau [735 872] and Afon Lluestgota, roads and tracks provide partial sections summarised below and correlated in Figure 35. The formation is comparatively thin (20 to 29 m) here; the beds of the *convolutus* Zone thicken northwards while those of the lower zones become thinner. The strata consist mainly of medium grey mudstone as thin rhythmites with thin siltstone interbeds mostly in the upper part. Dark grey graptolitic mudstone is not uncommon, and above the *triangulatus* Zone forms several graptolite bands. The lateral extent of these bands is variable; some, as for example the '*Monograptus-sedgwicki* shales' at the base of the Cwmsymlog Formation, are present in all sections; others, such as the *leptotheca* Band, are less certainly recognisable everywhere. In several sections there are two other bands between the '*Monograptus-sedgwicki* shales' and the *leptotheca* Band (as interpreted in Figure 35) but they are not present in all sections. To the SE around Ponterwyd five bands occur in this interval (Figure 32).

Arenites at the top of the *convolutus* Zone at Y Chwareli [736 879] (p.85 and Figure 41) are underlain by a thin graptolitic mudstone. These arenites mask the culmination of the thickening/coarsening-upwards sequence seen in nearby sections: they are also exposed in the east-facing scarp [7320 8949] just north of Bryniau Rhyddion and are again underlain by a thin graptolitic mudstone [7313 8930]. Here, however, the arenites are mere leaves of lenticular and cross-bedded siltstone up to 0.3 cm thick. Along the outcrop NNE from the eastern side of Moel Cyneiniog [7255 8807], the '*Monograptus-sedgwicki* shales' are absent, and the arenites may occupy part of the *sedgwickii* Zone as they do to the west between Bwlch-glas and Blaen-Ceulan (p.75). Clearly the outcrop lies across or near the NE margin of a body of more arenaceous turbidites of late *convolutus* Zone age.

Some of the recorded sections between Llechwedd Gwineu and Afon Lluestgota (Figure 35) and the contained faunas are as follows:

Section A. Roadside; [7393 8825]

Thickness
m

Derwenlas Formation

7. Mudstone, dark grey, shaly, with green stripe near top. *M. convolutus, M.* cf.*involutus, ?M. limatulus, M. lobiferus, ?Pristiograptus concinnus, P. regularis, Pseudoclimacograptus (C.) retroversus, P.(M.) undulatus. (convolutus* Zone)   at least 1.0

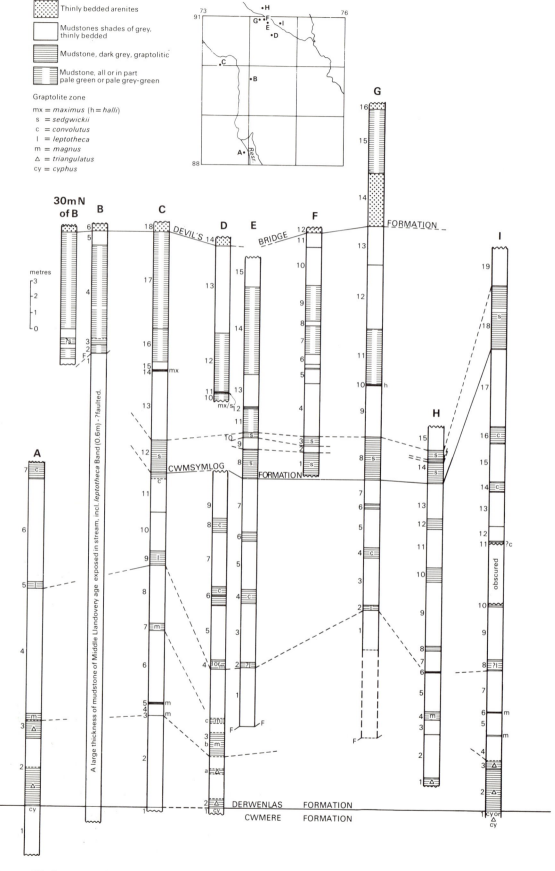

Figure 35 Comparative sections in the Derwenlas and Cwmsymlog formations in the area from Capel Tabor to the Afon Lluestgota

Thickness
m

6. Mudstone, grey-brown — 6.7
5. Mudstone, dark grey with white weathered stripes, green stripes near centre and top. *Glyptograptus?, M.* cf. *argenteus, M. capis, M. communis, M. denticulatus, ?M.triangulatus, P. leptotheca,* ?pernerograptids, indet. triangulate monograptids and biserials. (Probably *leptotheca* Zone) — 0.46
4. Mudstone, paler than below, banded, with pale bands bioturbated. Some thin interbeds of pale grey siltstone — 7.9
3. Mudstone, dark grey, shaly; top 0.46 m: *D. magnus, G. tamariscus, M.revolutus* s.l., *?M.triangulatus fimbriatus, M.t. major?, ?P.concinnus, Pseudoclimacograptus (M.)hughesi, R. longispinus, Rhaphidograptus toernquisti.* (*magnus* Zone); bottom 1.2 m: *G. tamariscus, G.*cf.*sinuatus, M.communis, M. triangulatus separatus, ?M.triangulatus triangulatus, P. (M.) hughesi, R. toernquisti.* (Probably *triangulatus* Zone) — 1.66
2. Mudstone, dark grey, blocky; bottom 2.4 m graptolitic, with *?A. atavus, Climacograptus sp., M.revolutus* s.l., *M. triangulatus triangulatus* and *R.toernquisti.* (*triangulatus* Zone) — 4.2

Cwmere Formation

1. Mudstone, dark grey cleaved and shaly. *A.atavus, Climacograptus sp., Coronograptus cyphus, Glyptograptus?, ?Monograptus difformis, M.revolutus* s.l., *Orthograptus* aff. *mutabilis, Pribylograptus incommodus* s.l., *Pseudoclimacograptus (M.) hughesi, R. toernquisti,* sponges.(*cyphus* Zone) — —

Section C. Ditch, SW side of road [7340 9000]

Derwenlas Formation

11. Mudstone, grey, softer than bed below 2.14–2.44 m above base: *Coronograptus gregarius, Glyptograptus sp., ?M. communis, M.convolutus, M. decipiens, M. lobiferus, Petalograptus ovatoelongatus, ?Pristiograptus regularis, Rastrites sp.* (Probably *convolutus* Zone); 1.83 m above base: *Climacograptus sp., ?M.capis, ?M.communis rostratus, ?M. convolutus, M. decipiens, ?M.denticulatus, M.limatulus, M.lobiferus, Pristiograptus* cf. *fragilis, ?P.regularis, ?Pseudoclimacograptus (C.) retroversus.* (Probably *convolutus* Zone) — 2.44
10. Mudstone, medium grey, with thin siltstone interbeds — 2.44
9. Mudstone, dark grey, soft, weathers to ash grey, only basal third seen. *Glyptograptus sp., M.* cf.*argenteus, M.lobiferus, Monograptus sp.* (triangulate), *P. (C.) retroversus, P.(M.)* cf. *hughesi, P.(M.) undulatus, R. longispinus?, R.* cf. *peregrinus, Rhaphidograptus toernquisti.* (Probably *leptotheca* Zone) — c.0.91
8. Mudstone, medium grey, slight green tinge — 3.66
7. Mudstone, very soft, weathered nearly white. *C.gregarius, ?Diplograptus spp., Glyptograptus sinuatus?, G.tamariscus* s.l., *Monograptus communis, M.* cf. *denticulatus, M. revolutus.* (possibly *magnus* Zone) — 0.46
6. Mudstone, medium grey, slightly brown weathering, becomes darker towards base — 4.57
5. Mudstone, dark grey. *Climacograptus sp., D.magnus, Glyptograptus sp., G. (Pseudoglyptograptus)* cf. *vas, G. (P.) sp., Monograptus triangulatus fimbriatus, M. pseudoplanus, Petalograptus ovatoelongatus, ?Pristiograptus sp., R. longispinus, R. sp., Rhaphidograptus toernquisti.*(*magnus* Zone) — 0.15

Thickness
m

4. Mudstone, medium grey — 0.69
3. Mudstone, dark grey. *D. magnus, Glyptograptus sp., ?G. (P.) sp., M. t. fimbriatus, M.pseudoplanus?,* triangulate monograptid, *M. triangulatus* s.l.(*magnus* Zone) — 0.04
2. Mudstone, medium to dark grey, blocky — 5.8

Cwmere Formation

1. Mudstone, dark grey, rusty weathered, cleaved — —

Section D. Forest track [7450 9062]

Derwenlas Formation

9. Mudstone, medium grey — 3.0
8. Mudstone, dark grey. *Climacograptus sp., G. tamariscus* s.l., *M.capis, M.convolutus, M.decipiens, M.*cf.*denticulatus, M.lobiferus, ?Petalograptus minor, P. ovatoelongatus, Pristiograptus regularis, Rastrites ?approximatus geinitzi, R. hybridus, R. spina.* (*convolutus* Zone) — 0.9
7. Mudstone, medium grey — 3.4
6. Mudstone, dark grey, graptolitic in top half yielding: *Climacograptus sp., Coronograptus gregarius, G.tamariscus linearis?, G.sp., M.communis, M.*cf.*convolutus, M.decipiens, M.involutus?, ?M.limatulus, M.lobiferus, P.fragilis, P. regularis, R.hybridus.*(*convolutus* Zone) — 1.22
5. Mudstone, medium grey — 3.4
4. Mudstone, dark grey. *Climacograptus sp., ?D. magnus, Glyptograptus sp., ?Monograptus cerastus, M.spp.* indet., *?R. toernquisti.* (Possibly *leptotheca* Zone) — c.0.61
3. Mudstone, dark grey, blocky. Graptolites from three rather shaly parts, see Figure 35
(c) *?C. gregarius, D. magnus, M.pseudoplanus, ?M. revolutus, P. ovatoelongatus.* (*magnus* Zone)
(b) *D. magnus, G. sinuatus, M. t. fimbriatus, M. t.* cf. *triangulatus, P. (M.) hughesi, R.toernquisti.*(*magnus* Zone)
(a) *C.gregarius, G.tamariscus* s.l., *M. triangulatus* s.l., *R. toernquisti.* (Probably *triangulatus* Zone) — 8.49
2. Mudstone, dark grey, rather shaly. *A. atavus, ?M.revolutus* s.l., *M. 't. triangulatus, M.t. separatus?, R. toernquisti.* (*triangulatus* Zone) — 0.61

Cwmere Formation

1. Mudstone, dark grey, cleaved and rusty weathering. *C.* cf. *cyphus, Cystograptus sp., M. revolutus* s.l., *R.toernquisti.* (Probably *cyphus* Zone) — —

Section E. Forest track [7440 9089]. Lithological characteristics as for Section D

6. *M. lobiferus, ?M.pragensis ruzickai, ?M. sedgwickii, M.* cf.*undulatus, P. regularis, Rastrites sp.,* triangulate monograptid
4. *M. convolutus, Rastrites?*
2. *Glyptograptus sp., Pribylograptus leptotheca?, Pristiograptus regularis?.* Probably the same bed [7440 9072] in the same section around a bend in the road yielded: *?M. decipiens, ?Pribylograptus leptotheca, Pristiograptus regularis?, Rastrites* aff. *fugax.* Some elements here are suggestive of high middle Llandovery horizons yet they lie below a band yielding: *M.* cf. *convolutus, M. lobiferus, P.* cf. *regularis, Pseudoclimacograptus (M.) undulatus, R.* cf. *hybridus,* triangulate monograptid, which might equate with Bed 4 of Section D.

	Thickness m
Section G. Forest track [7419 9094]	
Derwenlas Formation	
7. Mudstone, medium grey with pencil cleavage	1.52
6. Mudstone, dark grey. *?M. delicatulus,* *M.lobiferus,* triangulate monograptid	0.3
5. Mudstones, medium grey, massive	2.44
4. Mudstone, dark grey, shaly, rusty weathering. *M. lobiferus.*	0.61
3. Mudstone, medium grey, relatively massive and splintery	3.0
2. Mudstone, dark grey, shaly. *M.lobiferus, P. (M.) undulatus.* (*leptotheca* Zone or higher)	0.3
1. Mudstone, dark grey, blocky, 2.44 m visible to fault	c.7.9

	Thickness m
Section H. Track [7426 9115]	
Derwenlas Formation	
13. Mudstone, blue-grey, thin siltstone interbeds	2.13
12. Mudstone, dark grey, graptolitic	0.61
11. Mudstone, rather pale greenish grey becoming grey towards base	2.44
10. Mudstone, dark grey, graptolitic	0.9
9. Mudstone, greenish brown with irregular brown veinlets becoming paler and grey near base	4.0
8. Mudstone, dark grey, graptolitic	0.3
7. Mudstone, brown weathering	1.22
6. Mudstone, dark grey shaly	c.0.2
5. Mudstone, medium to dark grey	2.44
4. Mudstone, dark grey, rusty weathering with *Climacograptus sp.* and *D.cf.magnus*	0.46
3. Mudstone, dark blue-grey with thin siltstone interbeds	0.9
2. Mudstone, dark grey, shaly, rusty and white weathering with triangulate graptolites	2.75
?Cwmere Formation	
1. Mudstone, dark grey, cleaved and shaly, rusty weathering. The junction between the formations in this area is ill-defined. In Section I similar beds contain *M.triangulatus* in the top metre or so	—

	Thickness m
Section I. Roadside [7463 9083]	
Derwenlas Formation	
17. Mudstone, blue-grey, thin interbeds of pale grey siltstone	4.9
16. Mudstone, dark grey, rusty. *?Climacograptus scalaris, C. sp., M.* cf. *delicatulus, M.cf.lobiferus, P. regularis, R. longispinus.* (Probably *convolutus* Zone)	1.1
15. Mudstone, blue-grey, thin interbeds of pale grey siltstone	2.44
14. Mudstone, dark grey. *M.* cf. *convolutus, M. lobiferus, P. (C.) retroversus.* (Probably *convolutus* Zone)	0.61
13. Mudstone, grey-brown	2.31
12. Mudstone, blue-grey with a few thin sandstones	0.9
11. Mudstone, dark grey. *C.* cf. *scalaris, Coronograptus gregarius, Glyptograptus (P.) sp., ?Monoclimacis crenularis, Pristiograptus concinnus, ?P. regularis, ?Pseudoclimacograptus (M.) undulatus.* Loose: *P. leptotheca, ?Pristiograptus concinnus, Pseudoclimacograptus (M.) hughesi, Rastrites sp.* (Possibly *convolutus* Zone)	seen a few cm

	Thickness m
Obscured	c.4.0
10. Mudstone, dark grey, soft and graptolitic	seen a few cm
9. Mudstone, medium grey to grey-brown	3.35
8. Mudstone, dark grey, shaly. *Climacograptus sp., Coronograptus gregarius,* cf. *Monograptus argenteus, M. capis, M. communis* s.l., *P. leptotheca, Pseudoclimacograptus (M.) hughesi?.* (Probably *leptotheca* Zone)	0.61
7. Mudstone, blue-grey, paler than below, relatively massive	2.6
6. Mudstone, dark grey. cf. *D. magnus, M. capis, M.* cf. *denticulatus, Glyptograptus (P.) sp., R. toernquisti.* (Probably *magnus* Zone)	0.01
5. Mudstone, blue grey, relatively well cleaved	1.37
4. Mudstone, dark grey weathers brown and ash-grey, graptolitic at top and in basal 0.6 m (Bed 3) Top: *Climacograptus sp., D. magnus, ?G. (P.) vas, M. t. fimbriatus, M. t. major. P.* cf. *ovatoelongatus, R. longispinus, Rhaphidograptus toernquisti.* (*magnus* Zone) Bottom 0.6 m (= Bed 3): *Climacograptus rectangularis?, M. communis communis, M.cf. denticulatus, M.revolutus* s.l., *M.t. triangulatus, R. toernquisti.* (*triangulatus* or *magnus* Zone)	2.75
2. Mudstone, dark grey, crumbly, rusty and ash-grey weathering. *C. rectangularis, M. triangulatus, R. toernquisti.* (Probably *triangulatus* Zone, certainly post-*cyphus* Zone)	2.13
Cwmere Formation	
1. Mudstone, dark grey, crumbly and shaly, rusty weathering. Top 0.9 m: *A. atavus, C. gregarius, M. revolutus* s.l., *R. toernquisti.*(*cyphus* or *triangulatus* Zone); Remainder: *A. atavus, C. rectangularis, Coronograptus cyphus, Cystograptus vesiculosus, M. revolutus* s.l., *P. (M.) hughesi, R. toernquisti.* (Probably *cyphus* Zone)	seen 10.0

Siambr Traws-fynydd and Rhiw-gam area

There are several exposures near Siambr Traws-fynydd ranging in horizon from just beneath the Derwenlas Formation [7959 9283], where dark grey mudstones yielded *A. atavus, Climacograptus normalis, Coronograptus cyphus, ?G.sinuatus, M.revolutus, P. (M.) hughesi* and *R. toernquisti,* to just above it [7981 9302], where dark grey mudstones with rusty stains and white weathering-stripes yielded *M. sedgwickii.* Beds at a waterfall [7938 9269] contained *?C.gregarius, D. magnus, G.sinuatus?, M. communis, M. triangulatus separatus?, ?P. concinnus* and *Pseudoclimacograptus (M.) sp.,* while in the south bank, about 1.8 m lower, *D.* cf. *magnus, M.t.fimbriatus* and *R. toernquisti* were obtained, both faunas suggesting the *magnus* Zone.

Some 4.9 m of dark grey mudstones, which weather to produce white stripes, are visible in a gully [7961 9289]. They yielded *?D. magnus, G. tamariscus, M. triangulatus* aff. *separatus* and *?P. (M.) hughesi* near the base (*magnus* Zone). Nearby [7946 9271] more dark grey mudstone contained *G.* cf. *sinuatus, M.cf. argenteus, P. leptotheca* and *Pseudoclimacograptus (M.) hughesi,* probably of the *leptotheca* Zone, while at the eastern end of the gully [7964 9290] dark grey mudstone yielded *?C. scalaris* s.s., *Glyptograptus sp., M.* cf. *clingani, M.* cf. *denticulatus, Orthograptus insectiformis, ?P. concinnus, ?Pseudoclimacograptus (C.) retroversus,* and *R. toernquisti* indicating the *convolutus* Zone. A 5 cm sandstone in grey mudstone is visible [7990 9292], but in general these exposures contain a smaller arenite content than sections to the south, and an increase in thickness of the dominantly pelagic bands. The thickness of the formation remains, however, much the same, and it is still a coarsening-upwards sequence. In crags about 1 km farther NNE [8035 9417] the formation is c.36.8 m thick (measured).

Towards Rhiw-gam, most of the formation was measured in an old leat [8006 9433]:

Thickness
m

Cwmsymlog Formation

Mudstone, grey and, near base, pale green, structureless, no layering or banding. Cleavage uneven and phacoidal. Ramified by dark brown fine veinlets, subvertical and parallel with the cleavage. Forms marked topographical ridge — seen c.9

Mudstone, slightly silty, grey, blocky with abundant smoothly grooved dislocation surfaces, induced during soft deformation. Largely non-layered with the exception of a few thin layers of mudstone with a hackly fracture. Some veining present of the type common in overlying mudstone — 9

Mudstone, medium grey with pale banding; smoothly fluted fracture surfaces common — 1.8

Mudstone, dark grey, rusty weathering and graptolitic, poorly exposed; '*Monograptus-sedgwicki* shales', well developed and producing a marked topographical hollow — c.4.9

Derwenlas Formation

Mudstone, medium grey, non-banded; no arenite; top sharp — 2.1

Mudstone, dark grey, friable. *M. convolutus?*, *M. lobiferus*, *Orthograptus cyperoides*, triangulate monograptids — c.1.1

Mudstone, medium grey, massive, vaguely banded: no arenites — 6.7

Mudstone, dark grey, rusty weathering, friable. *M. lobiferus* and triangulate monograptid — 2.1

Mudstone, medium grey — 5.5

Mudstone, dark grey, friable. *M. lobiferus*, *P. (M.) hughesi* — 0.6

Mudstone, medium grey — seen 2.7

Leri Valley to Melindwr Valley

In the Pen y Castell Inlier [688 847] the following composite section towards the top of the formation is seen in Afon Stewy [6878 8464 – 6887 8464] and in nearby scattered exposures on the slopes of the hill fort.

Thickness
m

Mudstone, dark greenish grey and grey with thin (1 to 5 mm) paler current-laminated sandstone beds — seen 3

Sandstone, bluish grey, turbiditic; individual beds average 12 cm, but are commonly multiple; thin shale interbeds form 10% of the succession — 13

Mudstone, dark grey and bluish grey, silty; white siltstones (up to 5 mm) and sandstones (up to 4 cm) form 20% of the succession — 14

Sandstone, bluish grey, up to 30 cm thick; well graded distal turbidites commonly exhibiting a full Bouma *Ta – d* sequence (Figure 2a) with a characteristic interval of convolute and contorted laminae with small sandstone balls; the uppermost 3 m are similar but with sandstones only up to 5 cm thick. Pale grey and bluish grey intercalated mudstones and silty mudstones, commonly banded, occur throughout but only form 10% of the succession — 10

Thickness
m

Mudstone, pale bluish grey, silty, with thin (to 5 mm) sandstone beds — seen 3

Greenish to bluish grey mudstones with many thin siltstone beds and graptolite debris crop out in an old adit mouth [6789 8166] in an inlier in the Melindwr Valley.

Bwlch-glas – Blaen-Ceulan – Gweunbwll

Between Bwlch-glas and Blaen-Ceulan the formation is about 33 to 45 m thick, much thicker than to the east of Carn Owen. It is greatly inflated by arenitic detritus forming thin interbeds of siltstone, commonly cross-laminated. The mudstones are commonly wavy partings or drapes separating lenticular wavy bedded siltstone. Green and pale grey colours are common. Broadly the deposits are composed of the *c – d – e* intervals of the Bouma sequence (Figure 2a).

In places it is difficult to separate the Derwenlas and Cwmsymlog formations, for arenitic detritus has seemingly overwhelmed the marker bed of the '*Monograptus-sedgwicki* shales'. In general the arenites become thicker and coarser towards the top of the formation, with beds up to 0.3 m in Afon Cyneiniog (p.76). This suggests a prograding turbiditic body of rhythmites rather than channel-fill. A thinning- and fining-upwards sequence is visible in a lane [7091 8793] near Alltgochymynydd (p.76), so channel fills may be present within this prograding body. Graptolites are present, but hemipelagic dark grey mudstone is not obvious.

The following sections are typical:

Nant Bwlch-glâs [7031 8737 – 7035 8735]

Thickness
m

Mudstone, pale bluish grey, discontinuous exposure in upper part. Total thickness to base of Devil's Bridge Formation — c.12

Mudstone, bluish grey — 7.6

Mudstone, dark bluish grey, some black beds up to 5 cm thick with a fauna including *D. magnus*, *M.* cf. *denticulatus*, *P. ovatoelongatus*, *R. toernquisti*. (*magnus* Zone) — 3

Mudstone, bluish grey (discontinous exposure) — —

Afon Cynciniog [7021 8023]

Devil's Bridge Formation

Sandstone and siltstone, medium to pale grey, thinly bedded in parallel layers up to 6 cm thick — seen 1.2

Cwmsymlog Formation

Obscured — c.2.1

Siltstone grey, and mudstone, medium and pale grey banded, unevenly and thinly interbedded; some cross-lamination. Some arenite layers are coarse, gritty (with lithic grains) and graptolitic, the stipes orientated 065°, 110° and 200°. Graptolites from 1.83 m above base are *Acanthograptus?*, *Dendrograptus sp.*, *M.* cf. *pragensis ruzickai*, *M. sedgwickii*, *P. regularis*, *Pseudoclimacograptus (M.) undulatus*, retiolitid; also a scolecodont. (*sedgwickii* Zone) Graptolites from 1.22 m above base are *M. sedgwickii* and *P. regularis*. (*sedgwickii* Zone) — c.3.6

Mudstone, darker grey than below, colour banded and tending to rust, cleaved; top 1.83 m obscured. From c.1.5 m above base: *M. sedgwickii* and spiral monograptid — c.3.0

Thickness
m

Derwenlas Formation

Siltstone and sandstone, some beds 0.08 to 0.3 m
thick near top, mostly fairly clean siliceous detritus
but some dark lithic fragments are present.
Graptolites include *Cephalograptus cometa* cf. *extrema*,
Climacograptus cf. *scalaris*, *M. decipiens*, *M. lobiferus*,
?Pristiograptus jaculum, *P. regularis*,
Pseudoclimacograptus (M.) undulatus, *Retiolites perlatus*.
(*convolutus* Zone)   3.0

Mudstone, blue grey, massive banded, the paler
bands being more silty; thin darker graptolitic
mudstone stripes   5.5

Mudstone, green and olive grey. Thin siltstone
partings near the top   seen 3.0

Obscured   c.4.9

Mudstone, pale green, massive   seen c.0.4

Mudstone, medium grey, massive with paler
banding in parts and some thin silty layers   c.15

Mudstone, dark grey, shaly, some rusty weathering
(possibly basal beds of formation)   —

Laneside [7091 8793], near Alltgochymynydd

?Derwenlas Formation

Mudstone and arenite interbedded, grey; latter up
to 8 cm in lower part becoming thinner upwards   c.15.2

Derwenlas Formation

Mudstone, blue-grey, cleaved, with thin arenitic
interbeds. 4.6 m above base; *M. convolutus*, *M.*
lobiferus: 3.05–3.6 m above base; *?M. convolutus*,
M. lobiferus, *P. (C.) retroversus*: 3.0 m above base;
Glyptograptus spp., *Monoclimacis? crenularis*,
Monograptus convolutus, *?M. denticulatus*, *M. lobiferus*,
M. sedgwickii?, triangulate mongraptids, *P. (C.)*
retroversus. (Probably *convolutus* Zone)   6.1

Mudstone, dark grey, shaly, weathers rusty: *?C.*
scalaris, *?M. decipiens*, *M. limatulus*, *M.lobiferus*,
triangulate monograptids, *?P. (C.) retroversus*.
(Probably *convolutus* Zone)   0.3

Mudstone, darker greenish grey   0.61

Mudstone, pale grey-green massive, weathers fawn
or bronze at top, parts obscured   c.7.6

Mudstone, dark grey, massive   c.2.0

The thickness of some of the arenites and the thinning-upwards
sequence in the top 15.2 m make it uncertain to which formation
these beds belong but their general aspect and thickness favour the
Derwenlas Formation.

A section, along strike behind Cwm-byr [7108 8855] shows:

Thickness
m

Devil's Bridge Formation

Mudstone, medium to pale grey, well cleaved; pale
grey arenites, largely siltstone, up to 2 to 3 cm
thick, interbedded as rhythmites   27

Sandstone, in thick layers up to 16 cm, with minor
argillaceous partings   1.8

Derwenlas Formation

Mudstone, blue-grey, cleaved; thin siltstone
interbeds thickening upwards and including sandy
layers with dark lithic grains and graptolites.
From 5.5 m below the thicker sandstones; *?M.*
crenularis, *Monograptus lobiferus*. Basal 0.23 m darker
and rusting   6.4

Thickness
m

Mudstone, blue-grey, becoming paler below with
siltstone interbeds fairly common, to c.2 cm   0.6 to 0.9

Mudstone, rather massive, pale grey in places,
weathering to bronze hues   21 to 22

Mudstone, dark grey   c.18

Assuming there to be no fault in the section, the total thickness of
beds below the thicker sandstones is thus about 47 m; no part of
these can be certainly assigned to the Cwmsymlog Formation.

A trackside section [7068 8915] west of Moel-fferm reveals:

Thickness
m

Devil's Bridge Formation

5. Sandstone   0.15

4. Sandstone in beds up to 5 cm, cross-laminated with
   wispy uneven partings of pale grey mudstone   c.12

Derwenlas Formation

3. Mudstone, blue-grey, thinly interbedded with
   thinner pale grey siltstone up to 2 cm, cross- laminated
   wispy, and commonly lenticular 'ripple-
   bedded'. Some parts are rather coarse and sandy,
   and weathering gives rise to orange-rusty colours. In
   top 2 m the arenites are thicker. An 8 cm layer of
   blue grey porcellaneous mudstone is present about
   4.6 m above the base; *M. sedgwickii* occurs between
   0.3 and 0.6 m above this. Fauna 0.3 m below this
   mudstone is *Monograptus* cf. *proteus*, *M. sedgwickii*, *P.*
   regularis, *Pseudoclimacograptus (M.) undulatus*   7.6

2. Siltstone (mainly), in beds up to 2.5 cm thick,
   weathering to a very pale grey, with partings of blue-
   grey mudstone commonly showing rusty brown
   surfaces. Top 1.2 m consist predominantly of
   sandstones, some rather coarse, with argillaceous
   wisps and *?M. crenularis*, *Monograptus convolutus*, *?P.*
   regularis   7.9

1. Mudstone, mainly medium grey and vaguely silt
   striped. Darker mudstone at the base with triangulate
   mongraptids   c.24

It is evident that the characteristic lithologies of the Cwmsymlog
Formation are absent, though the *sedgwickii* Zone is plainly present
in the top 3.3 m of Bed 3. The remainder of Bed 3 together with
beds 1 and 2 provide a thickness of 36 m, equivalent to that of most
of the Derwenlas Formation of other areas.

A scarp [702 894] on the west side of Moel Fferm reveals in its
lower part, 22 m of thinly interbedded medium to dark grey silty
mudstone and siltstone. The siltstone beds are commonly 0.3 to
0.6 m thick but impersistent. Near the base some reach 4 cm,
showing convolute and cross-lamination while flute casts on at least
two levels indicate currents from 040°. Above, and up to the basal
sandstones of the Devil's Bridge Formation, are another 22 m or so of
similarly interbedded arenites and mudstones. In the middle and at
the top are layers with flute scours which indicate currents that came
from about 143°. The Cwmsymlog Formation is again absent, due
either to erosion or, more likely, facies change. In all, the Derwenlas
Formation here is 44 m thick; the previous section shows that some of
the top beds may belong to the *sedgwickii* Zone.

At the north end of Moel Fferm, a track yields a generalised section [7075 8968]:

	Thickness m
Mudstone, blue-grey, thinly interbedded with dark grey arenites up to 2.5 cm. Graptolites are present with *C. scalaris*, *M. communis*, *M. convolutus*, *M. lobiferus* and *Rastrites sp.* near base. (*convolutus* Zone)	c.14
Mudstone, medium grey, siltstone 'stripes' to 0.3 cm common	c. 4
Mudstone, grey, top half pale and greenish; a few very thin silty interbeds	5
Mudstone, dark grey and shaly, weathers very pale grey	0.76
Mudstone, blue-grey, thinly interbedded with dark grey arenite layers. Graptolites common in basal 1.5 m: *C. scalaris*, *?M. crenularis*, *Monograptus convolutus*, *M. communis*, *M.* cf. *lobiferus*, *?P. ovatoelongatus*, *Pristiograptus regularis*. (*convolutus* Zone)	c.11
Mudstone, blue-grey, weathering to bronze hues with siltstone interbeds up to 0.16 cm thick about 2.5 cm apart	c.18

The base of the formation is probably not reached in this section, and the centre portion clearly belongs to the *convolutus* Zone. At a minimum of 53 m the formation hereabouts is unusually thick.

Northwards, east of Moel-y-Llyn [712 917], the formation still retains a silty nature and is displayed at the end of a west facing scarp [7194 9116]:

	Thickness m
Mudstone, dark grey, with very numerous grey arenite interbeds up to 1.3 cm thick, wavy and lenticular with thin wispy argillaceous laminae	7.91
Mudstone, dark grey, shaly with thin siltstone partings. *Cephalograptus cometa*, *Climacograptus sp.*, *Diversograptus?*, *M. convolutus*, *?M. decipiens*, *M. lobiferus*, *P. jaculum*, *P. regularis*, *R. hybridus*. (*convolutus* Zone)	0.9
Mudstone, medium grey	c.1

The succession in the headwaters of Afon Einion, 1 km to the NE, is, however, different in that there is less siliciclastic debris. The Cwmsymlog Formation can be identified quite easily, the '*Monograptus sedgwicki* shales' clearly marking its base.

Forestry road [7270 9205], c 30 m west of anticlinal axis

	Thickness m
Devil's Bridge Formation	
Mudstone, medium grey-blue evenly colour-banded with rhythmically interbedded fine-grained arenites up to c.3 mm thick and c.5 to 7 cm apart near bottom, thickening upwards to c.6 mm spaced 2.5 to 5 cm apart near top; seen to centre of syncline	7 to 9
Arenite and mudstone, thinly interbedded. Bands of the former (fine sandstone and siltstone), up to 5 cm, are unevenly cross-bedded and laminated by darker grey argillaceous siltstone. Some layers coarser grained weathering brown	c.38
Cwmsymlog Formation	
Mudstone, banded blue-grey	8.5
Mudstone, dark grey, cleaved rusty weathering ('*Monograptus-sedgwicki* shales')	4.0

	Thickness m
Derwenlas Formation	
Mudstone, grey, weathers to bronze hues. Thin siltstone interbeds c.3 mm occur in top part	15.5
Mudstone, dark grey, friable. *?D. magnus*, *M. lobiferus*, *Petalograptus?*, *Pristiograptus?* and triangulate monograptids	0.3
Mudstone, medium grey	c.6.7
Mudstone, dark blue-grey. *C. gregarius*, *?D. magnus*, *G. tarmariscus*, *M. lobiferus*, *M. triangulatus* aff. *fimbriatus*, *R. toernquisti*. (probably *magnus* Zone)	0.9
Mudstone, medium grey, weathers to bronze hues, massive. Lowest few metres dark grey, weathering white or very pale grey	11.6
Cwmere Formation	
Mudstone, dark grey, rusty weathering, cleaved and friable. In top 0.3 m: *Climacograptus* aff. *alternis*, *Glyptograptus sp.*, *P.* cf. *ovatoelongatus*, *?R. toernquisti*. Just below top 0.3 m: *P.* cf. *incommodus*	33.5

Forestry road [7266 9231], where eastern limb of anticline crosses road

	Thickness m
Devil's Bridge Formation	
Sandstone, dark grey, thin-bedded, up to 5 cm thick and flaggy (i.e. parallel lamination prevalent). Argillaceous films are embodied within the laminations, and mudstone partings between the arenites are thin. The sandstone contains grains of argillite imparting the dark colour	1.5
Cwmsymlog Formation	
Mudstone, mainly grey, colour banded, with compactional or loading deformities. *?M. distans*, *M. sedgwickii* and *P. (M.) undulatus* form a dark grey layer in basal 1–2 m	c.9
Mudstone, dark grey, rusty and white weathering in parts and medium grey banded in parts. *M. sedgwickii*	6.0
Obscured	c.3.6
Derwenlas Formation	
Mudstone, medium grey, 'striped' by numerous siltstones up to 3 mm thick in top half. Bottom 2.0 m rather darker and rusty weathering with thinner siltstones and with dark shaly layer at top yielding *C. scalaris*, *?M. communis*, *M. convolutus*, *M. decipiens*, *M. lobiferus*, *?M. sedgwickii*, *P. regularis*, *Pseudoclimacograptus (C.) retroversus*, *P. (M.) undulatus*, *R. hybridus*. (*convolutus* Zone)	c.8.5
Mudstone, dark grey, banded, cleaved, weathers to pale and dark grey laminae. Base sharp. *M. convolutus*, *?M. decipiens*, *M. limatulus*, *?M. lobiferus*, *P. ovatoelongatus*, *Pristiograptus sp.*, *?Pseudoclimacograptus (C.) retroversus*, *R. hybridus*, triangulate monograptid	1.2
Mudstone, medium grey, fractures uneven, slightly reflective metallic lustre, bedding indistinct	4.8
Mudstone, dark grey, weathers white. *M.* cf. *argenteus* or *limatulus*, *M. lobiferus*, *Pseudoclimacograptus (M.) sp.*, *R.* cf. *hybridus subsp. nov.*, triangulate monograptids. (Probably *leptotheca* Zone)	1.2
Mudstone, medium grey, weathers to bronzy hues with rather metallic lustre. Dark grey for 0.3 m in middle part	7.3

Thickness
m

Mudstone, rather darker than above, fairly well
cleaved with slightly rusty stain 1.5
Mudstone, dark grey, rusty stained: *Climacograptus sp.,*
D. magnus, Glyptograptus sp., M. communis, M.
triangulatus fimbriatus, Orthograptus?, P. (M.) hughesi,
Rastrites cf. *longispinus, Rhaphidograputs toernquisti.*
(*magnus* Zone). 0.9
Mudstone, blue-grey 1.5
Mudstone, dark grey, weathers into white stripes.
?G. tamariscus, O. aff. *bellulus, M. revolutus, ?M.*
triangulatus major, M. t. separatus, M. t. cf.
triangulatus, R. longispinus, Rhaphidograptus toernquisti
(Probably *triangulatus* Zone) 1.8
Mudstone, dark blue-grey 1.5

Thickness
m

Cwmere Formation
Mudstone, dark grey, cleaved, rusty weathering.
Glyptograptus aff. *sinuatus, R. toernquisiti.* An
isolated exposure of mudstone, some 5 m to the
west yielded *Glyptograptus sp., P.* cf. *leptotheca* and
?Pristiograptus regularis seen 0.9

In the previous two sections, and in another [7291 9272] to the NE,
the Derwenlas Formation is 32 to 35 m thick, less than to the south.

Farther north this outcrop crosses three headwater tributaries
[7315 9304; 7341 9378; 7349 9415] where the beds dip very steeply
eastward. These sections show that the arenite content of the
formation is lower here, massive mudstone being dominant in a
sequence reduced to c.18 to 19 m.

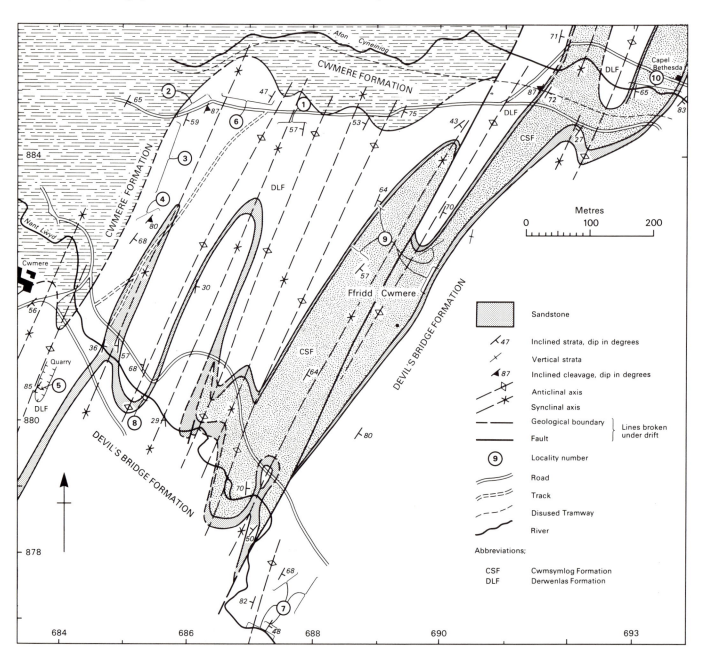

Figure 36 Sketch-map of the Cwmere area

Cwm Ceulan – Pen Dinas – Gwar-cwm

Thc formation is 50 to 60 m thick and dominantly mcdium to dark bluish grey mudstone, except in the middle where it is pale grey. Graptolite-rich 'bands' are commonest in the lower part. The upper beds throughout this area include thin siltstones and sandstone, but in the Leri valley west of Werndeg the proportion of arenaceous sediment is much greater with interlayered mudstones and distal turbidite sandstones throughout. This contrasts with the Ponterwyd area where the arenaceous beds are concentrated in a '*convolutus* sandstone' unit.

Colour-banded pale and bluish grey mudstone crops along two parallel ridges [6918 8858 – 6929 8894; 6932 8858 – 6947 8885] north of Afon Cyneiniog. The western ridge shows 38 m of bluish grey mudstone, with thin laminated siltstone beds 12 mm thick in its middle, overlain by the Cwmsymlog Formation.

Sections south of Afon Cyneiniog are shown in Figure 36 (locs.1 – 4,6). A 20 m section in an old quarry [6837 8805] (Figure 36, loc.5) shows dark bluish grey strongly cleaved mudstones with many siltstone laminae and graptolites including *D. magnus, G. (P.)* cf. *vas* and *M. triangulatus fimbriatus* (*magnus* Zone) passing up into paler and more flaggy mudstones. The full thickness of the formation (60 m) is seen [6775 8797 – 6767 8803] 250 m north of the Pen Dinas hill-fort:

	Thickness m
Devil's Bridge Formation	
Sandstone, grey, laminated; in 5 to 10 cm beds interlayered with mudstone	1.8
Derwenlas Formation	
Mudstone, bluish grey; many beds of laminated sandstone mostly about 1 cm thick, although some 10 cm in central part	6.3
Mudstone, dark grey, rusty brown weathering, some thin sandstone beds	1.8
Not exposed	5.5
Mudstone, dark bluish grey	3.7
Mudstone, pale grey; silty laminae up to 3 mm thick every 5 to 10 mm	9.1
Mudstone, dark bluish grey; darker and rusty brown weathering in middle of bed	0.9
Mudstone, pale to medium grey; silty laminae up to 3 mm thick every 5 to 10 mm, but less common in basal 3 m	12.4
Mudstone, dark bluish grey, a few silty laminae; dark grey and rusty brown weathering for 1.8 m at 12.5 m above base; unexposed for 4.3 m at 7.3 m above base	17.0
Mudstone, dark grey, many siltstone beds up to 5 mm thick	3.7

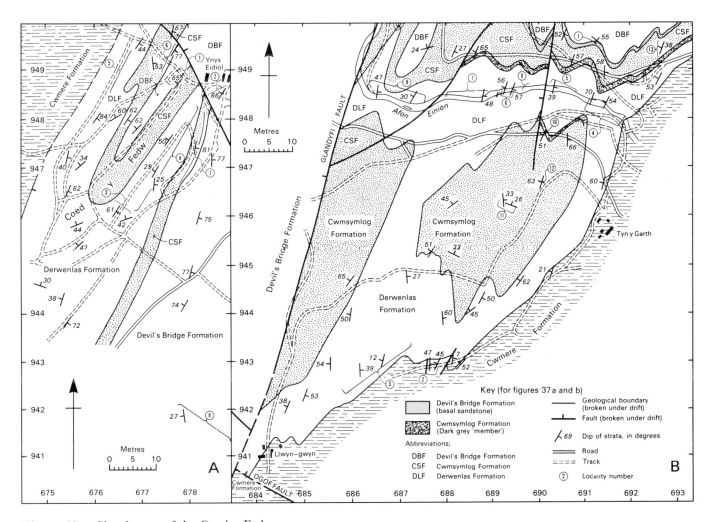

Figure 37a. Sketch-map of the Coed y Fedw area

Figure 37b. Sketch-map of the Tyn y Garth area

Cwmere Formation

	Thickness m
Mudstone, dark grey, many siltstone beds up to 5 mm thick	c.12.0

Sections on the southern side of Cwm Ceulan, SW of Fronlas, show a decrease in the arenaceous content of the strata. The best exposure is in crags [6807 8982] where 22 m of pale grey mudstone are seen, with abundant thin silty layers in the upper half of the section.

Tyn y Garth

The formation (Figure 37b) is 50 to 55 m thick: it comprises bluish grey and pale grey mudstones with a large number of thin fine-grained sandstone distal turbidites in its upper part.

An extensive section [6848 9424 – 6862 9434] (Figure 37b, loc.3) shows the following sequence in the lower part of the formation:

	Thickness m
Mudstone, grey, many sandstone beds up to 2.5 cm thick in basal 3.7 m, up to 1.2 cm thick above	8.2
Mudstone, dark grey, rusty brown weathering, sandy laminae	0.9
Mudstone, dark bluish grey	0.9
Unexposed	3.7
Mudstone, pale grey, brown weathering, thin sandstone beds in top 1.8 m	5.5

At loc. 4 of Figure 37b, 30 m of strata in the lower part of the formation are exposed. They comprise bluish grey mudstone, with thin siltstones and fine-grained sandstones in the highest beds at the west end of the section. Similar beds are seen at locs. 5, 6 and 7; at the last locality the lowest beds, lying in the core of an anticline [6874 9484], are dark grey rusty weathering mudstones with a graptolite fauna including *M.* cf. *convolutus*, *M.* cf. *lobiferus*, *P.* cf. *ovatoelongatus* and *Pseudoclimacograptus* (*Metaclimacograptus*) *undulatus* probably of *convolutus* Zone age. The upper part of the formation is seen in crags at loc. 9 (Figure 37b) and in a synclinal section at loc. 8, which shows 2.4 m of medium grey mudstone interlayered with sandstone beds up to 5 cm thick.

Derwenlas area

Derwenlas provides the type section of the formation and was accurately measured at about 37 m by Jones and Pugh, (1916, pp.353 – 358). A location plan of the exposures (Figure 38) shows, as near as possible, their bed numbers.

Erglodd – Ynys-hir

The sequence in this outcrop closely resembles that near Tyn y Garth (p.80). A quarry [6561 9105] shows cleavage with the unusually low inclination of 38° towards 310°. The basal beds of the formation are seen in an old quarry [6632 9205] where dark grey rusty brown weathering mudstone is overlain by grey mudstone with a few thin sandstone beds. Eastwards there is a discontinuous section [6637 9206 – 6655 9202] which displays a folded and faulted sequence of bluish grey mudstones with many fine-grained sandstone beds up to 2.5 cm thick. Within these mudstones [6643 9202] a 5.5 m sequence of sandstone in 15 cm beds with thinner mudstone interbeds is perhaps equivalent to the 'convolutus sandstones' of other areas.

To the north, a gorge [6690 9337 – 6693 9332] displays beds towards the top of the formation. They comprise 12 m of grey blocky mudstone at the west end, with sandstone laminae up to 1 cm thick in the upper part, overlain by 4.5 m of thinly interlayered mudstones and sandstones. The lowest beds in the section contain an abundant graptolite fauna including *M. denticulatus*, *M. limatulus*, *M. lobiferus*, *P. regularis* and *R. longispinus*, indicative of the *convolutus* Zone.

South-west of Ynys Eidiol (Figure 37a, loc.3) the upper part of the formation is exposed [6766 9460]:

	Thickness m
Mudstone, pale bluish grey; fine-grained sandstone laminae up to 5 mm thick every 10 mm; poorly exposed in basal metre	8.2
Mudstone, bluish grey; sandstone laminae 5 mm thick, 10 mm in top 0.6 m; some dark grey laminae	3.7
Mudstone, dark grey, some sandstone laminae, graptolite traces	0.9
Unexposed	c.1.2
Mudstone, bluish grey, a few 10 mm sandstone laminae in basal 2.4 m; abundant graptolites including *M. lobiferus* in a 2.5 cm bed 2.9 m above base	4.6

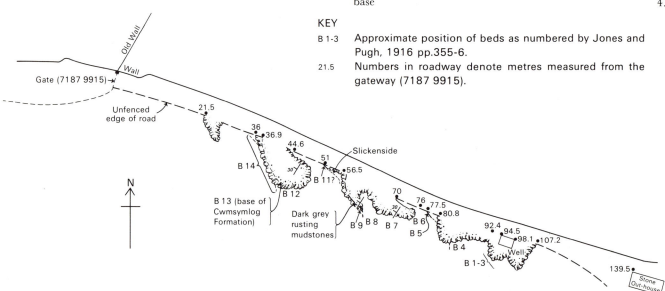

KEY

B 1-3   Approximate position of beds as numbered by Jones and Pugh, 1916 pp.355-6.

21.5   Numbers in roadway denote metres measured from the gateway (7187 9915).

Figure 38 Exposures along the old coach road, Derwenlas, type-section of the Derwenlas Formation

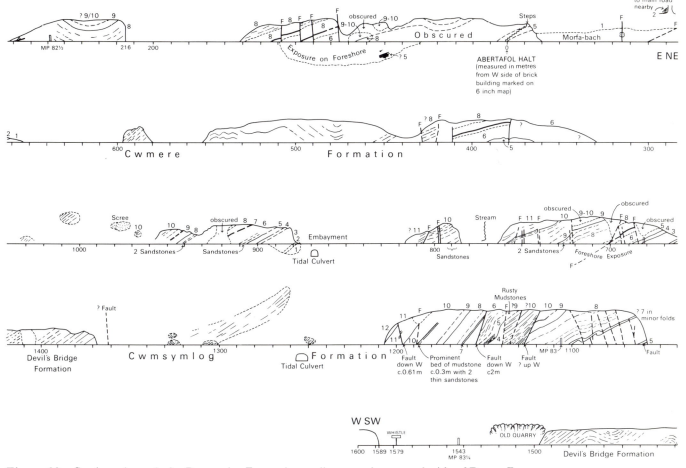

Figure 39   Sections through the Derwenlas Formation; railway cuttings, north side of Dovey Estuary

Part of this sequence can be seen farther NE (Figure 37a, loc. 4 and 5).

North of River Dovey

Exposures in several railway cuttings (Figure 39) between Abertafol Halt [6474 9685] and [6370 9636] can be linked to provide the following composite sequence through the entire formation:

	Thickness m
Cwmsymlog Formation	
12. Mudstone, dark grey, soft and friable, sharp base ('*Monograptus-sedgwicki* shales')	more than 2.00
Derwenlas Formation	
11. Mudstone, pale grey, massive. Thin layers of fine-grained arenite, less than 2.5 cm apart. Basal 2 m dark grey, becoming paler upwards	8.50 to 9.10
10. Mudstone, medium to dark grey, in layers 5 to 7 cm thick, cleaved and with rusty stain. A few prominent arenite layers up to about 2 cm thick	10.80
9. Mudstone, dark grey, friable and graptolitic, rusty weathering. Sharp base	2.13
8. Mudstone, medium grey, rhythmic banding distinct (cf. Devil's Bridge Formation). Layers are 5 to 7 cm thick	

with arenitic bases and pale fawn to grey bands at the tops. Black phosphate layers common as in Figure 2b. Occasional arenites up to 10 cm thick. Basal 1.8 m paler, becoming darker upwards as the banding becomes more pronounced — 6.40 to 6.70

	Thickness m
7. Mudstone, pale grey or fawn, unlayered but faintly graded (marker bed; see below)	0.20 to 0.23
6. Mudstone, medium grey, thinly bedded with numerous arenite laminae less than 2.5 cm apart. In several exposures an ochreous weathering layer occurs some 1.8 m below the top. In the basal 1 m are several arenites up to 3.5 cm thick which form useful marker beds. Base sharp	6.82 to 7.06
5. Mudstone, dark grey, graptolitic. Some interbeds up to 0.57 m thick of pale grey-green mudstone with dark streaks	2.44
4. Mudstone, medium grey, bedded and with dark grey streaks and mottles. Rusty weathering	0.76
3. Mudstone, medium to pale grey, massive; bedding indistinct. Brown weathering with a layer of rusty weathering concretions near middle	6.70

	Thickness m
2. Mudstone, dark grey, graptolitic. Pale grey layer c.7.5 cm thick 15 cm below top and two pale grey-green layers 1 cm and 0.5 cm thick 25 cm and 5 cm respectively from base	0.59 to 0.61
1. Mudstone, dark grey becoming paler upwards, massive. Weathers brown revealing bedding traces	seen 4.20

The beds dip generally westward and N–S faulting gives rise to repetition. The exposures can be correlated with one another by a few marker beds, in particular a layer of pale grey faintly graded bentonitic mudstone (Bed 7). The sequence reveals at least two coarsening-upward cycles. One embraces the *triangulatus*, *magnus* and *leptotheca* zones and ends in the *convolutus* Zone at the top of Bed 8. The other commences at the base of the succeeding graptolitic mudstone of Bed 9 and ends at the base of the '*Monograptus-sedgwicki* shales'. The cycles are interpreted as two prograding bodies of low energy, probably distal, turbidites that resulted from basin-wide pulses; they may, however, result from purely local causes. The marker bed (Bed 7) is unique to the sequence. Mr R. J. Merriman reported as follows on thin sections of samples from the railway cutting [6375 9638]:

'The matrix of the mudstone consists of an orientated intergrowth of muscovite and chlorite flakes up to 20μ m in length, and rutile needles ($<10\mu$ m). These minerals define a slaty cleavage which steeply cuts a poorly defined lamination. The sediment is conspicuously deficient in quartz or other orthoclastic material, but scattered books of chlorite probably represent ferromagnesian detritus. Rounded or elliptical areas of randomly intergrown muscovite and chlorite in the mudstone have a coarser texture than the matrix, and clearly deflect the cleavage. In the hand specimen these coarser aggregates form dark chloritic 'knots' up to 1 mm across, but although textural evidence shows they predate the deformation of the sediment, other petrographical evidence of their origin is lacking throughout much of the mudstone layer (E43767,8,9). However at the base of the mudstone (E43770), the proximity of an early quartz vein has prevented development of a strongly penetrative cleavage and aided preservation of original tuffaceous material. Here, rounded clasts of microcrystalline and devitrified glassy igneous material up to 0.5 mm across, are contained in a crudely cleaved matrix of chlorite and muscovite. Scattered chlorite aggregates commonly contain rutile needles arranged on an hexagonal grid, strongly reminiscent of exsolved Ti-oxides in biotite. Other chlorite aggregates appear to replace lamellar-twinned plagioclase fragments. Reddened opaque grains and rare apatite euhedra, 0.3 mm in length, are also present.

It seems likely therefore that the pale grey mudstone is a thin bentonite, and that the darker knots of phyllosilicate represent former mineral fragments (feldspar, biotite or other ferromagnesian) and pumice particles. Absence of quartz and other orthoclastic detritus indicates that ash accumulated directly from airfall material, rather than by reworking of nearby ash deposits. Argillisation of the ash probably took place soon after accumulation, and subsequently the bentonite was recrystallised and cleaved, erasing any evidence of vitroclastic texture. XRD analysis (chart DX 2911) shows that metamorphic recrystallisation has converted the bentonitic clay assemblage to chlorite and a $2M_1$ mica polytype with a crystallinity of $0.33° 2\theta$. The nature of the mica coupled with the sporadic occurrence of late granular epidote suggests that the grade of metamorphism was low-greenschist facies.'

Graptolites were obtained from the section (Figure 39) as follows; the metreage of the localities, WSW of an origin at Abertafol Halt is indicated:

Bed	m (approx)	
9	1175	*M. decipiens*, *?M. delicatulus*, *M. denticulatus*, *M. lobiferus*, *P. regularis*, *Pseudoclimacograptus* (*C.*) *retroversus*
	945	*Glyptograptus sp.*, *?M capis*, *?M. clingani*, *?M. convolutus*, *?M. decipiens*, *?M, limatulus*, *M.* cf. *lobiferus*, *?M. pragensis*, *Petalograptus sp.*, *Rastrites sp.*
	220–225	*?C. cometa* or large rastritid, *?M. convolutus*, *M. lobiferus*, triangulate monograptid.
9 (or top 8)	1117	*G. elegans*, *M.* cf. *delicatulus*, *?M. sedgwickii*, *M.* cf. *undulatus*, *?P. concinnus*, *P. regularis*, *Pseudoclimacograptus* (*M.*) *undulatus*, *R.* cf. *hybridus*
Basal 6	1067	*?M. halli*, *?M. proteus*, *M.* cf. *sedgwickii*
5 (top)	1151	*C. cometa*, *?M. clingani*, *M. lobiferus*, *?R. toernquisti*
5 (top)	1060	*?M. communis*, *M. halli* or *sedgwickii*, *M. proteus* (or *sp. nov.*), *M.* cf. *sedgwickii*
5 (lower)	1060	*M. communis?*, *M. halli*, *?Petalograptus sp.*, *Pseudoclimacograptus* (*M.*) *undulatus*
?5 (c.6 m below 1060)	c.1030	*C. simplex*, *M.* aff. *convolutus*, *?M. lobiferus*, *M. pragensis?*, *M.* cf. *sedgwickii*, *M. sedgwickii?* or *lobiferus?*, *Petalograptus sp.*, *R.* aff. *hybridus*, triangulate mongraptid
?5	70	*?M. convolutus*, triangulate mongraptid
5	0	*?M. convolutus*, *?M. denticulatus*, *M. limatulus*, *M. lobiferus*, *M. sedgwickii?* or *halli? M. triangulatus*, cf. *P. regularis* junction [6480 9693] of main road and lane to Aber-groes
2		*?M. argenteus*, *?Monoclimacis crenularis*, *?P. leptotheca*
1	70 (ENE)	*M. triangulatus* cf. *fimbriatus*, *P. leptotheca*

From 1060 m to 1117 m there appears to be a continuous upwards sequence, including Beds 5–10 of the Derwenlas Formation. A graptolite fauna from close to the junction between Beds 8 and 9 (at 1117 m), indicates, on balance, a mid-Llandovery age, and thus supports this interpretation. However, graptolite faunas recorded at 1060 m and 1067 m near the base of the sequence are interpreted as indicating the *sedgwickii* Zone, and they present an unresolved problem.

CWMSYMLOG FORMATION

Eisteddfa Gurig – Ponterwyd – Pen y Graig-ddu – Cymsymlog

The sequence between Eisteddfa Gurig and Llyn Blaenmelindwr broadly resembles that of the type area (p.84), although it is only

Figure 40 The Llyn Ieuan Cwmsymlog Formation inlier

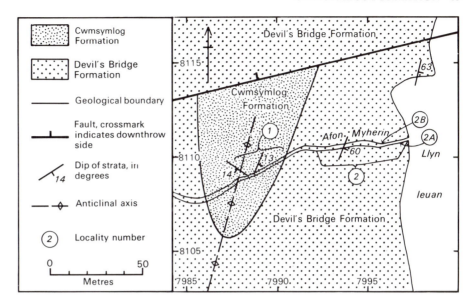

about 30 m thick over most of the area and reduces to 6 m near Disgwylfa Fâch (p.83). In crags [7974 8362] south of Eisteddfa Gurig the following section (Jones, 1909, pp.501–502) is seen.

	Thickness m
Mudstone, pale grey, some bioturbation	4.6
Mudstone, bluish grey, paler in top 3 m	7.6
Mudstone, dark bluish grey, some very dark laminae and silty laminae	3.0
Mudstone, dark grey, rusty brown weathering; a few 5 mm fine-grained sandstone beds; many graptolites including *M. sedgwickii*, *P. regularis* and *Pseudoclimacograptus (M.) undulatus*	9.0

Farther south there is a small inlier (Figure 40) in the core of an anticline [799 811] 1.5 km east of Y Glog. Bluish grey mudstones with some dark mudstone laminae and laminated sandstone beds up to 2.5 cm thick (Figure 40, loc. 1) contain a graptolite fauna (*M. distans?* or *lobiferus?*, *M. lobiferus?*, *M. sedgwickii?*) which suggests an horizon close to the boundary between the *convolutus* and *sedgwickii* zones. Westwards, in a line of crags [7768 8129 – 7772 8131] north of Nant Meirch, all but the very lowest beds of the formation are exposed as follows:

	Thickness m
Devil's Bridge Formation	
Sandstone, thin interbedded bluish grey mudstones	0.5
Cwmsymlog Formation	
Mudstone, pale grey, faint greenish colouration orange-brown weathering; 5 mm sandstones in basal 1.5 m, 12–25 mm sandstones above	9.1
Mudstone, pale grey, regularly striped with 5 mm fine-grained sandstones every 2.5 to 3.8 cm in basal 6.7 m	9.8
Mudstone, pale bluish grey	7.3
Mudstone, bluish grey, some thin siltstone beds	4.9
Mudstone, dark grey, rusty brown weathering, a few silty laminae ('*Monograptus-sedgwicki* shales')	—

On the eastern slopes of Brynbras, 24 m of bluish grey mudstones becoming paler upwards with some thin silty layers form crags [7480 7967] immediately above a slack containing the basal dark mudstones. The topmost beds are exposed, [7468 7962], and show a transition within 3 m from similar bluish grey mudstones to mudstones with interbedded sandstones up to 15 cm thick at the base of the Devil's Bridge Formation.

South of Disgwlfa Fâch the formation becomes much thinner, as is seen at loc. 3 (Figure 33), where it is about 6.4 m thick, only the top 4.9 m of bluish grey mudstone with thin siltstone beds being exposed. South and west of Llyn Blaenmelindwr the beds thicken; three divisions become mappable, these being the basal 'Dark grey mudstone member' ('*Monograptus-sedgwicki* shales'), the main undivided part of the formation, and the topmost 'Green mudstone member'. Their combined thickness increases from about 30 m at the lake, to 60 m on the flanks of the anticline near Carndolgau, and up to about 140 m on the western limb at Pen y Graig-ddu and northwestwards to Cwmsymlog. At Pen y Graig-ddu there is an inlier in the 'Dark grey mudstone member' and the undivided part of the formation, and in a forestry road there are complete sections through the formation on both limbs of the fold. The sequence on the western limb is:

	Thickness m
'Green mudstone member' (Figure 34 loc. 4)	
Mudstone, pale greenish grey, colour-banded; siltstone or fine-grained sandstone interbeds every 2.5 to 3.5 cm in the top 20 m gradually increasing in thickness upwards from 5 mm to 2.5 cm maximum	80.5
Undivided (Figure 34 loc. 5)	
Mudstone, bluish grey, faintly colour-banded, black laminae for 0.6 m at 3 m above base; 5 cm weathered nodular calcareous sandstone at 7.3 m above base	9.1
Mudstone, bluish grey, darker in basal 7.9 m; siltstone laminae, black laminae and scattered weathered calcareous nodules	35.1
'Dark grey mudstone member' (Figure 34 loc. 6)	
Mudstone, very dark bluish grey; greenish brown and orange-brown weathering, many thin black and pale laminae	14.0
Mudstone, bluish grey, many black laminae and some siltstone laminae	2.4

The sequence on the eastern limb (Figure 34 loc. 7) is similar, except that the 'Green member' is only 19.8 m thick. The thickness of the undivided beds (43.9 m) and 'Dark grey member' (17.5 m) are almost unchanged. On both limbs the 'Dark grey member' contains *M. sedgwickii* and *P. regularis*, confirming its position within the *sedgwickii* Zone. In these sections the undivided beds and 'Green member' both display abundant evidence of contemporary disturbance (see p.46).

Alltgwreiddyn – Afon Stewy

The three-fold division of the formation is pronounced along the southern extension of the Coed Dipws Anticline. The type section along a forestry road [6845 8408–6868 8415], NE of Pen-bont Rhydybeddau, is:

	Thickness m
'Green mudstone member' Mudstone, green and greyish green, well cleaved, dendritic 'pyrolusite' along cleavage planes	4.0
'Red mudstone member' Mudstone, purplish red and maroon, indistinct colour banding, well cleaved	7.3
'Green mudstone member' Mudstone, pale green, green and greenish grey, prominent colour banding and microfaulting; scattered thin (0.5 to 1 mm) decalcified sandy laminae; passing down into	38.1
Undivided Mudstone, dark greenish grey and bluish grey, silty, colour banded; sandy laminae (0.5 to 1 mm) give a pronounced striped effect; poorly bedded, blocky cleavage	22.0
Unexposed	10.0
Mudstone, bluish grey (paler than beds beneath), poor blocky cleavage; bluish grey and brown decalcified silt and sand laminae each up to 2 mm thick	12.2
Mudstone, bluish grey, silty; rusty weathering black mudstone partings, especially near the base	4.9
Mudstone, dark grey, rusty weathering. *M. decipiens?* and *M.* aff. *sedgwickii*	0.5
Mudstone, bluish grey, silty; some black mudstone partings to 5 mm thick, and thin beds of grey siltstone, or silty mudstone; prominent line of calcareous concretions, up to 0.22 m thick, at base	2.1
Mudstone, bluish grey, silty, many black mudstone partings to 5 mm thick	4.0
'Dark grey mudstone member' (total estimated thickness 23.8 m) Mudstone; thin black shaly beds (1 to 3 mm), bleached white, alternating with thicker pale grey beds; pyritous with intense rusty weathering; *M. sedgwickii*, *?P. jaculum* or *variabilis* and *Pseudoclimacograptus (M.) undulatus*	seen 7.0

Craigypistyll – Hafan Fault

Two inliers [706 854; 702 856] reveal pale greenish grey and bluish grey mudstones similar to the beds in the higher part of the formation to the south. The main part of the outcrop from Craigypistyll northwards, however, shows radical thinning and significant lithological changes (p.47). The basal 'Dark grey member' of the type area is replaced by greywacke sandstones, which in places still carry *M. sedgwickii*, and the overlying mudstones are much thinner and contain interbedded thin sandstones. Strong (1979) considered that these beds were deposited in a palaeochannel with SE–NW-trending axis through Bwlch yr Adwy [719 868] and Moel Golomen. The top of the formation is not clearly defined as there is an upward transition into the Devil's Bridge Formation. Though the top has now been taken at a rather lower level than by Strong (1979), this does not affect his conclusions.

The main cliff-face at Craigypistyll (Figure 29, loc.11) presents a good section. Similar beds at loc. 5, 12 and 13 show:

	Thickness m
Devil's Bridge Formation Mudstone, bluish grey, many fine-grained laminated sandstone beds to 2.5 cm thick, a few 7.5 cm thick	—
Cwmsymlog Formation Mudstone, dark bluish grey, smooth textures	1.8
Mudstone, dark bluish grey, smooth textures; fine-grained sandstone beds up to 2.5 cm thick	1.8
Mudstone, bluish grey; 5 to 10 mm laminated sandstone beds every 2.5 to 5 cm	4.6
Sandstone, fine to coarse-grained, graded, in units up to 30 cm thick with graptolites including *M. sedgwickii*; some mudstone laminae	3.7
Derwenlas Formation Mudstone, bluish grey, a few thin silstone laminae	—

Farther north the basal sandstones are exposed in a track [7148 8688]. They are at least 6.7 m thick, with graded units up to 30 cm thick in the top 5.0 m. The Bwlch yr Adwy section [719 869] (Strong, 1979) shows:

	Thickness m
Devil's Bridge Formation Sandstone in units up to 25 cm thick with thin mudstone interbeds	seen c.6.0
Cwmsymlog Formation Mudstone, dark grey, many thin sandstone beds	1.5
Mudstone, grey, many medium-grained sandstone beds up to 1.5 cm thick	5.2
Sandstone, grey, medium-grained with much dark mica, flaggy bedding, laminated; some thin dark mudstone interbeds	3.1
Derwenlas Formation Mudstone, bluish grey and pale grey	seen 13.7

Llechwedd Gwineu – Afon Lluestgota

Y Chwareli [7356 8782] (Figure 41) shows:

	Thickness m
Devil's Bridge Formation 9. Mudstone, pale grey and siltstone, medium grey, thinly interbedded. Some 4 m visible in small excavation [7359 8781] with proportions of mudstone to siltstone about equal; in overlying beds mudstones appear to be predominant	—
8. Sandstone and siltstone, medium grey, convoluted and cross-bedded and thinning upwards; forming up to 80% of the rock and interbedded with thin mudstones	17.0
Cwmsymlog Formation 7. Mudstone, pale greyish green, smooth textured; thin silt partings	2.4
6. Mudstone, medium grey with dark grey layers and dark grey mottles of *Chondrites* burrows. The layering	

reveals small scale penecontemporaneous load fractures with displacements of less than 1 cm, such features being characterstic of this formation. Siltstone partings up to 1 cm are present   **6.4**

5. Mudstone, darker grey, with a reddish stain, fissile, with very thin partings of siltstone   **4.0**

4. Mudstone, dark grey, fissile with some silty partings. *Climacograptus sp.* and *M. capis* (= 'Monograptus-sedgwicki shales')   **3.6**

Derwenlas Formation

3. Siltstone and sandstone in thin beds up to 12 cm thick with wavy lamination of thicker layers. Mudstone partings up to 1.5 cm thick. Top of a thickening/coarsening-upward sequence of thin turbidites   **c.3.0**

2. Mudstone, dark grey, friable, graptolitic   **0.13**

1. Mudstone, medium grey, thin siltstone partings   **2.4**

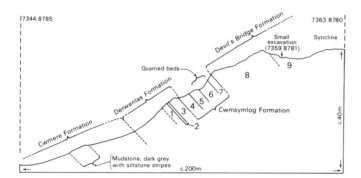

Figure 41 Sketch section through slope at Y Chwareli

The formation is 16.4 m thick, closely comparable with other nearby localities (Figure 35) where the 'Green mudstone member' is overlain directly by arenites of the Devil's Bridge Formation. Northwards the formation thins between Moel Cyneiniog and Bryniau Rhyddion, in the same direction as the underlying Derwenlas Formation loses much of its arenite. The green mudstone, 2.5 to 3 m thick, is visible in several places along Bryniau Rhyddion [7293 8878]. Farther north thickness comparisons become misleading, for the upper part of the formation seems to be diachronous (p.47) and the formation incorporates beds which are arenaceous in the south and have been included in the basal part of the Devil's Bridge Formation. The 'Green mudstone member' ceases to mark the top of the Cwmsymlog Formation, and the measures subjacent to the Devil's Bridge Formation are grey.

The Cwmsymlog Formation is exposed in the following sections shown graphically in Figure 35:

Section B: Stream NE of Dolrhyddlan [7403 8971]

Thickness
m

Devil's Bridge Formation

6. Mudstone, medium-grey, cleaved, thinly interbedded with arenites up to 2.5 cm thick in upper part but thicker at base   seen c.40

Cwmsymlog Formation

5. Mudstone, pale grey; thin parallel-laminated siltstone interbeds 0.3 cm thick at base, increasing to 2.0 cm at the top with convolutions   **0.93**

4. Mudstone, pale greenish grey   **6.10**

Thickness
m

3. Mudstone, dark grey, graptolitic   **c.0.3**

2. Mudstone, pale greenish grey   **0.6**

The section is terminated downwards by a fault. A thick mudstone succession below the fault contains graptolitic faunas. These do not appear to be in a meaningful sequence, and there may be undetected faulting.

A section some 30 m to the north, incorporating a small excavation, exposes the same beds in the upper part of the Cwmsymlog Formation. Bed 3 yielded the following graptolites, signifying the *sedgwickii* Zone but including the anomalous record of *R. maximus* and *M. halli*: *C. simplex, G. incertus, M. communis, ?M. delicatulus, M. halli, M. involutus, ?M. lobiferus, M. pragensis ruzickai?, M. sedgwickii,* triangulate monograptids, *Petalograptus sp., Pristiograptus regularis, Pseudoclimacograptus (C.) retroversus, P. (M.) undulatus, Pseudoplegmatograptus obesus, R.* aff. *hybridus, R. linnaei* and *R.* cf. *maximus.* If there is an exact correlation of graptolitic bands between nearby sections then this band should be the one noted from beds 14, 11 and 12 in Sections C, D and E respectively (Figure 35). However, in these latter three sections the graptolites indicate the subzone of *R. maximus,* and the bed may be the *Rastrites* Band of Jones and Pugh (1916, p.361).

Section C: Ditch, SW side of road [7340 9000]

Thickness
m

Devil's Bridge Formation

18. Siltstone and sandstone, thinly bedded with parallel and cross-lamination   —

Cwmsymlog Formation

17. Mudstone, pale grey and greenish grey, silty with closely spaced siltstone partings, micaceous and parallel-laminated, *M.* cf. *sedgwickii* in upper part   **6.1**

16. Mudstone, pale greyish green, smooth textured, with fine siltstone partings   **2.1**

15. Mudstone, dark grey, shaly   **0.46**

14. Mudstone, dark grey, fissile. *Glyptograptus sp., M.* cf. *knockensis, M. sedgwickii, M. undulatus?, Petalograptus sp., Pristiograptus regularis, Pseudoclimacograptus (M.) undulatus, R. fugax?, R. hybridus, R. maximus* (Base of *maximus* Subzone.)   **0.15**

13. Mudstone, dark grey, shaly   **4.27**

12. Mudstone, dark grey, shaly, with white weathered stripes: *Lagarograptus tenuis, M.* cf. *communis, ?M. involutus, M.* cf. *knockensis, M. sedgwickii, P.* cf. *jaculum, P. regularis, Pseudoclimacograptus (M.) undulatus*   **2.13**

Section D: Forest track, [7439 9067] on western limb of faulted syncline

Thickness
m

Devil's Bridge Formation

14. Sandstone, medium grey, in beds up to 10 cm with mudstone partings   **7.3**

Cwmsymlog Formation

13. Mudstone, grey, flaggy to shaly with thin siltstone laminae   **5.5**

12. Mudstone, green banded, smooth textures; abundant thin siltstone laminae   **3.7**

11. Mudstone, dark grey. *M. communis, ?M. involutus, M. lobiferus, M. sedgwickii, R. maximus?*   **0.05**

10. Mudstone, green banded, smooth textured; abundant thin siltstone laminae   —

Section E: Track; base [7441 9089]

	Thickness m
Cwmsymlog Formation	
15. Mudstone, grey	1.8
14. Mudstone, green	5.5
13. Mudstone, grey, smooth textured	1.8
12. Mudstone, dark grey, graptolitic	0.15
11. Mudstone, grey and green, smooth textured	1.5
10. Mudstone, dark grey, fissile	0.3
9. Mudstone, grey	0.6
8. Mudstone, dark grey; *M. sedgwickii*	1.83

Section F: Track; base [7428 9094]

	Thickness m
Devil's Bridge Formation	
12. Sandstone, medium, grey, up to 5 cm thick, cross-bedded and convolute	seen c.5.0
Cwmsymlog Formation	
11. Mudstone, strongly colour-banded, dark to medium grey and pale green, with siltstone interbeds up to 0.7 cm	0.91
10. Mudstone, strongly colour-banded, dark to medium grey and pale green	2.44
9. Mudstone, pale green, massive, unbanded, with anastomosing brown veinlets	2.44
8. Mudstone, medium to dark grey	0.30
7. Mudstone, pale green, massive, unbanded, with anastomosing brown veinlets	1.83
6. Mudstone, medium to dark grey, cleaved	0.61
5. Mudstone, as bed 4 with more dark grey banding and graptolites c.0.3 m below top	1.22
4. Mudstone, medium dark grey and paler grey wispy	
3. Mudstone, dark grey, rusty weathered, with *M. sedgwickii*	0.61
2. Mudstone, medium bluish grey non-rusty 'leaf' within the '*Monograptus-sedgwicki* shales'	0.46
1. Mudstone, dark grey, rusty weathered, with *M. sedgwickii*	1.37

This section, like others nearby (E, H and I), contains a non-graptolitic paler grey layer of mudstone dividing the '*Monograptus-sedgwicki* shales' above the middle.

Section G: Track; [7419 9094]

	Thickness m
Devil's Bridge Formation	
16. Sandstone in beds up to 10 cm thick	seen 2.0
15. Mudstone, colour banded, pale grey/pale green and darker grey, with bioturbation mottles (*Chondrites?*). A few arenites up to 4 cm thick	4.0
14. Sandstone/siltstone, medium grey, mostly c.2.5 cm thick but up to 5 cm, with mudstone partings	3.35
13. Mudstone, pale grey, interbedded siltstones and a few 2.5 cm sandstones	2.44
12. Mudstone, pale grey with dark grey wisps and mottles (*Chondrites?*) smooth-textured and shaly. Interbedded thin siltstones, thickening upwards	4.00
11. Mudstone, pale green, smooth-textured, fissile	3.35
10. Mudstone, dark grey, shaly: *M. communis*, *M. halli?*, *?M. sedgwickii*, *Petalograptus sp.*, *?Pristiograptus regularis*, *Pseudoclimacograptus (M.) undulatus*, *R. linnaei*	0.08

	Thickness m
9. Mudstone, pale grey to greenish grey and darker grey banded, smooth-textured and fissile	3.35
8. Mudstone, dark grey, rusty weathered: *?M. communis*, *?M. sedgwickii* from top. *M. lobiferus?*, *M. sedgwickii* from middle. *M. sedgwickii*, *?P. regularis*, *Pseudoclimacograptus (M.) undulatus* from bottom ('*Monograptus-sedgwicki* shales')	2.74

Nearby [7426 9115], Section H (Figure 35) reveals 2.1 m of '*Monograptus-sedgwicki* shales' with a 15 cm thick layer of grey non-graptolitic mudstone some 0.61 cm from the top. About 1.5 m of grey banded mudstone overlies the shales, the banding showing soft sediment deformations.

At the roadside in Section I [7464 9084], the '*Monograptus-sedgwicki* shales' are some 4.0 m thick and yielded: *Climacograptus sp.*, *Lagarograptus tenuis*, *M. lobiferus*, *M. planus?*, *?M. proteus*, *M. sedgwickii*, *P.* cf. *kurcki*, *Pseudoclimacograptus (M.) undulatus*. These beds are overlain by about 2.4 m of grey banded mudstones.

Anglers' Retreat to Rhiw-gam; Cwm Rhaiadr

In the NE of the district the Cwmsymlog Formation expands at the expense of the Devil's Bridge Formation. The 'Green mudstone member' provides a very distinctive marker band which enables this thickening to be determined, for just south of this area, as around Afon Lluestgota, it lies closely below the Devil's Bridge Formation while on the edge of Llyn Pen-rhaiadr [7531 9344] 17 m of beds separate the two, the section being:

	Thickness m
Devil's Bridge Formation	
Mudstone, banded grey, with thin arenite interbeds	—
Sandstone, in beds up to 8 cm thick, with interbeds of grey mudstone	3
Cwmsymlog Formation	
Mudstone, grey and greyish green, banded pale and dark, smooth-textured and well cleaved. Banding disrupted and streaked in parts by wet sediment deformation	c.17
'Green mudstone member': pale green to greyish green, massive, almost structureless. Banding indistinct or replaced by dark blebs and streaks. Pervaded by anastamosing brown veinlets coincident with a very uneven cleavage	9–10
Mudstone, dark and pale grey banded, well cleaved	8.5
Mudstone, dark grey, pyritous, weathering white or rusty. *M. sedgwickii?*	3.12
Derwenlas Formation	—

Nearby [7543 9335], dark grey mudstone laminae c.3 m below the 'Green mudstone member' yielded: *Glyptograptus sp.*, *M. sedgwickii*, *Petalograptus sp.*, *Pristiograptus regularis?*, *Pseudoclimacograptus (M.) undulatus*, *R. linnaei* (or *R. maximus*). This fauna is closely comparable with that found in hemipelagic dark grey shales just beneath the 'member' to the south; it is possibly of low *maximus* Subzone. North of Hyddgen, the 'Green mudstone member' comprises some 6 m of tough rather structureless mudstone. Its outcrop is readily recognisable by the ridge it produces. A section [7760 9148] displays its characteristic brown veinlets on a very uneven cleavage, as does another in a track [7818 9215] ESE of Ty Bwlch Hyddgen. Here again it is separated from the base of the Devil's Bridge Formation by banded grey mudstones; in a stream [7870 9260] these yielded: *Diversograptus runcinatus*, *M. gemmatus*, *M.* aff. *halli*, *M. planus*, *M. turriculatus?*, *P.* cf. *regularis solidus*, *R.* cf. *fugax* and *R. maximus*. These beds are still within the *maximus* Subzone, with a fauna indistinguishable from that of the Borth Mudstones and Aberystwyth Grits.

The interval between the 'Green mudstone member' [7998 9438] and the Devil's Bridge Formation in an old leat is c.60 m although the base of the latter is not distinctive.

A stream section in Cwm Rhaiadr [7527 9613 – 7526 9638] reveals most of the formation. The bottom part of the section is faulted and it is not certain whether the lowest beds belong to the Cwmsymlog Formation.

	Thickness m
Devil's Bridge Formation	
Mudstone, medium grey, thinly bedded, cleaved and of smooth texture. Many thin sandstones up to 2.5 cm thick; nearly vertical dips to synclinal axis	c.18
Mudstone, banded pale grey and medium to dark grey, well cleaved; the banded couplets are c.6 to 8 cm thick and many have arenite layers at the base which are about 1.5 cm thick at the base thickening upwards to over 2 cm	4.9
Thinly interbedded mudstone and sandstone, both up to c.2.5 cm in thickness, with the thicker sandstones c.15 cm apart on average	4.2
Cwmsymlog Formation	
Mudstone, in c.1 cm alternating bands of pale grey and medium grey, well cleaved; siltstone partings c.2 mm thick at the bases of many mudstone band couplets especially in top 2 m but not in bottom 7 m	12.2
'Green mudstone member': pale greyish green, structureless, weathers brown, pervaded by a fine plexus of thin brown veinlets following an uneven cleavage fabric, especially in the middle	c.6.5
Mudstone, medium grey, pale in parts, banding not evident but signs of bedding disturbances in places	12.8
Mudstone, medium to dark grey, alternating with thin sharp 'stripes' of pale grey mudstone 0.5 to 1.2 cm thick. Pyritous and rusty weathering. *Monograptus sp.* and *R. toernquisti*, the latter indicating sub-*maximus* Subzone	c.1.7
?Cwmsymlog Formation	
Mudstone, grey, well cleaved, thin laminae of pyrite at top; highly disturbed and wispy. Layer of ovoid concretions up to c.30 cm diameter in lowest 0.6 m	1.5
Fault	—
Mudstone, similar to basal bed of Cwmsymlog Formation, *R. toernquisti* at base	c.9.2
Mudstone, dark grey, abundant interbedded paler grey silty bands of up to c.2 cm thickness. Thin silty arenites of 2 to 4 mm thickness are present, on average c.1.5 cm apart. Pyritised graptolites occur but *M. sedgwickii* was not seen	c.9

Bwlch-glas – Blaen-Ceulan and Cwmere areas

Throughout the late middle Llandovery and early upper Llandovery (Figure 23) these areas consistently received a more energetic input of turbidites than most of the district: as a consequence arenites are much commoner and the diagnostic characteristics of the Cwmsymlog Formation are absent in places, and are replaced by those of the Devil's Bridge Formation. Outcrops of Moel Golomen and on Llechwedd Cwm-byr reveal respectively the SSW and NNE limits of this facies change. They suggest that the arenitic facies belt is aligned between SE–NW and ESE–WNW; in 1979 Strong recognised it is a channel-fill (p.48).

In the south, on Banc Ty-Newydd [6985 8708], there are coarser sandstones, individually up to 10 cm thick resting immediately on the Derwenlas Formation, and at South Moelglomen Mine [6983 8721] some bands reach 15 cm thick. They are commonly rusty-weathering, and have an abundance of feldspar, mica and dark grey lithic grains. Some are strongly convolute laminated, but parallel lamination makes them fissile and these laminae have yielded: *M. lobiferus*, *P.* cf. *regularis* and *Pseudoclimacograptus (C.) retroversus*. At the south end of Banc Ty-newydd, near Tre-boeth, these sandstones are visible below some 80 m of greenish and grey thinly banded mudstones and fine-grained arenites, which in turn are overlain by sandstones at the base of the Devil's Bridge Formation. This sequence of mudstones is clearly the Cwmsymlog Formation, with a thin packet of coarse basal sandstones separating it from the underlying Derwenlas Formation. Northward towards Moelglomen Mine the individual fine-grained arenite interbeds become thicker (up to 3 cm) and coarser, but the two sandstone-dominated sequences, one at the base of the Devil's Bridge and one at the base of the Cwmsymlog Formation, remain distinct and mappable. The intervening beds are thus still viewed as the Cwmsymlog Formation. On Moel Golomen, however, the whole formation passes into beds so like the bulk of the Devil's Bridge Formation that they have been placed within that formation.

Along Llechwedd Cwm-byr, some 2 m of thinly bedded bluish grey mudstones, siltstones and sandstones have been quarried for tiles [7103 8831]. Overlying them are sandstones which thicken upwards from c.5 cm in the bottom 1.2 m, to c.20 cm in the next 1.2 m. These sandstones are commonly graded, feldspathic, pyritous, and with dark grey lithic grains. They have been used to define the base of the Devil's Bridge Formation here. The quarried beds have yielded *M. convolutus* and *M. sedgwickii*, and are thus equivalent in age to the basal Cwmsymlog Formation elsewhere.

In another quarry [7131 8918] the middle of a sequence of 3–4 m of mudstones yielded: *?M. crenularis*, *?Monograptus planus*, *M. sedgwickii*, *P. regularis*, *Pseudoclimacograptus (M) undulatus* and, from the top few centimetres, *M. sedgwickii*. Thinly interbedded mudstones and flaggy sandstones up to 3 cm thick overlie these mudstones; the sandstones are cross- and parallel-laminated and viewed as part of the Devil's Bridge Formation. Upwards the bedding becomes thinner again, characteristic of the Devil's Bridge Formation, but the beds become green and are not unlike those in the Cwmsymlog Formation some 17 m above the base of the sandstones. About 1.5 km to the NNE [721 906], on the east side of the mountain road, beds much more typical of the Cwmsymlog Formation are exposed. They are mudstones, banded pale and medium grey with thin siltstone interbeds. The banding is disturbed by many mini-faults and attenuations indicative of soft sediment compactional deformation. On Moel Fferm and also farther west around Cwmere and Tyrhelig [6684 8907] the formation is again unrecognisable and the equivalent beds are either absent by overstep of the basal sandstones of the Devil's Bridge Formation, a distinct possibility around Cwmere (p.93), or have passed into the upper part of the Derwenlas Formation, as seems the case on the western side of Moel Fferm (p.76).

In the intervening ground between Cwm Ceulan [700 900] and Afon Cyneiniog the formation is about 35 m thick. The predominantly muddy upper part thins out towards the NW (Figures 36 and 42) and the basal sandstones unite with those at the base of the Devil's Bridge Formation to form a single arenaceous member. The basal sandstones are seen in Afon Cyneiniog (Figures 36 and 42, loc. 10). Southwards, on Fridd Cwmere, the mudstones in the upper part are already thinner (Figure 42, loc. 9) in a succession which has been built up from a number of adjacent exposures (Figure 36, loc. 9). North of Cefn-gwyn (Figure 42, locs. 6 and 7) the mudstones are even thinner, and sandstone interbeds from 2.5 to 10 cm thick are present in the top 10 to 12 m of the formation. From here to Tyrhelig and northwards to the Ceulan valley [666 898], there is no positive evidence of the presence of the formation.

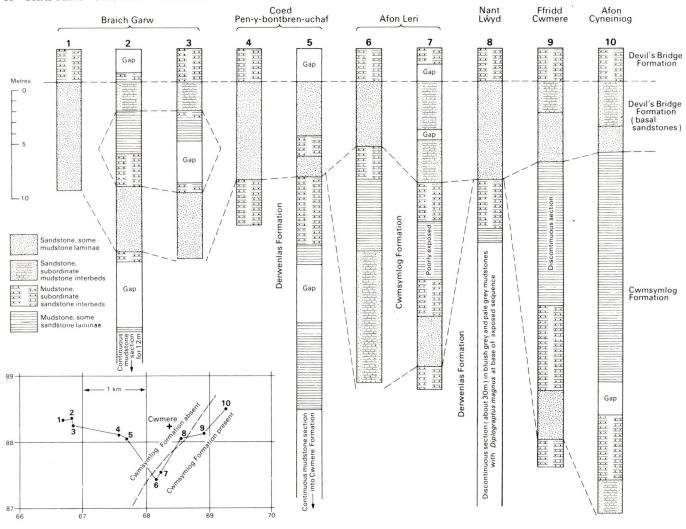

Figure 42 Comparative sections of the basal Devil's Bridge Formation, Cwmsymlog Formation and uppermost Derwenlas Formation in the Cwmere area

Afon Ceulan – Bedd Taliesin

Here there is a sequence of basal dark grey mudstone overlain by grey mudstone becoming paler upwards and with thin fine-grained sandstone interbeds increasingly abundant towards the top. There are no basal sandstones, and the total thickness is 25 to 30 m. An old quarry [6663 8976] shows all but the basal beds:

Thickness m

Devil's Bridge Formation
Interbedded mudstone and sandstone in 7.5 cm beds —

Cwsmymlog Formation
Mudstone, medium to pale grey, siltstone and fine-grained sandstone laminae 2.5 to 4 cm apart at base, gradually becoming more abundant upwards; pale grey in top 19 m, interlayered with fine-grained sandstone beds increasing in thickness upwards from 5 to 15 mm c.30

Talybont area (Glan-fred Borehole)

A borehole drilled at Glanfraid (Glan-fred) [6305 8812] (Cave, 1975a) provided data for this area. The borehole commenced in the Borth Mudstones Formation and ended in beds yielding *M*. aff. *sedgwickii*. The sequence was a continuum of interbedded fine arenites and mudstones, commonly banded, from top to bottom. The main distinctions between different parts of the core lay in the abundance and thickness of arenites, and in the colour. The base of the Devil's Bridge Formation was drawn rather arbitrarily at 280.97 m. The base of the Cwmsymlog Formation was not reached, though from the palaeontological evidence (Table 1), indicating the *sedgwickii* Zone, it is presumed to lie very closely below the bottom of the borehole. Its thickness here is at least 105 m, much greater than at Cwmere, only 5 km to the east, but similar to the type area (p.84). This suggests that a rapid thickening occurs south of the arenaceous facies and of a line from Borth [610 900] to Disgwylfa Fâch [736 838]. The bulk of the formation is similar to other areas, in that its top part (84 m) is pale grey and greenish grey while its lower part is darker grey. In general the turbidites display the Bouma sequence *Tcde* (Figure 2a), but some have a basal *Ta* which is dark grey, coarse, and contains much feldspathic and lithic grain detritus. Binary banding in the mudstone interval is well displayed (Figure 2b), and secondary phosphatisation at the interface is also well displayed.

Blaeneinion area

The formation is more uniform and thinner here than to the south (sections p.83). In stream sections [7327 9303; 7343 9378; 7353 9404] the beds are nearly vertical and the thickness of the formation is c.9 m. At the base the dark grey '*Monograptus-sedgwicki* shales' are about 3.5 m thick and a 15 cm sandstone [7343 9378] occurs in the middle. No sandstone of this thickness occurs in other nearby sections. Green mudstones appear 5.3 m above the base, and a dark grey graptolitic band 20 cm thick and 6.17 m above the base probably correlates with that seen in sections south of Afon Lluestgota (Figure 35). The mudstones above are still pale but more greyish-green. In the top 0.9 m, thin sandstones appear, and become thicker upwards probably representing a transition into the Devil's Bridge Formation, the base of which has been taken arbitrarily where the sandstones are c.2.5 cm thick.

Tyn y Garth – Pont Llyfnant and Derwenlas

The formation is 25 to 30 m thick and lithologically similar to that in the Ceulan valley. The basal beds are dark grey mudstone with *Lagarograptus* cf. *tenuis* and *Pristiograptus regularis* (Figure 37b, loc.10). Higher beds are bluish grey mudstone, with sandstone beds up to 1 cm thick; they are exposed farther south (Figure 37b, loc.11). Pale grey mudstones with many sandstone beds up to 2 cm thick, probably near the top of the formation, are seen at loc.12 (Figure 37b). North of Afon Einion the most complete section (Figure 37b, loc.13) is:

	Thickness m
Devil's Bridge Formation	
Interlayered sandstone and mudstone in 5 cm beds	—
Cwmsymlog Formation	
Mudstone, bluish grey; fine-grained laminated sandstone beds 1 cm thick, about 5 cm apart	c.7.3
Mudstone, bluish grey; sandstone beds 3 to 5 mm thick about 1 cm apart	3.7

	Thickness m
Mudstone, dark bluish grey, slight orange-brown weathering; some 5 mm sandstone beds	3.7
Gap	3.0
'*Monograptus-sedgwicki* shales' represented by topographical slack	—

The top of the sequence is exposed in an old quarry face [6978 9622] SSE of Cymerau. The base of the Devil's Bridge Formation is taken at the base of c.1.5 m of beds which contain sandstones up to 15 cm thick. These overlie 3.66 m of grey mudstones with thin, lenticular and cross laminated siltstones, mainly c.0.5 cm but up to 1 cm thick, and between 0.5 to 1.5 cm apart. Beneath are c.2.0 m of pale greyish green mudstone with siltstone interbeds up to 0.3 cm thick. This is the 'Green mudstone member'. It is estimated that another 18 to 19 m of grey mudstone with thin silty partings are present in the slope below.

In the eastern face of Morben Quarry [7152 9918] (Jones and Pugh, 1916, p.262), the beds are overturned. Flute moulds indicate strong currents from an easterly direction. These moulds are in a mudstone only some 3 to 4 mm thick, and the scour was limited by the underlying thin siltstone, so that the moulds are broad but shallow (Plate 15). The underside of the thin siltstone shows a cuspate interference pattern of beds of arcuate narrow and subdued ridges, formed by coarser foreset laminae of small scale linguoid current ripples that are impressed into the underlying mud. The directions of the traction current responsible for them and of the subsequent scouring turbidity are identical.

Plate 15 Flute moulds, Cwmsymlog Formation in Morben Quarry [7155 9921]. The arrow indicates the direction of the current, MN26231

Erglodd – Ynys-hir

The succession is similar to that at Tyn y Garth (p.89). In the south, a section [6628 9172] shows:

	Thickness m
Interlayered pale grey mudstone and fine-grained laminated sandstone in 5 to 10 mm beds	13.7
Obscured	c.3.0
Mudstone, bluish grey, a few thin sandstone laminae	7.3
Mudstone, dark bluish grey	3.7

Farther north [6694 9312] the lower part of the formation includes beds of weathered calcareous nodules in a colour-banded bluish grey mudstone similar to those at Pen y Graig-ddu (p.83). On the Coed y Fedw ridge the topmost 15 m of the formation (Figure 37a, loc.1) comprise interlayered bluish grey mudstone and fine-grained laminated sandstone in 5 mm beds. Localities 6 and 7 of Figure 37a display a similar sequence.

A quarry [6821 9615], together with a small nearby exposure revealed:

	Thickness m
Cwmsymlog Formation	
Mudstone, pale grey, thinly interbedded with siltstone laminated which increase in thickness upwards to layers c.2 cm thick	c.13.0
Mudstone, very pale grey with thin siltstone laminae. *M.* cf. *involutus*, *M. sedgwickii* and *R.* cf. *maximus* c.0.9 m above base	1.5
Mudstone, darker grey and banded with lenticular laminae of siltstone c.2 mm thick and 0.5 to 1.0 cm apart. *M.* cf. *distans*, *M. sedgwickii*, *M.* aff. *toernquisti*, *R. fugax* and *R. maximus*	0.15
Mudstone, very pale grey with thin siltstone laminae	0.6
Obscured	c.3.0
Shales, dark grey, pyritous. *Climacograptus sp.*, *Glyptograptus sp.*, *M.* cf. *involutus*, *M. sedgwickii* (very common), *M. spiralis* and *P. jaculum*	0.30
Derwenlas Formation	
Mudstone, medium to pale grey	seen c.1.0

Similar beds are exposed in another quarry [6801 9567] SW of Ynys-hir, where 8.2 m of pale grey mudstones, thinly interbedded with siltstone laminae up to 0.5 cm thick, are visible. About 3.65 m from the top of the section, some thin laminae of dark grey mudstone yielded *Glyptograptus sp.*, *M. sedgwickii*, *M. spiralis*, *M. toern- quisti?* s.l., *P. jaculum*, *P. nudus*, *P. regularis* and *R. maximus*. Some 2.89 m higher the mudstones become paler grey, and a 7 cm dark shale band at the top of the section yielded *P.* cf. *jaculum*.

Approximately 0.6 m below the lowest beds of the quarry, dark grey shale was excavated in the turf and yielded *D.* cf. *runcinatus*, *M. halli*, *M.* cf. *involutus* and *M. sedgwickii*: in view of the *M. halli* this assemblage appears to lie above the '*Monograptus-sedgwicki* shales'. A further 3 m lower, more dark shales yielded *M. sedgwickii* and these probably belong to the '*Monograptus-sedgwicki* shales'.

North of the Dovey Estuary

At the NE end of a forestry road [6584 9787 – 6580 9775] some 2 m of dark grey mudstone, typical of the '*Monograptus-sedgwicki* shales' are exposed above medium grey mudstone of the Derwenlas For- mation. They yielded *M.* aff. *sedgwickii* and *R. longispinus?*, and are overlain by a further 2 m of dark grey rusty weathering mudstone showing white weathered laminae which yielded *M.* aff. *sedgwickii* and *M. sedgwickii* or *marri*.

Across a fold-pair the section continues:

	Thickness m
Devil's Bridge Formation	
Mudstone, thinly interbedded with sandstone, the latter up to 8 cm thick in basal 3 m, thinning upwards to 4 cm or less and 3 to 6 cm apart	8 to 9
Mudstone, as below but shattered and quartz-veined	c.3
Mudstone and sandstone, thinly interbedded. Sandstone beds abundant and 2.5 to 8 cm thick except towards base where they are thinner	c.9
Cwmsymlog Formation	
Mudstone, pale grey, smooth-textured and cleaved. A few sandstones up to 2.5 cm thick: one at base is 5 cm thick with basal scour casts. Top 0.6 m pale greyish green (= 'Green mudstone member')	2
Mudstone, medium grey and pale grey bands in couplets 2 to 3 cm thick, with cross-laminated siltstone	15 to 18

Another exposure of the '*Monograptus-sedgwicki* shales' [6600 8907] yielded: *M. sedgwickii*, *?P. regularis* and *Pseudoclimacograptus (M.) undulatus*; yet another [6598 9810] yielded *M. sedgwickii*; a third [6591 9798] produced *M. communis?*

In this area the 'Green mudstone member' closely underlies the Devil's Bridge Formation. It crops out [6422 9686] about 18 m above the '*Monograptus-sedgwicki* shales'. A crag [6407 9673] is in some 4 m of massive pale grey mudstone above c.12.8 m of banded grey mudstone. A 25 cm graptolite band occurs 11.9 m up this suc- cession and yielded: *M.* cf. *sedgwickii*, *?P. regularis*, *R. maximus* or *lin- naei*. These forms indicate the *maximus* Subzone, and the band may correlate with the one known in this position in many other sections.

Farther north [6400 9754] the 'Green mudstone member' is separated from the Devil's Bridge Formation by 3 m of medium grey, banded mudstone. The roadside east of Braichycelyn Lodge [6356 9636] exposes imperfectly:

	Thickness m
Devil's Bridge Formation	
Mudstone, medium grey, cleaved and thinly interbedded with abundant sandstones commonly c.1.3 cm thick but up to c.7.5 cm	c.4
Mudstone, grey, cleaved, and sandstones up to c.5 cm thick, interbedded. Flute casts (currents from NE). In lower part sandstones are thinner and more widely spaced, mudstones are colour- banded in couplets 5 to 7 cm thick	6.1
Obscured	6.0 to 7.5
Cwmsymlog Formation	
Mudstone; grey, in bands 5 to 16 cm thick, dark phosphatic concretions in layers near tops of bands. Some mudstone bands have thin cross- bedded arenite bases	2.4
Mudstone, medium and pale grey streaky, poorly cleaved. 1.8 m in middle is pale grey, probably the equivalent of the 'Green mudstone member'	5
Obscured	c.6
Mudstone, pale and medium grey banded in couplets 1 to 2.5 cm thick, well cleaved	1.5
Mudstone, as above, less well cleaved	1.6
Mudstone, dark grey, pyritous and banded. *M. sedgwickii* ('*Monograptus-sedgwicki* shales')	seen 1

In Aberdyfi the formation crops in the core of an anticline. Two good, though partial, exposures have been noted. The first [6127 9600] is in a quarried yard to the rear of a shop frontage and east of a mine entrance. Here some 10 m of grey banded mudstones with interbedded very thin arenites and layers of phosphatic and siliceous concretions are well displayed. The concretions are each tilted by minor displacements on the cleavage planes. The second [6122 9612] is in a quarry exposing some 12 m of similar mudstone in bands 2.5 to 7.5 cm thick, each having a pale and dark grey layer. At the base of the road cutting at the quarry entrance graptolitic layers occur with *Glyptograptus sp.*, *Monoclimacis?*, *?Monograptus involutus*, *M. marri?*, *M. sedgwickii* or *halli*, *M.* cf. *sedgwickii*, *?M. undulatus*, *Petalograptus kurcki?*, *Pristiograptus regularis* and *Pseudo-climacograptus* (*M.*) *undulatus*. These forms indicate the *sedgwickii* Zone.

DEVIL'S BRIDGE FORMATION

Eisteddfa Gurig – Afon Myherin – Tynyffordd

The basal beds, with a maximum thickness of about 18 m, are more arenaceous than the main mass of the formation, the sandstone beds being locally up to 30 cm thick and commonly the predominant lithology. There is a rapid transition into these sandy beds from the mudstones of the Cwmsymlog Formation, usually within about 2 m: the basal sandstones are well exposed and commonly form a strong positive topographic feature.

North of the Castell Fault these basal sandstones are only about 4.5 m thick, as in crags [7970 8343] 0.6 km south of Eisteddfa Gurig: individual beds are fine grained and 10 cm, rarely 20 cm, thick. *Ta* and *b* divisions of Bouma (Figure 2a) are absent. On the southern side of the Castell Fault exposure is almost entirely limited to the easterly dipping limbs of folds, as in crags [7807 8150 – 7808 8158] SW of Fagwr-fawr and on both sides of Nant Meirch [7759 8105; 777 813] and Nant Fuches-wen [771 805; 773 809 – 773 811]. The total thickness of the sandstones is here about 18 m, but individual beds are rarely more than 15 cm thick. To the SW the sandstones maintain this thickness, SE [757 796] of Tynyffordd and and on Mynydd yr Ychen [761 796 – 765 794], where individual beds reach 30 cm in thickness and the mudstone beds are thin and subordinate. Here the *Ta* and *b* divisions of Bouma (1962) are commonly present, the basal parts of the sandstones being medium to coarse grained and graded.

The remainder of the formation in this area consists of rather monotonous interbedded mudstones and sandstones, the mudstones generally tending to become darker grey higher in the sequence. There are many open folds, making estimates of thickness difficult and unreliable. The best section is in Nant Rhŷs [7874 8158 – 7920 8148]. It is almost continuous and shows three cycles in each of which there is an upward increase in the thickness and decrease in the spacing of sandstone beds. The lowest cycle, some 110 m thick, has very thin (5 mm or less) widely spaced sandstone beds in its basal part which gradually thicken to 150 mm at the top and become predominant over the mudstone interbeds. It is succeeded by the second cycle, only 25 m thick, in which individual sandstone beds rapidly increase in thickness from 25 mm at the base to a maximum of 225 mm at the top. A fault of unknown throw separates these beds from the third cycle. This is less clearly defined, the lowest 100 m containing sandstones increasing in thickness from 25 to 35 mm and the remaining 75 m with beds up to 75 mm thick. The sandstones are almost all parallel laminated (Bouma *Td*, Figure 2a) with convolute lamination (*Tc*) also present in some of the thicker beds. In this section there appears to be a steady easterly dip, apart from two small folds, and the full thickness exposed is at least 310 m.

There is also an excellent section in the gorge of Afon Myherin

[7972 8078 – 7992 8043]. The lowest 18 m of beds, in the core of a broad anticline, are bluish grey mudstones with local reddish brown weathering, and with laminated and, in places, graded sandstone beds up to 25 cm thick. These beds contain *D.* cf. *runcinatus*, *M. halli* or *sedgwickii*, *M.* cf. *marri*, *M.* cf. *proteus*, *M.* cf. *tullbergi* and *Pristiograptus regularis solidus?* suggestive of *maximus* Subzone or possibly *sedgwickii* Zone. They are overlain by 35 m of bluish grey mudstones with many sandstone beds up to 5 cm and locally 7.5 cm thick. Southwards at least a further 30 m of massive bluish grey mudstones with fewer sandstones underlie the Dolwen Mudstones. Only 83 m of beds are seen here, yet the graptolite fauna near the base is suggestive of a low horizon. It is possible, therefore, that the formation is thinner than to the NW. The presence of an inlier of beds apparently referable to the Cwmsymlog Formation some 400 m to the north along the main anticline (p.83) supports this suggestion.

There is an adjacent section (Figure 40, loc. 2) near the base of the formation:

	Thickness m
Mudstone, bluish grey, many 2.5 cm laminated sandstone beds	11.6
Mudstone, dark bluish grey, rusty brown weathering, a few thin laminated sandstones (locality 2A)	3.7
Mudstone, bluish grey, interbedded sandstones up to 7.5 cm thick (locality 2B 6 m from top)	22.0

Graptolites from loc. 2A include *M. lobiferus* and *M. proteus*; they may indicate a position near the boundary between the *sedgwickii* and *turriculatus* zones. A similar zonal position is suggested by a rich shelly fauna, in lenticular sandstones at loc. 2B, including *Aegiria grayi*, *Dicoelosia alticavata*, *Eoplectodonta penkillensis*, *Leangella scissa*, *Skenidioides lewisii* and *Visbyella pygmaea*.

Brynbras – Goginan – Cwmere – Glandyfi Fault

Here the binary or (around Goginan) tripartite subdivision of the formation (p.49) has been mapped. From Brynbras to Pen y Graig-ddu the combined lower and middle divisions are about 80 m thick, including the basal sandstones which are up to 15 m thick with in-dividual sandstone beds reaching 30 cm in places and commonly predominant; the upper division is at least 310 m thick.

The basal beds are well exposed in crags along the eastern slopes of Brynbras [e.g. 7462 7936]:

	Thickness m
Devil's Bridge Formation	
Mudstone, bluish grey, interbedded with sandstones up to 2.5 cm thick	12.0
Basal Sandstone	
Mudstone, bluish grey, interbedded with sandstones up to 15 cm thick	5.5
Mudstone, bluish grey, some thin sandstone beds	1.2
Mudstone, bluish grey, interbedded with sandstones up to 15 cm thick	1.8
Cwmsymlog Formation	
Mudstone, bluish grey, many sandy laminae, some 2.5 cm sandstones towards top	3.4

The most extensive outcrops in the main part of the lower and middle division are seen [7427 7974 – 7409 7923] east of the road to Ystumtuen where about 38 m of grey mudstone, with many sandstone beds generally up to 2.5 cm thick, but a few 10 cm thick,

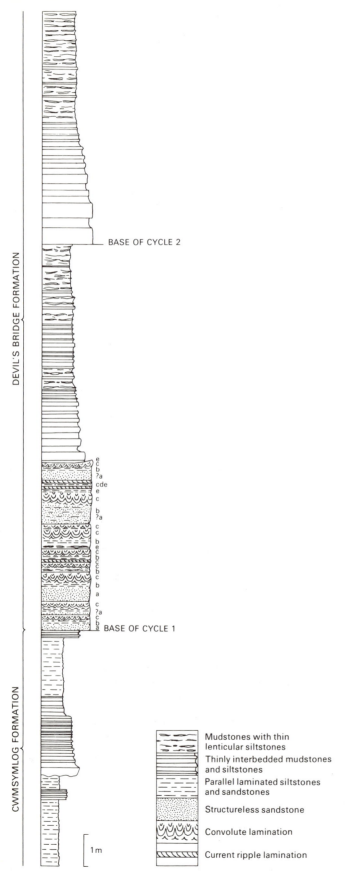

DEVIL'S BRIDGE FORMATION

CWMSYMLOG FORMATION

BASE OF CYCLE 2

BASE OF CYCLE 1

Mudstones with thin
lenticular siltstones

Thinly interbedded mudstones
and siltstones

Parallel laminated siltstones
and sandstones

Structureless sandstone

Convolute lamination

Current ripple lamination

1 m

Figure 43 Lower quarry, Craig yr Hesg; lower part of the Devil's Bridge Formation

are seen. The thinner sandstones are mainly parallel laminated but the thicker beds display a graded unit at the base. The basal beds are exposed in a quarry [7405 8090] (Fitches, 1972, pp.149–151) and nearby outcrops [7410 8076: 7410 8072], where about 15 m of mudstone interbedded with sandstones up to 30 cm thick are seen.

The complete combined lower and middle divisions and part of the upper division are exposed along the forestry road [7096 8182–7119 8177] (Figure 34) SE of Pen y Graig-ddu:

	Thickness m
Upper division (loc.1)	
Mudstone, dark bluish grey, interbedded with grey to dark grey laminated fine-grained sandstones mostly 2.5 to 4 cm thick, some up to 7.5 cm thick; slight orange-brown weathering in basal 28 m; sandstones predominate for 60 cm at 10 m above base	c.70
Lower and Middle Division (loc.2)	
Mudstone, dark bluish grey, orange-brown weathering; many thin siltstone beds for 1.8 m at 3.7 m above base; 10 cm sandstone bed 0.9 m above base	c.13.4
Mudstone, bluish grey, orange-brown weathering; regularly interbedded with 6 to 12 mm pale grey fine-grained sandstones	13.4
Mudstone, bluish grey; thin siltstones increasing in abundance upwards; a few laminated sandstone beds up to 2.5 cm thick	c.42.7
Mudstone, pale bluish grey, with 2.5 cm siltstones and fine-grained sandstones	c.7.3
Lower and Middle Division; basal sandstone	
Mudstone, pale bluish or greenish grey, interbedded with sandstones 5 to 15 cm thick	c.6.1
Sandstone, medium-grained, deep brown weathering, many small pyrite cubes; resting on	0.15 to 0.22
Cwmsymlog Formation	—

Dark grey shaly mudstones with deep rusty brown weathering, belonging to the upper division, occur in a few places along a forestry track, and at one spot [7050 7949] have yielded an extensive graptolite fauna including *?Climacograptus nebula*, *D. runcinatus*, *Glyptograptus* aff. *tamariscus fastigans*, *M. halli?*, *M.* cf. *marri*, *M. proteus*, *M.* cf. *pseudobecki*, *M. rickardsi?*, *M. turriculatus?*, *Petalograptus* cf. *wilsoni*, *Pristiograptus nudus* and *P.* cf. *regularis solidus*. The fauna indicates an horizon in the upper part of the *turriculatus* Zone.

Approximately 138 m of the upper division are exposed in an almost continuous road section [7146 8103–7187 8128] west of Bwlch Nant-yr-arian:

	Thickness m
Mudstone, grey to dark grey, interbedded with sandstones up to 7.5 cm thick	25.9
Mudstone, dark grey, rusty brown weathering in places, sandstone beds up to 5 cm thick	21.3
Mudstone, medium to dark grey, closely interbedded with sandstones mostly up to 2.5 cm thick, but some to 7.5 cm; especially about 6 m from the top	19.8
Mudstone, grey, closely interbedded with sandstones, mostly up to 5 cm thick but some 10 cm thick near base (small break in section, possibly some overlap with beds below)	30.5

	Thickness m
Mudstone, grey, interbedded with sandstones up to 15 cm thick (these sandstones form the top of prominent crags to the south of the road)	21.3
Mudstone, grey, closely interbedded with sandstones up to 2.5 cm thick	12.2
Mudstone, grey to dark grey, closely interbedded with sandstones up to 15 cm thick	6.7

The selective exposure of eastward dipping beds (see p.105) is clearly demonstrated at the eastern end of this section.

Around Goginan, where the formation has its maximum of about 600 m, all three divisions have been mapped separately. The persistence of the greenish grey colouration that is typical of the mudstones high in the Cwmsymlog Formation into the Devil's Bridge Formation is demonstrated in a quarry [6864 8080] at Goginan where 18 m of pale green and green mudstone, with 5 mm brown sandstones at the top, lie low in the latter. The middle division is exposed [6886 8176] at Old Goginan where dark greenish grey reddish brown weathering mudstones, with sandstone beds, have yielded the following *maximus* Subzone fauna: *D. runcinatus, Glyptograptus sp., ?Monoclimacis galaensis, Monograptus gemmatus, ?M. halli, M. planus, M.* cf. *pragensis ruzickai, M.* cf. *proteus, M. pseudobecki, M. turriculatus, P. regularis, ?P. variabilis, Pseudoplegmatograptus obesus, Rastrites* cf. *fugax* and *R. maximus.* Just north [6831 8100 – 6830 8111] of Penbryn, 2.5 m of pale bluish grey, reddish brown weathering mudstone with sandstone beds (middle division) is overlain by 17 m of dark bluish grey mudstone with parallel-laminated sandstones up to 5 cm thick (upper division). In the upper division shelly faunas have been obtained from two localities, south of Goginan almost along strike from one another. A stream section [6977 7967] in Coed Ty-llwyd shows dark grey mudstones with a few sandstones up to 50 mm thick; in their decalcified lower parts, these have yielded *Aegiria grayi, Dicoelosia* cf. *alticavata, Eocoelia?* and *Leangella?* as well as crinoid columnals, a fauna of high *sedgwickii* or *turriculatus* Zone. To the south in a track [6993 7922], some sandstones have basal decalcified zones up to 20 mm thick with *Eocoelia?, Hyattidina?, Mendacella?* and *?P. regularis.*

North and west of Goginan it is possible to distinguish only a binary subdivision of the formation. Between Craigypistyll and Bwlch-glas the combined lower and middle division is about 90 m thick, the best section being at Craigypistyll [7124 8556 – 7125 8565] (Figure 29, loc. 9). The basal arenaceous beds are not clearly defined here, and the base of the formation is taken at the first appearance of sandstone beds above mudstones of the Cwmsymlog Formation.

	Thickness m
Devil's Bridge Formation (lower and middle division)	
Mudstone, grey, a few thin sandstone beds	seen c.6.0
Mudstone, bluish grey, orange-brown weathering, a few thin laminated sandstone beds: 1.8 m of grey fine-grained, thinly bedded sandstone interlayered with thin mudstone at 4.3 m above base and in topmost 1.2 m	8.5
Mudstone, bluish grey, orange-brown weathering, interlayered with laminated sandstones generally 1 to 2 cm thick, some up to 7.5 cm thick in basal 1.8 m	23.4
Cwmsymlog Formation (Figure 29, loc.11, see p.84)	—

Many of the basal sandstones in crags [7186 8691] display a graded interval (*Ta*) as well as the usual laminated and cross-laminated or convolute-laminated divisions. Crags [7049 8749 – 7045 8738] east of Bwlch-glas show about 14 m of beds near the base of the sequence. The main part of the lower and middle division in the Bwlch-glas area comprises grey and pale grey mudstone interlayered with pale grey fine-grained parallel-laminated sandstones.

Between Bwlch-glas and the Glandyfi Fault there is again a dominantly arenaceous member at the base. Additionally, as the Cwmsymlog Formation thins westward (p.87), the arenaceous beds at the bases of the two formations amalgamate (Figure 42). The thickness of the combined lower and middle division, including the arenaceous beds ranges between 190 and 240 m. A section [6854 8804] (Figures 36 and 42, loc. 8) shows 9 m of massive fine- to medium-grained sandstone in beds up to 30 cm thick with thin mudstone beds and partings, overlain and underlain by grey mudstones interbedded with fine-grained sandstones up to 2.5 cm thick. South-eastwards the sandstone packet splits, allowing mudstones of the Cwmsymlog Formation to crop out on Ffridd Cwmere and in Afon Cyneiniog and Afon Leri (Figure 42). Figure 42 also displays sections farther west (locs. 1 – 5).

Two quarries in Craig yr Hesg reveal the basal beds of the formation in three broadly thinning- and fining-upwards cycles. In the lower quarry [6770 8850] the thickest sandstones occur at the base of the lowest cycle, and appear in the bottom of the quarry as three massive, but composite, beds nearly 3 m thick in total (Figure 43). The beds are of uneven thickness. They show well developed convolute lamination in their higher parts with some parallel lamination below or, in the case of the bottom massive bed, structureless mildly graded sandstone. *Ta?bcd* intervals appear to be represented; *Te* is very thin. The base of the uppermost of the three massive beds is fluted, with flute-casts splaying towards 285°. The top of this sequence consists of mudstones, with thin, commonly cross-laminated, siltstones up to 0.7 cm thick. The base of the second thinning-upwards sequence is marked in the main quarry face by two 25 cm sandstones. This second sequence thins upwards to where the beds of arenite are between 2.5 cm and 5 cm thick, but its top is not exposed. In the upper quarry [6772 8845] the bottom cycle is very similar, but the second cycle is complete showing a thickness of c.6.1 m but containing, at its top, 61 cm of mudstones with sandstones up to 12 cm thick, which may be a subsidiary cycle. Overlying these beds are more thick sandstones representing the base of the third cycle. There are four beds of sandstone, the thickest basal one being 40 cm thick, and the others being between 10 cm and 15 cm thick separated by mudstone partings (*Te*) of 2.5 to 7 cm thick. Flute casts on one of these beds indicate a palaeocurrent from the east. The highest beds seen are c.1 m of interbedded arenites and mudstones. From the debris in this quarry it is clear that flute casts are fairly common, and some reveal soft-sediment deformational shortening on their long axes. These beds are interpreted as the deposits of a shallow depression, or channel, on the sea bed. The successive thinning upward cycles are looked upon as the product of resurgences of deposition in this depression, rather than the prograding or lateral migration of sand-dominated bodies: both the latter processes tend to produce coarsening- and thickening-upward cycles.

Farther north the basal arenaceous beds are less well developed [6668 8996]:

	Thickness m
Mudstone, grey to pale bluish grey with interbedded sandstones 2.5 to 4 cm thick	4.6
Obscured	2.4
Basal sandstones: sandstone, grey, fine-grained, commonly with graded bedding; in 10 cm beds interbedded with thin grey mudstones	5.5

Cwmsymlog Formation

	Thickness m
Mudstone, grey, interbedded with grey sandstone in 2.5 cm thick beds	3.7
Mudstone, bluish grey, regularly interbedded with fine-grained sandstone in 5 mm beds	13.7

Westward, crags [6655 8994 – 6662 8994] in slightly higher beds show 16.5 m of grey mudstones with interbedded sandstones 0.6 to 2.5 cm (rarely 7.5 cm) thick overlying 8.2 m of pale grey mudstone with very abundant 6 mm sandstone beds and with several beds of calcareous nodules. These lower beds are unusual, resembling the closely interbedded mudstone and sandstone facies of the Cwmsymlog Formation noted at the previous locality.

Sections in the main part of the lower and middle division in Afon Cyneiniog [6983 8833 – 6954 8845], Nant Lŵyd [6869 8768 – 6880 8777] (Figure 36, loc.7) and Afon Leri [6802 8658 – 6807 8719] reveal monotonous grey to pale grey mudstone with 1 to 2.5 cm thick (rarely 7.5 cm) fine-grained sandstones, many with parallel lamination, at 2.5 to 8 cm intervals. At the western end of crags on Braich Garw [6621 8851] there is an upward transition into the darker mudstone and sandstones of the upper division.

North and NW [694 880] of Moelglomen many strike ridges in the upper division reveal medium to dark grey mudstone with dark grey argillaceous fine-grained sandstones, usually parallel laminated, up to 5 mm (rarely 10 mm) thick, and about 2.5 to 5 cm apart. Slightly paler sandstones and some orange-brown weathering mudstones in the lowest beds are exposed [6942 8757]; they suggest there is a downward transition into the lower and middle division. This passage is exposed also [6621 8851] on the south side of the Leri valley, where dark grey mudstone with some dark grey sandstone beds up to 5 cm thick passes down into pale to medium grey mudstone with many grey sandstones up to 2.5 cm thick.

Llyn Conach and Bwlch Hyddgen

In the area around Llyn Conach [740 930] the trace fossils *Nereites sp.* and *Dictyodora sp.* are usually preserved in the parallel laminae of the Bouma *Td* division (Figure 2a). The approximate thickness of the formation is calculated as 225 m [7312 9264 – 7330 9259].

Creigiau Bwlch Hyddgen exposes pencil-cleaved, dark grey mudstones with rusty weathering surfaces. They are pyritous and graptolitic, and possess bands, some over 1.5 m thick, representing pelagic deposition in a distal turbidite sequence of mudstones and thin fine-grained sandstones. Stratigraphically the beds are high in the formation, and the graptolites from 1.5 m of dark grey mudstone in a small anticline in the stream [7680 9368] and from similar mudstones nearby [7711 9324] include: *?D. runcinatus, Monograptus barrandei?, M.* cf. *marri, M.* cf. *planus, M.* cf. *proteus, M. pseudobecki, M. turriculatus, ?P. wilsoni, Pristiograptus nudus, P. regularis* (transient to *P. nudus*) and *Rastrites sp.*, indicating the *turriculatus* Zone.

Talybont – Glandyfi

Although the full sequence is present it is not possible to subdivide it or to make an accurate estimate of its thickness because of repeated folding. Sandstones are commonly up to 10 cm thick, and flute casts were noted [6394 8827], indicating a palaeocurrent from between 060° and 070°. Sections [6504 8902 – 6517 8903; 6522 8905 – 6530 8911] near the top of the formation show grey to blue-grey mudstones with many fine-grained argillaceous laminated sandstones up to 1.5 cm thick, and with a few up to 7.5 cm thick in the western section.

North-east of Taliesin are three small outliers [664 918; 667 924; 668 929]. In the northernmost the basal beds are seen [6686 9295 – 6687 9291]; at least 9 m of massive fine- to medium-grained sandstone, with convolute and parallel lamination in beds up to 20 cm thick, with thin mudstone interbeds are exposed. The sandstones pass upwards into grey mudstones with many sandstones up to 2.5 cm thick.

The basal beds are present at the western end of a forestry road [6721 9383 – 6727 9385]:

	Thickness m
Devil's Bridge Formation	
Mudstone, dark grey, slight orange-brown weathering in places; interbedded with grey laminated sandstone beds up to 2.5 cm thick	14.0
Mudstone, grey and sandstone, grey laminated, in 5 cm beds; some thicker sandstones including a 15 cm bed at base	4.8
Cwmsymlog Formation	
Mudstone, grey, silty, and sandstone, fine-grained, laminated, in 1 to 2.5 cm beds	1.2

Here the basal arenaceous beds are thinner than around Taliesin. Another section [6731 9385 – 6734 9384] just to the east shows the range of thickness of the sandstone units:

	Thickness m
Mudstone, dark grey, orange-brown weathering; interbedded laminated fine-grained sandstones mostly up to 5 mm thick, a few to 25 mm	c.8.2
Mudstone, grey, interbedded with laminated fine- grained sandstones up to 12 mm thick, some to 50 mm in top metre	c.3.4
Sandstone, fine grained, massive, convolute-laminated	0.15
Mudstone, grey, interbedded with laminated fine- grained sandstones mostly up to 12 mm thick, a few to 25 mm	c.9.8
Mudstone, bluish grey, slight orange-brown weathering, many fine-grained sandstones up to 5 mm thick, a few to 25 mm	c.4.0

The basal arenaceous beds are also well exposed on Foel Fawr [6903 9496] (Figure 37b, loc.1), where 3 to 4.5 m of interbedded sandstone and mudstone in 7 to 10 cm beds are seen in an anticline with a steep plunge of 55° towards 40°.

The Cwmsymlog Formation is exposed in the base of the sequence [6974 9622] at the SE end of the rock-face just above a path near Cymerau. The base of the Devil's Bridge Formation is sharp, and succeeded by c.1.5 m of interbedded arenites and mudstones, the former reaching thicknesses of c.15 cm. Flute casts occur with splay towards 290°. These beds are overlain by a further 50 m of thinly interbedded arenites and mudstones. In the basal 13.5 m of these, the arenites, fine sandstones and siltstones are less than 2.5 cm thick; in the following 36.5 m they are commonly c.5 cm thick (but up to 7 cm) and c.2.5 cm or less apart. Above this sequence, in the core of a syncline at the NW end of the section, well cleaved mudstones with much thinner arenites appear.

A road cutting [6481 9177], west of Neuadd-yr-ynys, shows 7.6 m of grey silty mudstone with many 5 mm beds of fine-grained laminated sandstone and some thicker sandstones. These thicker beds, up to 25 mm thick in the lower part of the section and up to 75 mm thick towards the top, show both graded bedding and convolute lamination. An unusual feature is a bed of rotten ferruginous nodules (probably originally calcareous) in the central part of the section. Immediately NE of Pen-y-graig [6533 9261], a quarry and adjacent sections show about 36.5 m of grey silty mudstones with many 10 to 30 mm beds of graded fine-grained sandstones. Two 75-mm-thick sandstone beds in the central part of the quarry display convolute lamination and box-type ferruginous weathering.

North of the Dovey Estuary

The basal part of the sequence has been described above (p.90). Higher parts are displayed near Fron-gôch [664 973] and are generally thinly bedded mudstones and arenites, with mudstone in beds 10 to 40 cm thick dominant over cross bedded arenites, mostly 1 to 2 cm thick. The rhythms are very typical of the more argillaceous parts of the formation, and clearly show the 3ii division with phosphatic concentrations below. Beds in a quarry [6643 9732] were sufficiently cleaved to work for flagstones.

One or two packages of thicker sandstones occur in the sequence in this area. One was traced from the eastern end of Pant Eidal Wood [6568 9726] to a nearby stream [6602 9740]. At its western end one exposure reveals 2.4 m of beds consisting mainly of fine- to medium-grained sandstones between 20 cm and 40 cm thick, with largely convolute and parallel lamination, separated by mudstone layers between 5 cm and 76 cm thick. Sandstones of this thickness are very uncommon within this district except at the base of the formation: their presence may imply an independent source to the ENE, but no current indicators were seen. However, several exposures [e.g. 6377 9668; 6374 9664] ENE of Braichycelyn reveal sandstones with flute casts indicating palaeocurrents from between east and ENE.

In the railway cutting [6356 9632] near Braichycelyn Lodge, 30 m of thinly bedded fine arenites (up to 10 cm thick) and mudstones are overlain by a similar thickness of thinly bedded mudstones with thin arenites (up to c.6 cm thick).

To the west the formation thins, but is still present around Aberdyfi. It is exposed at the roadside [6096 9603] where the beds dip SE overturned at 87°. The sequence is thinning upwards at the top, and reflects a waning in the energy of transportation of sediment and a transition into the Borth Mudstones facies.

	Thickness m
Mudstone, thinly bedded, with siltstones/fine sandstones thinning upwards from c.2.5 cm to 0.3 cm at the top	c.8.0
Arenites, fine-grained, up to 8 cm thick, bedding convolute and parallel at top (*Tcd*), interbedded with mudstones up to 1 cm thick	0.76
Arenites, up to 2.5 cm thick, commonly with parallel lamination (*Td*) and some with coarser feldspathic sandstone bases (*T?a*). Other layers are strongly cross-laminated (*Tc*) and in the thicker layers the lamination is convolute. Mudstone interbeds (*Te*) less than 2 cm thick	5.5

BORTH MUDSTONES FORMATION

Capel Bangor – Penrhyncoch

A quarry [6434 7958] exposes 6 m of poorly cleaved dark grey mudstones with several paler grey siltstones up to 2.5 cm thick and three laminated sandstones up to c.5 cm thick. A bed of Harp Rock Type (p.55), 23 cm thick, is also present consisting of grey mudstones with very small shale clasts and a 3 cm layer of fine-grained sandstone at the base. Sections in Afon Peithyll [6553 8246 to 6562 8254] reveal dark grey mudstones, with sandstones or faintly laminated bluish grey siltstones interbedded at c.3 m intervals. The latter increase in frequency towards the base of the formation.

Thin layers of black graptolitic mudstone occur, and in one place [6532 8243] yielded *D. runcinatus*. The lane east of Garth Penrhyncoch [6577 8390 to 6522 8408] shows medium grey, poorly cleaved mudstones, sparsely interbedded with thin siltstones and sandstones from 2 to 5 cm thick, which have sharp bases and graded bedding. In Nant Silo [6512 8377 – 6509 8380], dark grey mudstones with thin siltstone interbeds and sporadic sandstones up to 15 cm thick occur.

Llandre and Talybont

A quarry [6407 8629] east of Dolau exhibits 9 m of massive poorly cleaved medium grey mudstone in beds up to c.30 cm thick with a few interbedded sandstones up to 1.5 cm thick. At Cwarel [6303 8627], another quarry shows 15 m of pencil-cleaved dark grey mudstones.

An excavated face behind a barn [6169 8781] at Cil-olwg exposed c.25 m of mudstones containing several structural discontinuities; the thickest unbroken succession was c.8 m thick. The mudstone is medium grey; parts show bronze weathered surfaces, other parts are dark grey tending to rust when weathered. Cleavage intensity is not uniform. Sandstones are sparse, generally c.30 cm apart and c.2 cm thick, but some up to 5 cm thick. In another excavation [6303 8832] at Henllys, Dolybont, fracture-cleaved, pencilly, medium grey mudstones occur in beds 30 to 60 cm thick which contain basal sandstones up to 2.5 cm thick. A long continuous section in Afon Leri [6589 8900 – 6599 5894] shows dark bluish grey mudstone with a few thin paler mudstones and siltstones and some widely spaced structureless fine-grained sandstones up to 7.5 cm thick.

Upper Borth to Harp Rock (Craig y Delyn)

The foreshore WSW from Upper Borth to Careg Milfran [5995 8848] and from there SSW to Harp Rock [5962 8768] provides almost continuous exposure of many metres of strata, so folded and faulted that the total thickness is uncertain: it may be well over 100 m. The wave-smoothed cliff, especially under Craigyrwylfa [602 886], reveals bedding and secondary banding of turbidite mudstones with Division 3ii oxic hemipelagite (Figure 2b). South of Careg Milfran there is also a well exposed package [6000 8850] of mud turbidites with anoxic hemipelagic intervals of Division 3i. The black mudstones yielded some 80 graptolites including *D. cf. runcinatus*, *Glyptograptus sp.*, *M. exiguus*, *?M. pseudobecki*, *M. turriculatus?*, *P. regularis* and *P. regularis solidus?* indicating the *turriculatus* Zone.

Farther SSW, in the 400 m of foreshore north of Harp Rock, sandstones are more numerous and thicker. Mudstone layers are commonly c.22 cm (and up to 35 cm) thick, with sandstone (c.2 cm thick and up to 8 cm) at the bases of some. Cone-in-cone sideritic concretions, up to 30 cm long and 6 cm thick, occur in some mudstones, usually in the top quarter of the bed. A few beds of mudstone are very silty, containing wisps of silt, and these are probably early precursors of the Harp Rock type sandstones (p.55). The section across the junction into the Aberystwyth Grits is shown in Figure 44.

An overgrown dingle 600 m WNW of Dolybont exposes medium grey mudstones, thickly bedded and homogeneous. A single 15 cm sandstone [6187 8822] possesses flute casts indicating currents from the west. No bed recording this turbidity direction has been observed on the forshore to the west though the horizon may be concealed under Borth beach.

Cors Fochno and adjacent areas

The Borth Mudstones probably underlie Cors Fochno (Borth Bog); they form a number of 'islands' projecting through the Holocene

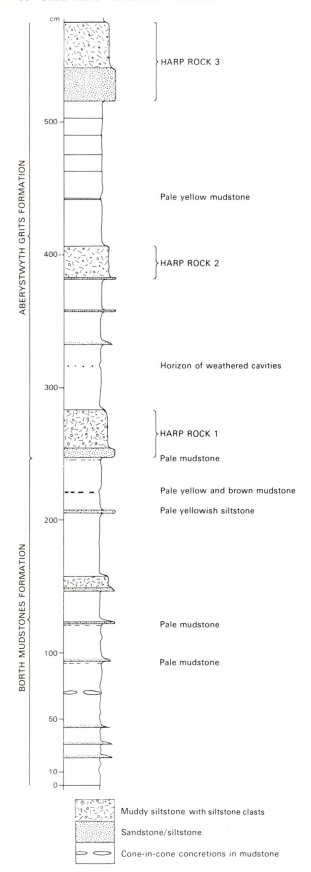

Figure 44 Section through the base of the Aberystwyth Grits Formation at Harp Rock

deposits [e.g. 629 926]. A quarry and road cutting [6279 9268 – 6306 9275] at Ty Mawr show about 16 m of grey to blue-grey cleaved silty mudstone with widely spaced discrete fine-grained sandstone beds up to 7.5 cm thick, and also thin sandy layers some 30 to 90 cm apart which grade upwards into the overlying mudstone. An impersistent bed with cone-in-cone structure is present about 2 m from the base of the section.

Aberdyfi area

The thickness of beds here is large, perhaps exceeding the 300 m estimated near Borth, but intense folding and faulting make precision impossible. The rocks are no different from those around Borth. Some of the better exposures are given below:

A gateway [6268 9629] and drive, WNW of Penhelig, shows 20 m of mudstones, thickly bedded (up to 50 cm), moderately cleaved and with approximately ten sandstones up to 8 cm thick, in the bottom 5 m of beds. At the base of one mudstone, the Division 1 arenite (Figure 2b) is replaced by lenticular siliceous concretions. Division 3ii mudstones are common, with the underlying phosphatic concretions also present. Flute casts on two sandstones indicate palaeocurrent directions from 060° to 080°.

A laneside [6306 9656] NNW of Trefri, shows some 2.5 m of medium grey moderately cleaved mudstones in beds commonly c.60 cm thick. Sandstones up to 10 cm occur as Division 1. They are moderately dark grey, argillaceous and micaceous, and their tops are not sharply demarcated. Flute casts indicate a palaeocurrent direction from 060°. This palaeocurrent indicator and those of the previous locality do not differ from those in the Devil's Bridge Formation nearby.

Near Penhelig there are three entrances to stone mines [6202 9636; 6199 9642; 6196 9645], the last of which has been developed into a quarry. The first two are unusually large, about 2 m square, and lead into large underground caverns. The mudstone is soft, medium grey, and in beds some 60 cm thick. Thin sandstones, up to 2 cm thick, occur as Division 1 in some beds, while Division 3ii is commonly present. Cleavage is usually poor.

Just south of the railway cutting at Penhelig, a site clearance excavation [6200 9614] exposed c.10 m of mudstone, mainly medium grey, homogenous Division 2, in beds 15 to 30 cm thick, usually with Division 3ii mudstone at the top, and the arenite of Division 1 commonly absent or a mere film. There are, however, a few sandstones, up to 10 cm thick, with tops grading into the Division 2 mudstone. In some beds Division 1 is replaced by a layer of ovoid siliceous concretions.

ABERYSTWYTH GRITS FORMATION

Craig y Delyn (Harp Rock) to Wallog

The base of the formation is exposed on the coast at Craig y Delyn [5958 8761] at the base of the three prominent Harp Rock Type 2 deposits lying in the westward dipping limb of a syncline on the foreshore. The structure is relatively simple. The beds are turbidite mudstones with subordinate sandstone ribs (less than 3 cm thick) in packets up to 3 m thick, separated by thicker turbidite sandstones (up to 0.2 m thick) which are less common. Type 2 deposits are plentiful, but subordinate to the latter: they are up to 0.3 m thick. Very thin black graptolitic mudstone beds are common beneath sandstones, especially north of a NW – SE normal fault [5918 8640]. Graptolites were obtained from large fallen blocks along the foreshore mainly from the thin graptolitic pelagic mudstone beds. Details of the graptolite localities are listed in Table 2.

Plate 16 Aberystwyth Grits. Evenly bedded bi-modal turbiditic sandstones [5828 8262]. Cliffs below Constitution Hill. A12963

Wallog to Clarach Bay

In the coastal section the structure is fairly complex, the folding being complicated by numerous thrusts and faults. The beds are turbidite mudstones with many sandstone ribs (less than 25 mm thick) in packets up to 1.2 m thick, and thicker sandstones, up to 0.3 m thick but on average 75 mm to 0.15 m thick (see also p.55). Type 1 and 2 Harp Rock deposits (see p.55) are common, and some are up to 0.38 m thick. A group of four Type 1 deposits in a sequence of normal sandstones and mudstones some 7 m thick, seen 900 m south of Wallog [5865 8485], is repeated three times to the south [5858 8640; 5855 8432; 5856 8430] as far as Clarach Bay. Cone-in-cone nodules are common in the turbidite mudstones and can be seen spectacularly in plan-view on foreshore reefs [5860 8470].

Dark graptolitic interturbidite mudstones are up to 3 mm thick beneath thin turbidite sandstones; they are common north of northing 8490 but far less abundant to the south (Table 2).

Clarach Bay to Constitution Hill

Apart from the northern end, this section of the coast has a relatively simple structure. The formation consists of turbidite mudstones with numerous sandstones. The latter are bimodal (Plate 16), of thicknesses up to 1 cm, in packets of up to 1.2 m, and of thickness between 1 and 7.5 cm, also in packets. There are also thicker, more prominent, sandstones (7.5 cm to 0.22 m thick) which occur scattered throughout (see also p.55). The highest Harp Rock deposit (of Type 1) in the succession is seen in the foreshore immediately south of Clarach Bay [5862 8352], and is 0.175 m thick. In the cliffs at the same locality a 25 mm white to creamy grey ash bed is interbedded in mudstones.

Graptolites have not been collected in this section, and graptolitic interturbidite mudstones are extremely uncommon. A brachiopod fauna has been recorded (Bates, 1982) from a sandstone in the mouth of a cave [5826 8270] in the first cove to the north of the breakwater at the southern end of the section. The trace fossil *Palaeodictyon* is common at the base of the cliff [5830 8280].

Aberystwyth to Allt-Wen

South of Constitution Hill only isolated foreshore sections are available, the main ones being situated around the pier [580 816] and in the south side of the harbour [5821 8106 to 5796 8068]. Sandstone beds exceeding 0.2 m are uncommon, the thickest being 0.3 m thick: the average thickness is 5 cm to 0.15 m (see also p.55). Graptolite localities are listed in Table 2.

Inland sections

These show features similar to those in the coastal section. The main fossiliferous localities of significance are listed in Table 2. The better sections are described briefly below.

Quarries [5982 8343 – 5975 8349] on the south side of the Clarach valley (loc.15, Table 2) expose some 100 m of interbedded turbidite sandstones and mudstones, the forming being on average individually 5 – 10 cm thick. Another disused quarry [6090 8185], SW of Cefnhendre Farm (loc.16, Table 2) is in c.45 m of interbedded sandstones and mudstones. Yet another [6354 8285], NE of Peithyll (loc.17, Table 2) is within 130 m of the base of the formation. About 18 m of rusty weathering mudstones with black pyritous graptolitic shale beds and scattered sandstones up to 0.15 m thick are visible; there are also deposits of Harp Rock Type (type 2). Lastly, a disused roadside quarry [6111 7954], beside the A4120 road, is mainly in the inverted limb of a gently inclined minor fold. Its west face shows inverted turbidite soles dipping at 49° to 130°, and displaying the full spectrum of sole structures, especially flutes and groove casts.

DOLWEN MUDSTONES FORMATION

The lower part of the formation is mainly forested and poorly exposed, though discontinuous sections in blocky, bluish grey, brown-weathering mudstone, with sandstone and siltstone beds up to

5 mm thick, occur in the two minor tributaries [7906 7916 – 7908 7905; 7994 8042 – 7999 8047] of Afon Myherin. The highest beds are well exposed in a number of open southward plunging folds [7955 7915 – 7985 7925] south of the river. The beds are grey to bluish grey mudstones, orange-brown weathering towards the top of the exposed strata, with sandstones up to 25 mm, and rarely 50 mm, thick. Decalcified shell fragments were found at two localities (Appendix 3, locs.19, 20).

CHAPTER 6

Structure and geophysics

Throughout the Ordovician and Silurian the Aberystwyth district was situated in an unstable marginal basin to the SE of the closing Iapetus Ocean. This Welsh Basin was initiated by Cambrian times, the rocks were deformed during the Ordovician and Silurian and the basin was eliminated in the Downtonian in the final phase of the Caledonian earth movements. There is no evidence within the district of deformation which post-dates that period (currently placed at c. 408 Ma). Even during the limited period of deposition represented by the rocks at outcrop in the Aberystwyth district, deformation was active; to the east the Towy Anticline had been initiated and northward of it, passing near to Welshpool, there was an active lineament, which for a time controlled the lower Llandovery marine transgression. Beyond this line the Ordovician rocks of Shelve and Builth had been folded already. It is thus clear that there was a 'continuum' of deformation, including the early part of the Silurian, a period normally considered docile. Inevitably this deformation acted upon sedimentary piles that were at various stages of dewatering and lithification. It also produced unevennesses of basin profiles, thus promoting gravitational stresses and strains further to those indigenous to differentially accumulating sedimentary and volcanic systems.

The structure of the basinal Lower Palaeozoic sedimentary/volcanic accumulation is thus of two types; that with a regional tectonic origin and that with a gravitational (soft-sediment) origin. In the Aberystwyth district both types are evident (Davies and Cave, 1976). However, the distinction between them is blurred, and many structures, including some folds, cannot be assigned with certainty to either. It does not seem possible, for instance, to discriminate between folding produced in a thin skin overriding a décollement and that produced at depth in compression above a foreshortened basement. However, excess pore-water is an essential ingredient to much gravitationally induced sediment/volcaniclastic deformation, so an association of particular folds with water-escape structures is suggestive of this type of deformation.

It has been noted already (Jones and Pugh, 1935a, p.284; 1:50 000 Geological Map sheet 163 (Solid Edition), 1984) that there are differences between the fold patterns above and below certain formations. Locally these include the Nod Glas (of Caradoc age) and the Cwmere Formation at the base of the Silurian. Both formations consist largely of dark grey pyritous mudstone, and it is evident that these layers exerted an influence over the distribution of stress within the sedimentary pile. Small-scale regionally abnormal folds and faults are present at places within the Cwmere Formation. They can be seen, for instance, in a track [7004 8741] near Bwlch-glas and at a locality [7185 8678] 1.2 km east of Nant-perfedd. In the former the outcrop is intensively fractured, largely along strike, and displays extensional movements.

Low-angle glide planes, small step faults and boudined sandstone layers are common. A linear exposure of this type inhibits full appreciation of relative movement of the faulting but displacements appear to be down to the SW. At the latter locality the formation is strongly folded. Small folds plunge northward at up to 80°; they are quite incongruous with the larger folds here. Fracture surfaces are tight and detectable mainly by the bedding discordances they impart. No part of this structure carries mineralization; neither quartz nor carbonate is present. Such structure may have arisen in rocks which retained a comparatively high content of pore water, thus producing ductility contrasts with neighbouring formations and perhaps facilitating a measure of décollement at this stratigraphical level.

FOLDS

The dominant folds are the major periclines of the Plynlimon and Machynlleth inliers, both with a NNE trend. Another major anticline with the same trend enters the district in the north near Carn March Arthur [651 981], and is separated from the major periclines by a southwesterly plunging curvilinear synclinal area trending through Gogarth [674 982] on the Dovey estuary.

These major folds have a wave-length of about 10 km, but upon their limbs are several orders of smaller folds. The immediately subordinate folds have wave-lengths of about 1.5 km within which are the anticlines and periclines of Cefnyresgair [748 889], Carn Owen and Banc Lletty Evan-hen [717 852], Moel-y-Llyn [713 916] and the train of periclines between Moel y Garn [692 912] and Cwmere [682 883]. The third order of folds, formed upon the limbs of the second, vary in size between two to four per km in the south-east, where they are open, low amplitude folds, and up to 7 per km in the centre of the district. Below this order, the frequency and size of folds varies considerably with the lithologies and horizon. Thinly multilayer sequences in the Silurian are commonly pervaded by small folds, yet similar lithologies in the Ordovician are little affected at this scale.

East of the Glandyfi Tract (Figure 45) the trends of fold axes of the lower order folds are, in general, the same as those of the major folds. West of the Glandyfi Tract this is not so, and north of the Dovey estuary minor folds trending c.050° cut across the major Carn March Arthur Anticline. The disparity of trend here affects both limbs of the major fold and thus appears not to be a function of plunge.

The Glandyfi Tract also marks a divide in fold (and cleavage) vergence. To the east the fold style is mainly upright, or with easterly vergence. To the west, including the area north of the Dovey estuary, there is commonly a strong westward or north-westward vergence.

FAULTS

Transverse faults

Most the large faults of the district trend approximately ENE–WSW in the south and nearer to E–W in the north. They dip steeply, usually 75° or over, and are normal faults with downthrows inconsistent in both direction and amount. Some reveal lateral tear movements, and thus in strongly folded rocks estimates of the amounts of apparent vertical throw can be spurious, but in places there are throws of more than 100 m. Many of these faults have rather sinuous courses, commonly branching or anastamosing, but in the north they are straighter and less complex e.g. Llyfnant Fault. A set of NNW–SSE faults complements the WSW–ENE set. It is very subordinate both in the number of faults and in the movements along them, though it is more strongly developed in the NW of the district. Both sets of transverse faults carry metalliferous sulphides in places together with quartz and iron-rich dolomites.

A major zone of faulting occurs at the northern limit of the district (Figure 45). It has a heading approximately 080°, roughly in line with the other major transverse faults, but some aspects of the faults in this zone make it unique. It is composed of many sub-parallel faults, of which the Cwm Sylwi Fault and the Pennal Fault are prominent components. They are straight and, although subvertical and of normal throws, their vertical displacements are small, showing change from down-north in the east to down-south in the west. In places a wide zone of brecciated country rock lies adjacent to the fractures and is sparsely veined by quartz, as for example in an Ordovician sandstone [6510 9860] on the south side of Llyn Barfog. Although many of the faults comprising the zone have small throws, their influence on the topography is marked; indeed, between Aberdovey and Trefeddian, only the topography determines the position of many of the closely spaced fractures.

Strike faults

Major faults parallel with the fold axes are numerous and commonly long. They appear to be normal faults with inconsistent directions of downthrow of up to 30 m or so. Characteristically they are unevenly distributed and form linear packs. They do not carry metalliferous sulphides, but bear irregular masses of opaque quartz commonly enclosing clasts of country rock together with some carbonate mineralisation. One such pack of strike faults bounds the Machynlleth Inlier on the east side. At its northern extremity this appears to pass into a thrust (the Cascade and Forge thrusts), while at its southern end it is intimately linked to the very straight line of minor periclines which trail SSW from the SE corner of the Machynlleth Inlier. Another pack of strike faults lies immediately to the east of the Carn Owen Pericline.

Thrust faults

Two major thrusts are present, both within the Machynlleth Inlier. The larger is the Brwyno Overthrust which courses along the western side of the inlier repeating the western limb of the pericline, and bringing different stratigraphies into juxtaposition. It has a dip westward of c.55°-60°, only a little steeper than the overlying beds. In the north it is transected by the Llyfnant Fault (Figure 45) and continues on the north side of this fault as the Gelli Goch Overthrust. Beneath the main thrust-plane are several subsidiary thrusts which, especially just south of the Llyfnant Fault, cause considerable disturbance of up to 100 m of strata.

The smaller thrust occurs in the NE of the inlier. This too has been affected by the Llyfnant Fault; south of the fault it appears as the Cascade Overthrust, while to the north it has been displaced eastward and continues as the Forge Overthrust.

CLEAVAGE

A single cleavage affects the rocks of the district. In general the cleavage planes are spaced several millimetres apart, and are even and parallel. However, in the mudstones of the Drosgol Formation, the Bryn-glâs Formation and the 'Green mudstone member' of the Cwmsymlog Formation, cleavage planes are uneven, curved and anastomosing. This distinction is considered to stem from the sedimentational history of the rocks affected. In the case of the Drosgol Formation and the Bryn-glâs Formation, at least, this history included considerable synsedimentary disturbance which may have affected the fabric of their clay minerals at an early stage.

Cleavage within the district is commonly not axial planar, but normally deviates dextrally. Its disposition is not uniform; one of the main variants is the direction of dip. In general the Glandyfi Tract (Figure 45) divides an area to the east where the cleavage is upright or dips steeply westward from an area to the west where the cleavage dips eastward, commonly at low angles. This cleavage facing-divide is considered to be a fundamental factor in the structure of the district and possibly of western mid-Wales.

MINOR STRUCTURES

Cusp and furrow structures

This consists of a ripple-form regular pattern of anticlinal and synclinal flexures affecting only the thin arenite layers (Plate 17), in particular their interfaces with mudstone interbeds. The amplitudes of the flexures are commonly little more than a centimetre, and they have a wave length between 10 cm and 25 cm. The anticlinal crests are asymmetric and rather acute; on bed surfaces their traces anastomose to form a regular elongate reticulation which encloses eye-shaped synclinal furrows or flat areas which are commonly comparatively smooth and open. The scale of this pattern is constant on any one bed, but differs from sandstone to sandstone; such differences probably relate to grain size, bed thickness and separation. It may be allied to cusp and dome structure.

The structure occurs in formations containing thinly multi-layer couplets of arenite and mudstone. It is common, for instance, in the Nant-y-Môch, Cwmere, Derwenlas, Cwmsymlog and Devil's Bridge formations, and occurs widely throughout basinal mid-Wales, west of the Towy Anticline.

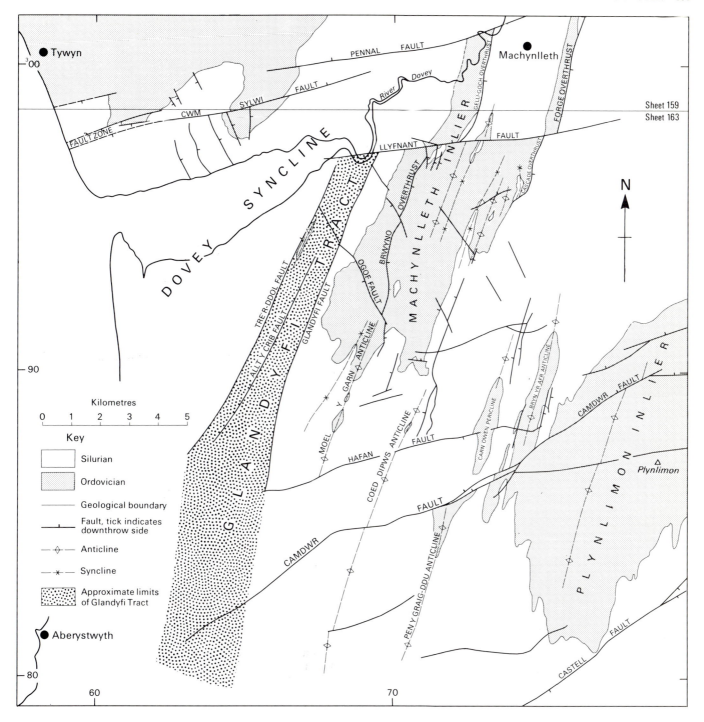

Figure 45 Main structural features of the Aberystwyth district

In section the anticlines normally are ruptured along their axes by one or more uneven planes of fracture that strike subparallel with the cleavage. In many cases the cusped arenitic layer has been parted on the fracture, and the mudstone below penetrates through the rupture making continuity with the mudstone above. The parted ends of the arenite layers are invariably rounded and adjacent internal laminae are usually distorted. In some places carbonate veinlets accompany the fractures and commonly extend through the mudstone above and below, where they are thinner, passing through the next arenite along the axis of a similar ruptured cusp.

The association between cusp and furrow structure and cleavage is always so close as to infer a genetic relationship. Burial and lateral compression are required for both to form, but cusp and furrow structure is considered to be the result of the lateral compression of a sheet of sediments still to some degree hydroplastic, in which argillaceous layers yielded by

Plate 17 Cusp-and-furrow structure, Devil's Bridge Formation. Ynys Edwin [6780 9630]

loss of water and mineral diagenesis, while the grain-supported arenaceous layers, with their pore spaces perhaps already clogged, behaved competently, buckling and rupturing, and in some cases disaggregating to produce sand dykes. Later regional compression may merely have tightened up these early structures.

Minor faults

There is a host of minor normal faults throughout the district. These may be divided into two broad categories.

Normal faults unrelated to cleavage direction

The more steeply dipping of these, commonly little more than a centimetre apart, usually form step-faults or complementary sets forming mini-horsts and graben. The lower angle ones are commonly curved and in places become bedding-parallel; they are also associated with very disturbed bedding fabrics in which the layered lithologies have become mixed and streaked in a plastic fashion. The fractures are sharp, and not obviously aligned with any other fabric such as the cleavage; they are not mineralised, and thus hanging and footwalls are in tight contact. The surfaces of these walls are smooth, some polished, but commonly uneven, forming broad grooves or flutes aligned in the direction of movement.

This type of mini-faulting has been observed in three formations within the district: the Cymsymlog Formation (Cave, 1971, pl.1, p.80) and the glacial lake clays of Hengwm (Cave, 1974, pl.1, figure 2, p.20), in both of which low-angle faulting and sediment disturbance has occurred; and the Borth Mudstones, together with the Grogal Sandstones (Anketell and Lovell, 1976), in which the faults are mainly steep step-faults.

The host beds are all thinly multilayer, mainly colour-banded, mudstones. They are also thought to have accumulated rapidly, and are overlain by coarser sedimentary rocks of even more rapid accumulation. The structures appear thus to have been generated by loading and compaction.

Normal faults along cleavage direction

These faults are very similar. They are closely spaced, usually a few centimetres apart, but are aligned with the cleavage folia. They show generally small normal displacements, usually of less than 1 cm, that are commonly westward in mid-Wales. Such movements could, in total, comprise a significant element in the post-depositional dynamics of the sedimentary pile.

Again, they are generally non-mineralised with foot- and hanging-walls in tight contact along smooth, but uneven surfaces. The unevenness is of broad grooves or flutes, aligned in the direction of displacement, there are also concordant ridges and depressions that reflect the heterogeneity of the thin beds in the fault-walls. The ridges are formed by the more competent sandy laminae which standing proud of the much thicker mud interbeds; the depressions are the images of the same layers impressed from the opposite fault-wall after displacement (Davies and Cave, 1976, pp.103 and 104).

These mini-faults occur commonly in the Bryn-glâs Formation in those places where it has thinly multilayer bedding in a largely mudrock sequence and where the beds have clearly undergone disturbance soon after deposition [700 958]. They are present widely throughout mid-Wales in thinly multilayer rocks of high mud content, e.g. the Nant-y-Môch Formation [770 882], which have no obvious symptoms of early post-depositional disturbance such as slumping or rapid differential loading. In some cases at least, there appears to be an association between these faults and cusp and furrow structures, and they are considered as yet another element in a suite of minor structures that was impressed upon the sedimentary pile during the early stages of lithification.

CURIOUS MARKINGS (OR HIEROGLYPHS)

These structures were observed on several bedding surfaces within the Aberystwyth Grits by Challinor and Williams (1926). Challinor elaborated the descriptions in 1928a, 1929, 1949 and 1978, and his observations have been repeated during the survey and by Dr D. E. Bates of University College, Aberystwyth. They also occur in the Devil's Bridge Formation near Forge in the extreme north of the district, demonstrating that they are not peculiar to the Aberystwyth area or to the Aberystwyth Grits.

The bedding surfaces are of polished and finely striated mudstone. The marks are fine rounded grooves on these surfaces and clearly were caused by small 'clasts' which lie along the bedding and which acted as engraving tools during shear movements across the bedding plane. In the Forge example, the tools are minute oblate pellets of concretionary sediment, probably diagenetic, and all of the same size and shape. Some remain embedded in the hanging wall and some in the footwall, so that there are moulds and casts of the tool marks on both surfaces. The casts are proud of the surface, and are made of mudstone not a secondary mineral filling.

The lengths of these grooves vary from under 1 cm to 3 – 4 cm, and usually they possess several geniculations. The movements which gave rise to them were, therefore, small and changed direction abruptly after each pause. Detachment along the bedding plane was followed by motion of the overburden in random jerks in different directions. On any single bedding surface the hieroglyphs are very similar, but over several centimetres the 'signature' changes slightly, and shows that movement was in part rotational.

This variety of bedding-plane slip is presumably similar to the more common one that produced uniformly linear grooves. The obvious differences in the type of grooves sets it apart but, as contended by Challinor, the generative movements probably occurred in ductile sediments at depths of burial perhaps of only tens of metres. The cause of the movements is not known; perhaps earthquakes were a contributary factor.

SUGGESTED SEQUENCE OF STRUCTURAL EVENTS

Two distinct points of view of the genesis of most of the structures of this district persist; one is the traditional view of an end-Silurian convulsion linked to a collision of lithospheric plates, the other is that a great deal of the structure developed synbasinally and synsedimentarily. In the west, if the minor disruptive structure is excluded, the folding and cleavage can be viewed as having had the same upright style that is still evident in the east, but which has suffered a westward collapse: the variegated pattern of minor structures may be local adjustments within the collapsed sedimentary pile. If this assessment is correct then the more orderly upright folds in the east were an earlier product than the confused pattern in the west; moreover, if the sediments in the west were in a rather wet, incompetent state at the time of folding, then the same was probably true for the sediments elsewhere in the district.

The following sequence of events is thought to have affected the deposits at any fixed level:

1. Diagenesis commenced and concretions formed including phosphatic concretions within the latest turbidite.
2. Dilation by excess pore water of certain bedding planes and cumulative movement westward of the sedimentary pile by gliding along these planes in response to gravity. This produced interface grooves and ?ripples, ramping, and very acute discordance in middle and distal positions on slide planes with some crumpling in distal positions.
3. Much of the folding and cleavage was produced by compressional tectonics within a thin skin (or successively within several skins) riding on one or more supra-basement décollements. Sediment dewatering proceeded.
4. Possibly concomitantly, axially-aligned basement faulting, or a thin-skin compressional up-arch developed along the Glandyfi Tract, creating a sedimentary pile unbutressed on the west. This pile collapsed westward and then vertically upon itself (Figure 46).
5. This collapse tightened folds, distorted both folds and cleavage, and generated thrusts and tear faults as seen in the Aberystwyth Grits.
6. Vertical 'settling' of the displaced pile flattened the cleavage, and produced some flat topped (box) folds in the Glandyfi Tract and many medium ($40° – 60°$) small normal faults (step-faults) mostly throwing down east, but some down west. Bedding-parallel slide surfaces were created (or reactivated) in beds with steep eastward dips and took up some of the eastward component of this collapse.
7. Major transverse faults formed across the district, and became mineralised with ferruginous dolomite from the turbiditic sands, some metallic sulphides from hemipelagic muds, and some baryte, most of which subsequently emigrated leaving pseudomorphs.

Figure 46 A generalised structural model of the area

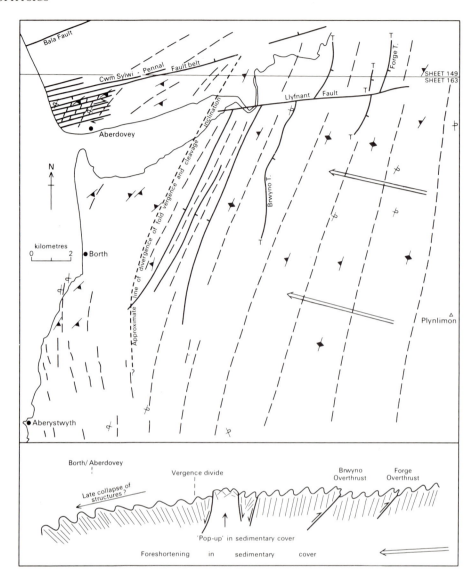

Surface trace of fault; T=overthrust

Direction of alignment of fold axes

The approximate divide of fold facing and direction of cleavage dip

Supposed direction of bulk movement of sedimentary cover above basement

Cleavage strike-bar; tick indicates direction of dip, double tick indicates near verticality

Occurrences of overturned strata; tick indicates direction of dip

Folding

Cleavage

GEOPHYSICS

The Bouguer gravity anomaly map shows a steep negative gravity anomaly gradient of around 1 mGal per km falling eastwards away from the highest anomaly values of around + 34 mGal near Twywn and + 32 mGal near Aberystwyth to around + 12 mGal in the south-eastern corner of the district. The main cause of this gradient is likely to be deeply buried. Griffiths and Gibb (1965) attributed the regional anomaly in Central Wales to lateral density variations occurring at some considerable depth in the crust, with a maximum depth to the top of the structure of 28 km. This would explain the general lack of correlation between the local Lower Palaeozoic structures and the regional gravity variations. Griffiths and Gibb removed the main regional field and found positive residual anomaly highs remained near Twywn and south of Aberystwyth which they suggested might define areas of maximum thickness of Lower Palaeozoic rocks with a higher density than the Precambrian. The average densities they measured of the Precambrian rocks in Pembrokeshire were lower than the Ordovician and Silurian rocks in Central Wales.

Powell (1956) estimated that about 4500 m of Lower Palaeozoic rocks would be required under Aberystwyth to account for this anomaly if these rocks were 0.1 g/cm³ denser than the Precambrian. Alternatively he suggested that this anomaly may be due to dense igneous rocks at depth, possibly similar to the basic igneous rocks in the Ordovician of the Cader Idris succession.

The aeromagnetic anomaly map shows generally weak magnetic anomaly gradients of only a few nanoteslas per kilometre, with smooth contours which indicate a deeply buried magnetic basement. There is a general positive gradient towards the east possibly suggesting a shallowing magnetic basement in that direction. Superimposed on this gradient in the north-east of the district is a small anomaly of about 10 nT in amplitude which suggests a local structure with a north–south strike.

Geothermal measurements in the Glan-fred Borehole (Richardson and Oxburgh, 1978) revealed a mean temperature gradient of 21.2°C/km calculated from an equilibrium temperature of 19.9°C at a depth of 397 m. The heat flow was calculated at 59 ± 14 mW/m², which is about average for the British land mass and indicates a normal crust as regards its heat production properties.

DETAILS

East of the Glandyfi Tract

FOLDS

The fold style is generally upright or with a small ESE vergence. Major culminations occur widely in the fold pattern of the Welsh basin; the Plynlimon Inlier (Jones, 1909) and the Machynlleth Inlier (Jones and Pugh, 1916) are two of these. Smaller culminations occur at Carn Owen, Cwmere and Bwlch-glas. Fold axes are linear with a consistent trend NNE–SSW. Southwards the axial trend converges acutely with the Glandyfi Tract (Figure 46) and there appear to be no folds on this trend to the west of the latter.

Two factors appear to have influenced the intensity of the folding. First is the position within the sedimentary pile, for the folding is generally tighter in the Silurian than in the Ordovician. The second is the bedding; thinly multilayer formations like the Devil's Bridge Formation exhibit more intensive folding than more massive sequences like the Derwenlas Formation.

Many of the folds, both small and large, are non-cylindrical, commonly being replaced en-échelon over short distances. Interlimb angles place them broadly within the categories 'open' and 'close' (Fleuty, 1964). Easily accessible examples of this style of folding, in beds of the Devil's Bridge Formation, occur along the A44 road at Bwlch Nant-yr-arian [718 812] (p.92 and Plate 18), where interlimb angles range from 60° to 70°. The section also demonstrates the selective nature of the natural exposures in revealing mainly the easterly-dipping beds, the factor which led to the Plynlimon area being mistaken for a major syncline containing the highest strata of the district (Keeping, 1878, 1881). In another section (Plate 19), in a cutting beside the A487 [739 004] north of Wylfa, the folds tighten and verge eastwards against a crushed easterly dipping limb.

FAULTS

The fault pattern is dominated by ENE-trending transverse faults, many of which are mineralised; they are all normal faults with only local and subsidiary tear components. There are also a few NW-trending faults with similar characteristics, the most important being the Ogof Fault in the Machynlleth Inlier. The transverse faults displace the fold axes and other structures such as strike faults and overthrusts. Raybould (1976, p.637) has noted that these transverse faults have their maximum throw near their mid-point, and steepen and become subvertical breccia zones with little vertical movement towards their extremities. The transverse faults are visible in a number of old opencast mineral workings. One of them, the Camdwr Fault, is particularly well displayed in the gorge at Craigypistyll (Figure 29). Another is exposed [7358 9137] at Esgair-hir, revealing much quartz-veined breccia.

It has been claimed (Jones, 1922, p.17) that the Llyfnant Fault, together with the Ystwyth Fault just south of the district, is of a different type, being without sulphide minerals though also trending between east and ENE. Chalcopyrite is, however, very evident in workings [7196 9755] on small WNW–ESE faults which appear to be associated with the main Llyfnant Fault. Movements on the latter were analysed (Jones and Pugh, 1916) using the relative position of thrusts, fold axes and geological boundaries on each side of the fault. They deduced that the vertical displacement on the fault in its central section from Alltddu [715 973] to the Forge Overthrust [758 978] varied between 24 m and 188 m; using the same criteria they deduced that there was a dextral tear component of between 518 m and 975 m. While their matching of individual thrusts across the fault may be correct, their matching of fold axes is not, for they misplaced the position of the major anticlinal axis of Tarren Tyn-y-maen by some 200 m. Their conclusions are, therefore, suspect.

There are many strike faults in this eastern area, but most fall within two narrow belts parallel with the fold trends. The faults in these belts are rather sinuous, forming anastomosing complexes as for instance north of Alltgochymynydd [708 881]. They are normal and usually throw down west, though easterly downthrows occur on

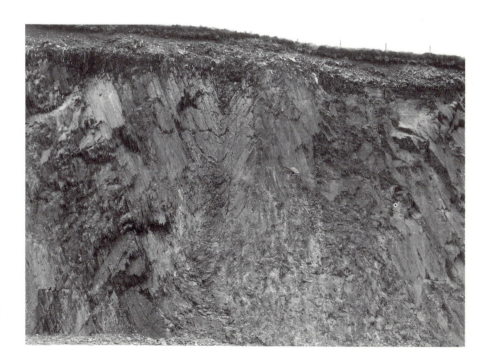

Plate 18 Fold, east of the Glandyfi Tract. Roadside (A44) [718 812], Cwmbrwyno, Goginan

Plate 19 Fold, east of the Glandyfi
Tract. Roadside (A487) [SH 7388
0037], Ogo-fach, near Machynlleth

some in the anastomosing branches; the direction of downthrow is, however, commonly difficult to discern. In places the strike faults carry large mineral bodies of white quartz containing mudstone clasts. There are thus areas along them that were under tension, perhaps at places of unevenness on the fault-plane. Carbonate and chlorite are also present as part of this mineralisation, giving it the same aspect as that noted farther west on bedding-parallel slides and strike-zones of disruption (Davies and Cave, 1976, p.101). In these later cases the mineralisation is considered to be relatively early (p.131) and, by analogy, so are the strike faults in this eastern area. They are believed to be the product of extension in the sediment cover, but could be penetrative upwards from basement fractures. Their outcrop provides no obvious evidence to suggest that they are listric, but this is their most likely form, connecting in depth with concordant décollement at the base of, or within, the sediment cover: they might thus have acted as channels for escaping pore-fluids (Cave, 1975b, p.26). It is noteworthy that the two major bunches of these faults lie eastwards of marked anticlinal tracts at Carn Owen and Alltgochymynydd.

In the Machynlleth Inlier there are major overthrusts with westward dips of 40° to 60° (Jones and Pugh, 1916). Of these, the Brwyno Overthrust has been traced southwards almost to Cwm Ceulan. The existence of the postulated parallel Wern-deg reverse fault (James, 1972, fig. 2) has not been confirmed by the present survey, but disturbance immediately subjacent to the Brwyno Overthrust is intensive in places. In addition to the major strike faults and thrusts, some of the minor folds possess faulted or overthrust axial planes. In most cases the throw is less than 1 m, and they are commonest in the western part of the area towards the Glandyfi Tract (Figure 45). Examples can be seen in crags [667 884] at Braich Garw and on the hillside [7781 8139] SE of the Dyffryn Castell Hotel.

CLEAVAGE

Cleavage dips are steep, on either side of the vertical, but predominantly to the west. Cleavage strike commonly deviates from the trend of fold axes, usually dextrally and by as much as 30° (Davies and Cave, 1976).

BEDDING-PARALLEL SLIDE-PLANES

These are very common, not only in this eastern area, but generally in mid-Wales wherever the sequences are thinly interbedded arenites and mudstones (Davies and Cave, 1976, pp.97 – 101; Cave, 1978b). They usually lie at the tops of the turbidite units, commonly in the pale grey mudstone layer. Their surfaces are covered with a thin film of quartz, dolomite or chlorite, in some cases as fibrous crystalline growths, or by pyrite cubes. Cataclasis on the slide planes is present in places, but commonly is not evident (Strong, internal BGS report). The resultant striations are aligned approximately down the dip of the beds, but usually with a slight dextral deviation. In many cases there is but little mineralisation, and the striations in some of these cases appear to be gouges in the sediment rather than fibrous mineral growths. Some of the latter have revealed gouges with rectangular section, and these appear to be marks caused by drag of early diagenetic pyrite which formed along the plane of slide but has since been removed at outcrop by weathering. There are cases in which several layers of mineral growth have built up on a slide-plane to form a 'vein' up to 2 cm thick. In adjacent areas of Wenlock rocks the 'vein' minerals are largely quartz and calcite (Wedd and others, 1927, p.91). Whilst the striations of each of these layers have the same general trend, in detail the trends differ from layer to layer. It would seem in these cases that, although sliding was located in one sedimentary position, on many occasions the original glide-plane was not reactivated, and subsequent sliding was initiated immediately adjacent to its already mineralised predecessor.

In folded strata these bedding-parallel slide-planes and their veins occur throughout the geometry of the fold—in the limbs and over the axes. The slide movements thus predate the folding. In most cases it is not possible to quantify the movements, but there are instances where the groove-forming tools are still visible, as well as the resultant grooves. In such cases the amount of travel of the superincumbent strata was but a few centimetres, while the general direction was towards the west or WNW. In one example [7551 0032] slip has occurred on a plane occupied by a layer of quartzose sideritic ovoid concretions. The latter behaved as tools, already lithified, within a less competent argillaceous medium, and produced linear scars on the slide-plane some 25 cm long. Other slide-planes may represent similar or even smaller amounts of travel, but within the sedimentary pile they are innumerable and together must represent a large movement of structural significance.

Many bedding-parallel slide-planes and their veins are now intensively puckered or rippled (Lewis, 1946) and these are the tectonic ripples of other authors (e.g. Warren and others, 1970; see also Nettle, 1964; and Nicholson, 1966, 1978). The crests of these ripples show close alignment with the cleavage, but they are commonly ruptured and thrust, and show asymmetry with a preferred westward vergence. It is assumed that this puckering and disruption of the slide-planes and veins resulted largely as a response to the lateral compression of an incompetent, largely argillaceous, sedimentary pile within which lay very thin, but lithified, bedding-parallel veins. Again this points to an early, pre-cleavage origin for both the sliding and the veins.

Some exposures reveal minor discordance along slide-planes, where the top few centimetres of the underlying strata have been truncated. Examples occur in Forge [762 001] (Plate 20), to the north of the district, and in a laneside near Llanwrin. They show that minor folding and ramping occurred during, and probably because of, the sliding. These movements presumably result in a 'want' in the continuity of the stratigraphy 'upslope', and a duplication in the stratigraphy 'downslope'. There may, however, be no volumetric increase downslope if partial dewatering of the sediments during diagenesis and, possibly, cleavage production were concomitant with the movement.

Such sliding is thought to have been facilitated by the rupture and dilation of certain layers within the sediment pile due to pressure and ingress of excess pore-water. A low-friction cushion could thus have been inserted from time to time. Any lateral movement induced by gravity down a slope would create tensile conditions upslope, and thus possibly initiate fractures there, to allow escape of the pressurised fluids. Movement would then cease, at least until the process was repeated in the same layer or a higher one.

Plate 20 Bedding-parallel slide, Devil's Bridge Formation, Forge [SH 7621 0007]. Diameter of penny is 2 cm

Plate 21 Fold, west of the Glandyfi Tract. Anticline, cusp-shaped and axially ruptured, in the Aberystwyth Grits [5896 8518]. A12959

West of the Glandyfi Tract

FOLDS

The folds in this area have a strong westward vergence, and in general the west-facing limbs become progressively steeper westwards towards the coast, where they are overturned; this applies also to the western part of the area north of the Dovey estuary. The structure may be related to the distribution of the Aberystwyth Grits and the Borth Mudstones formations. As in the eastern area, the intensity of folding is greater in the heterogeneous, thinly multilayer Aberystwyth Grits than in the more massive Borth Mudstones, but in other respects the fold-style is very different from that in the eastern area (Plate 21); there is no simple pattern with a constant trend. Boat-shaped folds are very common, disposed en échelon, and commonly with axes that are non-parallel, curved and faulted (Davies and Cave, 1976, fig. 4). A good example of such a fold occurs in, and south of, Careg Milfran. It has been concluded (Price, 1962) that the major folds here are a set of *en-échelon* asymmetric periclines up to 8 miles long which, together with the minor folds, most of the faults (including the thrusts), and the cleavage, can be ascribed to the Caledonian Orogeny. It is however, difficult to view the overall structure as the product of a regionally ubiquitous, end-Silurian, crustal compression.

FAULTS

The major transverse faults of the eastern area fail to cross the Glandyfi Tract. In the western area faults are small and varied, many being related to local stresses.

There are low angle eastward dipping thrusts, some of which can be seen to be related intimately to folds, e.g. on the cliff west of Constitution Hill, Aberystwyth. Other faults follow the axial planes of anticlines and, in some of these, e.g. the anticline through Careg Milfran [5997 8854], the ruptured sediments have been rewelded, making the sutures difficult to locate with precision. There are also thrust faults which are directed eastwards, like a high-angle fault in the eastern limb of an anticline c.1 km north of Clarach. These are possibly local responses to the secondary stresses that resulted from the early gravity movements.

Normal faults, also of low dip (less than 60°), dipping eastward and aligned approximately along strike, are numerous. They are commonly of such small throw that marker beds can be matched in either wall. For instance, in the roadside outcrop due south of Aberystwyth Castle there are at least six of them, with dips eastward of about 50° and throws of less than 1 m. Their strike is similar to that of the cleavage, about 010°, which also dips eastwards but at 40°; the beds are horizontal.

A third set of faults are tears. They are steeply dipping and have two main trends, NW and WSW, where they are well displayed on the foreshore between Wallog and Borth. Drag of the beds and the displacements of fold axes indicate sinistral movements along them. Some parts of the foreshore, e.g. south of the pier at Aberystwyth and just south of Clarach, are transected by a plexus of high angle faults.

CLEAVAGE

This is also very inconsistent, in its amount and direction of dip. Usually, however, dips are low, even as low as 30°, and commonly between east and SE but not uncommonly SE to south. Disparity between the cleavage-strike and the traces of the fold axes is very large in places and the disparity is usually dextral.

Variations in the cleavage occur abruptly in places, as across some faults, including bedding-parallel faults. In other places there are gradual changes, for instance in cleavage dips along folds with curved axial-planes.

BEDDING-PARALLEL SLIDE-PLANES

These are as common in the west as elsewhere and have the same characteristics. They are folded like the bedding e.g. the small anticline on the foreshore just south of Allt-wen (Cave, 1978b) where movement is generally in a westward direction. In some instances, however, usually in eastward-dipping beds, the opposite sense of movement is indicated by displacements of up to several centimetres in discordant quartz veins (displacements of up to a few metres occur in other places, but the fractures responsible for those movements are more complex). In these examples the quartz veins are not fractured by the slides but appear to continue along the slide-planes between the displaced portions, suggesting that the tension which opened up the bedding-normal fractures was synchronous with movement on the bedding planes.

In this western area, bedding-parallel slide-planes are very well displayed in coastal exposures. They are numerous, and persist laterally generally adhering to one horizon. Good examples occur at the foot of Allt-wen in the south and between Borth and Careg Milfran in the north. At the foot of Allt-wen several examples depart slightly from the bedding, while the strata between some others is contorted and ruptured like the deformation in a slump (Plates 22 and 23). At the base of the cliff these slide-planes form a narrow zone, the position of which can be traced up the cliff. This zone underlies chaotically disturbed strata in the westward dipping,

Plate 22 Bedding-parallel slide, Aberystwyth Grits; foot of Allt-wen [5757 7952] looking N. Footwall to the right with even bedding below a narrow disturbed zone; hanging wall overlain by much disturbance. Ruler = 30.5 cm (lower centre, left hand edge)

downslope (or down the pressure gradient). The magnitude of this particular example necessitates a 'down-slope' movement far greater than those known in the eastern area.

As in other areas these bedding-parallel slide-planes have been folded with the containing beds; generally, therefore, they are considered to predate the regional folding, but small-scale folding and ramping that takes place in the toe of sliding masses or at obstructions anywhere on their paths must be synchronous with sliding. An example occurs on the foreshore beneath Craigyrwylfa, where the slide plane and super-incumbent beds are repeated three times as three rotated underthrusts. Several other examples where bedding-parallel slides have resulted in repetition are visible in quarries between Aberdovey and Trefeddian (Figure 47). Here the cleavage is also deformed, and shows similarities to some of the deformed cleavage near Borth.

Plate 23 Bedding-parallel slide, Aberystwyth Grits; foot of Allt-wen [5757 7952] looking S.

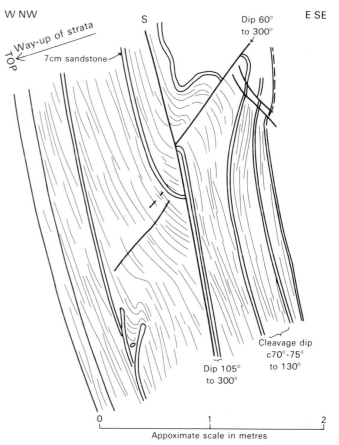

Figure 47 Small ramp of strata produced by movement on a bedding-parallel slide, between Aberdovey and Trefeddian

steep limb of a major fold and could be construed as a gravitational collapse of part of the limb on these slide planes, which rise up-sequence westwards across the foreshore. Near Careg Milfran, beneath the northern end of Craigyrwylfa [6027 8867], a compound bedding-parallel quartz vein some 2 cm thick, consisting of several striated quartz/mudstone laminae, is readily accessible in the cliff base and foreshore. Other less obvious bedding-parallel dislocation planes are present nearby. One reveals a slight, but distinct, discordance with the beds below, and perhaps represents the end of the slide-sheet, for such a body must eventually terminate either by the glide-plane rising or ending in a steep shear zone. Another reveals a similar, but more abrupt, condition where the slide-plane steps sharply up to a succeeding bedding-plane along strike. The bend in the incumbent strata produces a detachment prism, presumably initially occupied and supported by excess pore water, but now filled with quartz and mudstone clasts (Davies and Cave, 1976, fig. 5); these bends are possibly also side-wall components, aligned with the movement. The duplication of strata consequent upon termination of a slide-sheet is also evident on the foreshore. Although the Aberystwyth Grits is a monotonously repetitive alternation of mudstones and sandstones, the patterns of some sandstone/mudstone sequences are sufficiently unique to show that small packages of strata are duplicated above and below bedding-parallel slide planes. Two such are identifiable at Allt-wen, while a much larger package just south of Clarach Bay was noted by Wood (written communication). The lower half of such repetitions of strata represent ramps over which the repeated package has ridden

Glandyfi Tract

This tract separates the two above areas and shows a mix of their distinctive structural styles.

FOLDS

Upright, westward-verging and eastward-verging folds are all present in this tract. It is not uncommon for beds to display low dips or to be almost horizontal. The beds in the Glan-fred Borehole, near the western edge of the Glandyfi Tract, had a low dip but turned abruptly into the subvertical in parts: the hole probably intersected open folds verging westward. Along the eastern side of the tract, near Glandyfi, Furnace and Talybont, folds verge eastward.

FAULTS

Three long linear strike-faults dominate the tract (Figure 45). The eastern pair bound a narrow graben, and the western pair a narrow horst which brings up an inlier of Ordovician rocks at Penrhyngerwyn. It is along and westward of this horst that the contrast of structural style is greatest. There are several small normal faults crossing parts of the tract in a generally E–W or SE–NW direction, notably near Llancynfelin and just west of Talybont. Both sets carry sulphides. There are also very small listric faults, which are probably the contemporaries of bedding-parallel slides in that they curve from a subvertical to a bedding-parallel disposition, some amalgamating with bedding-parallel slides, and some being truncated at their upper end by a higher bedding-parallel slide e.g. one [6791 9647] near Ynys-Edwin. On a forest track [6476 8906] near Talybont similar listric faults, with a dip to the SW, occur in a small quarry. They are considered to represent the lateral margins or back walls of small, generally westward directed, slides.

CLEAVAGE

This is as varied as is the style of folding. In places along the east of the tract, the cleavage is of low dip to the WNW, e.g. 35° along the roadside just west of Glandyfi, east of Eglwysfach and again just SW of Taliesin. There are, however, places on the western side of the tract where cleavages have low dips uo the west or NW, e.g. in the stoping [6459 8932] SE of Staylittle. Nearby, another exposure [6492 8941] reveals cleavage, rather poor and well spaced, which dips at 25° NW at the eastern end and 20° SE only 20 metres away at the western end. This exposure also shows several normal strike faults (with displacements of only a few centimetres and dips slightly greater than that of the cleavage) comparable with the faults near Aberystwyth Castle (p.108). At the eastern end of the section these faults throw down west; at the western end they throw down east. The beds are almost horizontal with a very slight anticlinal arch; perhaps they represent a collapsed anticline.

BEDDING-PARALLEL SLIDE PLANES

Good examples are seen in the Ynys-Edwin exposure (p.110), where beds above and below one slide-plane have slightly different dips (Plates 24 and 25).

Plate 24   Two bedding-parallel slides (arrowed) with a slightly discordant package of beds between; Devil's Bridge Formation [6791 9647]. Near Ynys Edwin. Scale = 15.2 cm

Plate 25 Two small listric normal faults dipping northward into the lower of the bedding-parallel slide planes of Pl.24

CHAPTER 7

Quaternary – general account

Superficial or Drift deposits of Quaternary age are widespread (Figure 48). The unconformity between them and the underlying Solid rocks spans about 420 m years from the late Llandovery (Aberystwyth Grits) to the late-Pleistocene. There is no positive evidence that sediments were laid down in the district during this period, though they were near Harlech (Woodland, 1971) and west of Aberdovey (Penn and Evans, 1976). Some authorities have, however, suggested that Mesozoic rocks were deposited over much of mid-Wales, but were eroded following uplift during the late

Cretaceous or early Tertiary (Challinor, 1969, pp.26–27).

The glacial history of the Aberystwyth district has been the subject of controversy during recent years. Some authors (Mitchell, 1960, 1972; Synge, 1963; Watson, 1970) considered that eastern Cardigan Bay and the adjoining Welsh hills were largely ice-free during the last glacial phase of the Devensian, the most recent of the Quaternary stages, and that the great majority of the drift deposits in the district have resulted from periglacial action. On the other hand, evidence has been amassed (Bowen, 1973, 1974) from

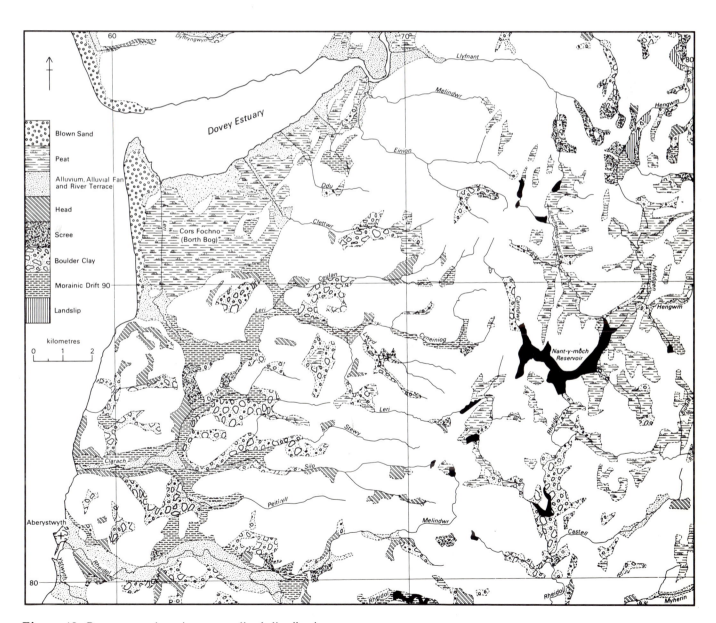

Figure 48 Quaternary deposits; generalised distribution

numerous workers to show that mid-Wales underwent extensive glaciation during the late Devensian, with glaciers extending eastwards into the Welsh Borders and westwards into Cardigan Bay, though much of the till deposited during this glaciation has since been modified by periglacial action. Results obtained during the present survey support this latter theory, but there is no positive dating of any of the till or morainic drift though the freshness of many of the glacial landsforms and deposits strengthens the case for their being Devensian in age.

The most widespread deposits are Boulder Clay (till), Morainic Drift and Head, with smaller areas of Glacial Lake Deposits, Scree and Glacial sand and Gravel.

BOULDER CLAY

Although widespread, Boulder Clay is not a major deposit. It occurs in patches where topographical depressions have preserved it from degradation, and as valley-fill in the upland areas such as that around and to the north and west of Plynlimon: the hills are usually free from it. Away from the uplands it is present on valley sides, beneath morainic drift in places, and on interfluves like those SE of Talybont. The deposit is commonly modified by some later degree of down-slope translation.

It consists of stiff clay and silt containing liberal admixture of ill-sorted clasts ranging from boulder to granule size and from rounded to sub-angular in shape; beneath it, striated surfaces are common. The clasts are of rock types indistinguishable from the 'local' mid-Wales sediments, i.e., coarse greywackes to laminated fine siltstones and mudstones. Vein quartz is also common. In the north and west there are some more exotic clasts that can usually be matched with volcanic 'felsites' from Meirionnydd. They indicate a more northerly source for the ice. Beach stones around Ynyslas and Aberdovey are very varied and include flints, limestones, sandstones, and igneous intrusive and volcanic rocks, all exotic and possibly winnowed from Irish Sea Boulder Clay not evident on-shore. The provenance of these stones is northern Britain, the Irish Sea and possibly Northern Ireland.

In the upland valleys much of the deposit forms gentle, laterally sloping 'terraces' and is commonly deeply dissected by fluvial action. Downslope orientation of the included stones and crude bedding parallel with the local hillslope make it clear that the material in these 'terraces' has undergone solifluction. It was concluded (Watson, 1968; 1970) from studies in the upper Rheidol basin and elsewhere that the 'terrace' deposits are periglacial head and that the area was largely ice-free during the Devensian. However, a study (Potts, 1971) of a wide area of Central Wales concluded that the stony clay is a Devensian till, which has been modified during periglacial conditions at the end of the glacial episode. Grain size distribution, the morphometric features of the stony clay, and the abundance of subrounded to subangular striated stones were used to support this case. The present work has shown there is till outside the solifluction 'terraces': in certain cases, as in the upper part of the Nant-y-moch valley, it extends well up the valley sides.

MORAINIC DRIFT

The valley-fill in the major valleys has been called Morainic Drift, a term used in this account for a heterogeneous deposit of clay, silt, sand, pebbles, cobbles and boulders. Parts of the deposit are bedded and show aqueous sedimentary structure; others are structureless and unbedded. Small pods of boulder clay are commonly present within the deposit. The main tracts are in the bottoms of valleys, mainly in the lower reaches of the Leri, the Rheidol and the Ystwyth valleys, where these are deep and broadening and where they merge into the subdued topography of the tract between Talybont and Rhydyfelin. It is a deposit of wide occurrence in the valleys of mid-Wales. In general, the rivers follow the northern boundary of the terrace-like valley fill. Morainic Drift is also found at the heads of some valleys in the NE of the district where the deposit forms small patches with a fresh morphology.

Bedding within the sand and gravel, and imbrication of the gravels clearly indicate aqueous transportation in places, while the wide range of grain sizes and the abrupt lithological differences between beds indicate rapid changes in the energy of the water. Abrupt contortions of the layering record local collapses. The ratio of bedded to unbedded deposits differs down the length of a valley and, more obviously, between different valleys. Bedded deposits are more common in the longer valleys, due presumably to their larger volume of melt-water.

In the south-western valleys Morainic Drift infill reaches levels over 30 m above the rivers in places; little is known about the configuration of its base though this is likely to be very uneven. The surface of the deposits is distinctively gentle; it is uneven, very dissected, and has a general down-valley gradient. Where incised only by the main river, Morainic Drift clings to the valley sides in the general form of a high terrace. In the higher reaches of some valleys it is clear that the main stream has incised a meandering channel through the soft deposits very rapidly and with much greater flow than it now has. Small benches are preserved at different levels on the sides of the main incision, and represent particular levels of the river bed during the erosive process. They simulate constructional river terraces, but possess no obvious aggraded materials and do not conform to a single thalweg. They are particularly obvious at the confluence of the rivers Leri and Cyneiniog, but are also discernible on the sides of the more mature valley at Clarach.

It would seem, therefore, that since their inception at the recession of the ice, the present rivers have been eroding this deposit, first rapidly, then less energetically. This suggests that these rivers at no time constructed the valley fill, but only removed it; it was the legacy of an earlier regime. Such a deduction is compatible with the internal structure and fabric of the Morainic Drift which are indicative of ice-contact deposition during a period of ice-wasting. Most Morainic Drift is thus considered to be a form of Kame Terrace, composed of materials both released from a valley glacier and imported on to and into it by copious melt-waters, which then partially sorted, reworked, and redeposited them as a pile of englacial debris and flow-till. The surface of the detritus was presumably controlled by the level of the valley glacier, but was partially aggraded as the ice wasted. In the larger valleys like the Rheidol and Ystwyth, the deposit, particularly in its upper parts, has been almost totally sorted and relain by aqueous processes.

In the NE extremity of the district there are several S–N 'blind' valleys which end headwards in steep walls. Some of these valley-heads are corries, e.g. Rhiw-gam and Cwm-byr; of them some now possess a catchment area above the head-wall in which a V-shaped gorge has been eroded, which now carries a rapid or waterfall, e.g. Hengwm and Cwm Rhaiadr. In all of these corries the steepest slopes, sur-mounted by rocky crags, are north- or east-facing. At the heads of all these valleys there are small patches of gravelly Drift mapped as Morainic Drift. In constitution these pat-ches closely resemble the Morainic Drift in the SW of the district.

The clasts within the Morainic Drift of the southern valleys are almost entirely sandstones, siltstones, mudstones and vein-quartz from rocks that occur at outcrop to the east in the hilly hinterland of Talybont and Bow Street. An ex-ception is the Morainic Drift of the Lower Leri Valley which also contains blocks of fine-grained volcanic rock ('felsite') comparable with outcrops in Meirionnydd. These 'exotic' rock fragments enter the Leri Valley at Talybont. They are presumed to have been contributed directly from Dovey Valley ice standing over Borth Bog, though it is equally possible that they were derived from till deposited on the slopes just north or west of Talybont by earlier Dovey Valley ice.

HEAD AND SCREE

Head is a variable deposit usually comprising angular rock debris of local derivation set in a clay or silty clay matrix. Some of these deposits may have resulted from periglacial solifluction, but they also include recent slope deposits, hillwash and alluvium. It fills hollows and occurs in patches along most valley sides. It is particularly extensive along the valley sides in the south-western extremity of the district where an apron of head commonly obscures the upper edge of the valley fill of Morainic Drift.

Two types of scree are present in the district. Mudstone scree (Watson, 1965b, p.23; Potts, 1971, p.66) consists of angular mudstone debris of variable size with little or no matrix. The deposit is virtually uncemented. Because of lack of sections some areas mapped as Head may be of this type. It formed by periglacial action at the end of the last glacial period (Devensian), and post-dates glacial channels and other glacial features. Screes in several localities exhibit a three-fold subdivision, with interbedded layers of coarse and fine debris ('stratified screes') lying between two generations of unsorted scree. It has been suggested (Watson, 1965b, p.23; Potts 1971, p.66) that this three-fold division may be related to climatic variations (Older Dryas–Allerød–Younger Dryas) of the Late Glacial, but there is no direct evidence to support this suggestion.

Screes of sandstone blocks are present beneath sandstone crags in a number of places, e.g. the western face of Plynlimon [788 866 – 789 871].

GLACIAL SAND AND GRAVEL

Small areas of sands and gravels of indeterminate origin have been classified as Glacial Sand and Gravel.

GLACIAL LAKE DEPOSITS

Pale grey, smooth and very soft clay is present in a few places, most notably in Hengwm, c.10 km SSE of Machynlleth (Figure 49) but also in Afon Leri [6785 8841]. It is thinly laminated (?varved), the laminae varying be-tween c.1 mm and 1 cm in thickness. Some layers, usually thin, are coarse, silty and darker grey. No dropstones are evident, though small chips of mudstone occur in laminae otherwise of very fine grain and are unlikely to have been in-troduced by current traction. The laminae are disrupted by many minor step faults and low angle glide planes. Such disturbances are common in glacial lake clays, resulting from compaction, loading and possible slope instability. These disturbances closely compare with those in the Cwmsymlog Formation (Cave, 1971, pl.1; and 1974, pl.1, fig. 2).

The clays appear to be barren of both fauna and flora presumably indicating a very cold climate. Mr B.R.Young comments that a sample of the clay from Hengwm (Film X6385)contains abundant chlorite and clay-grade mica, ap-preciable quartz (?c.20 per cent) and a little feldspar. The clay-grade mica is probably close to muscovite in composi-tion, unlike the typical clay-mica or illite in most clays which has been degraded by loss of interlayer potassium during weathering and transportation. The chlorite is probably more abundant than the mica, and it is unusual to find a clay containing as much chlorite. According to Perrin (1971) chlorite has not previously been reported from the Pleistocene in this country.

In Hengwm the deposit thins against the valley sides and passes into thinly bedded silts and sands. The fluvial input to the lake appears to be of low energy. Lenticles of nivation scree gravel occur in the clays in places near Hengwm-cyfeilog [7807 9481]. Probably these are of contemporaneous inclusion and, if so, this indicates that the age of the clay is probably early Dryas.

Intervals of well stratified fine silt and clay occur beneath Morainic Drift and above Boulder Clay in Cwm Rhaiadr [7560 9478] and in Cwm-byr [7858 9488]. These may be correlatives of the Hengwm clay.

The lake-clay in Afon Leri lies beneath Morainic Drift and is probably slightly older, possibly correlating with clays also lying beneath ?Morainic Drift in one of the Clarach boreholes (Heyworth and others, in press).

DIATOMACEOUS EARTH (DIATOMITE)

The only record of Diatomite is from a ditch [7368 9791] near Garthgwynion, Glaspwll, (p.126; Figures 50 and 51). Mr C.J.Wood reported that it comprises organic (plant) material, unspecified silt-size sand grains, and large numbers of whole and fragmentary diatoms. The date of this

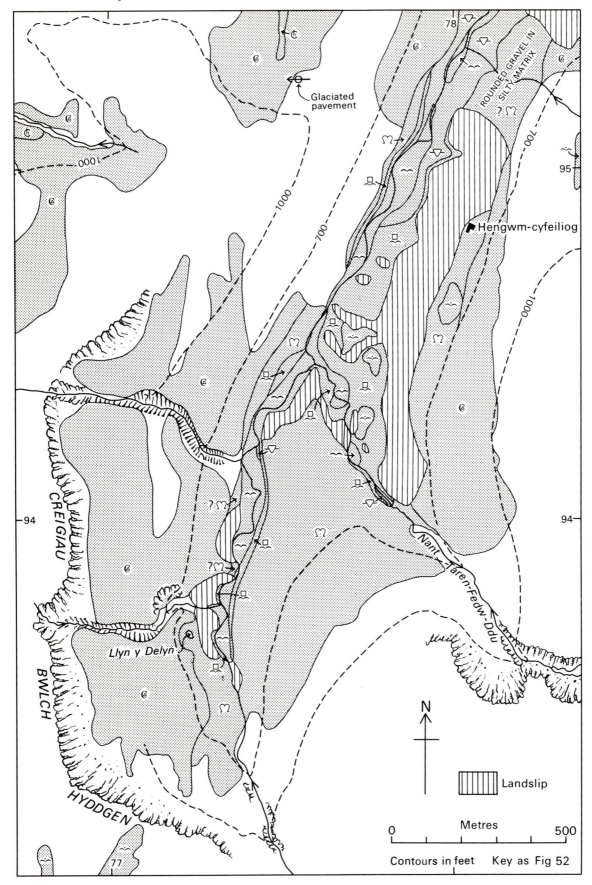

Figure 49 Drift deposits in Hengwm

deposit is 11 430 ± 160 BP at youngest. Its thickness has not been proved.

North of the Dovey, diatomaceous earth may well be present below recent deposits (cf. Thomas, 1972) at Llyn Barfog [653 988] and at an adjacent lake site now infilled with peat [655 989].

RIVER TERRACES, ALLUVIAL FANS AND ALLUVIUM

Small discontinuous patches of River Terrace have been mapped in several valleys. They are at various levels above their rivers, commonly aggraded to transient cross-valley thresholds. Others, in the lower part of the Rheidol valley (pp.128), are more significant. Most of the deposits are of pebble gravel, or in places cobble to boulder gravel, commonly with a thin capping of silt or peat.

There are numerous Alluvial Fans within the district. Most occur where a small tributary stream joins a larger valley. The deposits range from pebbly silt to boulder gravel, depending on the gradient and original size of the stream concerned. The fans are probably largely formed at the end of the glacial period when stream flow was much greater than at present. Few, if any, are accumulating now and in some places, as at point [7753 8153] SE of the Dyffryn Castell Hotel, the present day alluvium truncates fan deposits.

Broad alluvial floodplains, up to 0.8 km wide, are present only along Afon Ystwyth near the south-western limit of the district, and along the lower reaches of the Rheidol (west of Capel Bangor), the Clarach (west of Gogerddan) and the Leri (NW of Glan-Lerry). Narrow discontinuous strips of

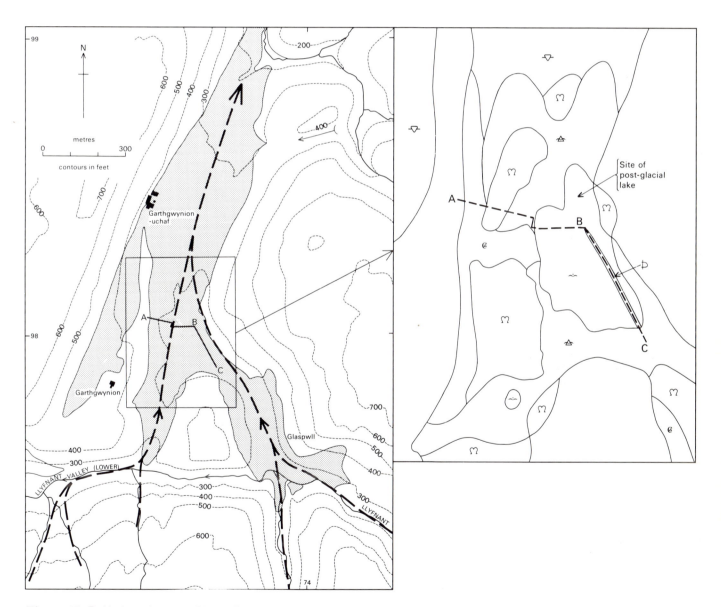

Figure 50 Drift deposits near Glaspwll

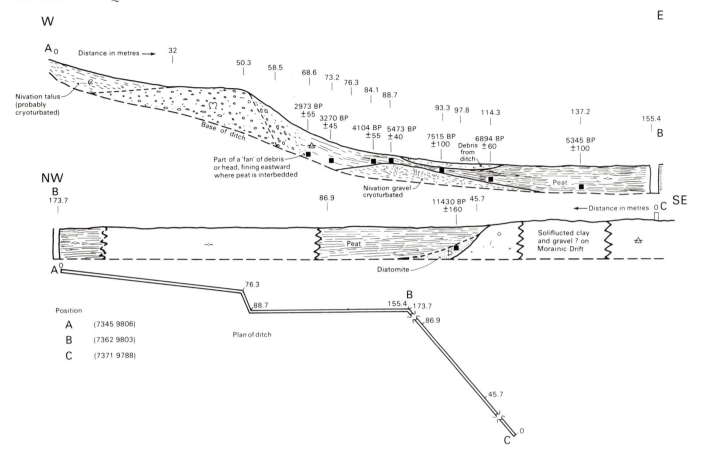

Figure 51 Ditch section through the Drift deposits near Glaspwll (see p.123)

PEAT

Peat has formed in several environments:

a. High-level peat forms a blanket up to 2 m thick over many mountain areas, for example east of Hyddgen. It rests on a thin weathered zone (soil), in which tree stumps and roots are common, at the surface of the solid rocks. The age of the base of this peat in the Hyddgen area [8001 9163] has been established as 5570 ± 100 BP (Welin, Engstrand and Vaczy, 1974, p.98). This supports the thesis that 'Welsh blanket bogs started to form during Pollen Zone VIIa and are to be attributed to the rise in precipitation at that time' (Bowen, 1974, p.412).

b. Peat, commonly less than 1 m thick, covers spreads of Boulder Clay in most of the high valleys such as that of the Afon Hyddgen and Afon Lluestgota. The low angle of slope of the Boulder Clay 'terraces' together with the impervious nature of the deposit has led to extensive development of thin peat. In some areas peat extends beyond the outcrop of the

Alluvium, rarely exceeding 100 m in width, are associated with the higher reaches of most of the rivers and streams. The alluvial deposits range in grain-size from silt to coarse gravel.

glacial drift, most notably when the valley sides are gentle, as in the upper reaches of the Nant-y-moch valley. Its distribution accords with the findings of Taylor and Tucker (1970).

c. Local patches occur at low levels such as at Garthgwynion [7355 9803] (p.126) where its base is much older than that of the high level peat, being some 7515 ± 100 BP at least (Welin, Engstrand and Vaczy, 1974, p.102). Wood from another low level peat [6657 9873] near Cefn-cynhafal gave a date of 4225 ± 55 BP (Harkness and Wilson, 1974, p.241). Hollows, caused by landslip, in the uneven surface of the Boulder Clay between Coed Pant-y-bedw and Dolcaradog contain thin peat. Wood (*Quercus*) at the base of this peat [7925 9752–7976 9715] has given dates of about 3300 to 3600 BP. A further date of 1835 ± 40 BP (Harkness and Wilson, 1974, p.243) was obtained from wood in the top 61 cm of the peat [7925 9752].

d. In the alluvium of aggraded estuaries as at Clarach and Aberystwyth (p.129).

e. In the submerged forest along parts of the coast (p.127) and in the ombrogenous raised bog of Borth (Cors Fochno). This latter area of peat is the largest in the district (p.117) and, at least in the central part of the bog, peat growth has continued from about 4700 BP, when it replaced the earlier forest cover, to the present day. The base of the peat is approximately at sea-level, and it has a maximum thickness of some 6.7 m (Rudeforth, 1970, p.62).

MARINE AND ESTUARINE DEPOSITS

Cors Fochno (Borth Bog) and the Dovey Estuary

These deposits cover a roughly triangular area of some 40 km² extending from Borth to Aberdyfi and eastwards to Glandyfi. Together with any underlying glacial or fluvial deposits they fill an irregular buried valley. Geophysical work (Blundell and others, 1969, pp. 163–167) suggests that the depth to bedrock is between 49 and 85 m under the Dovey Estuary and between 61 and 107 m at the south-western corner of Borth Bog; Jones (in Yapp and others, 1916, p.29) records that the rock floor is more than 45 m below sea-level at Dovey Junction. The irregular nature of the rock floor is clearly shown on geophysical profiles (Blundell and others, 1969, figs. 3 and 4), and is emphasised by the several small 'islands' of solid rock which project through the Holocene sediments, as for example at Foel-ynys [607 930], Ty Mawr [630 927] and Ynys-fergi [614 897].

After the last glacial episode, as the sea-level began to rise, a coastal barrier started to build northwards by longshore drift from the headland south of Borth. The date of its initiation and the rate of its growth is not known, but in time the area behind it became protected from the open sea and started to silt up. Foraminiferal evidence from a borehole at Borth [c.609 901] (Haynes and Dobson, 1969, fig. 17 and p.254), suggests that from about 8000 BP to nearly 6000 BP sedimentation there kept pace with the sea-level rise, allowing the deposition of grey silty sands of a low marsh environment (i.e. flooded by almost all high tides). The borehole proved these sands to about − 25 m OD. Just before 6000 BP sedimentation began to overtake the rise in sea-level, possibly due to a decrease in the rate of that rise (Wilks, 1979, p.33; Heyworth and Kidson, 1982, p.110). The low marsh sands were succeeded by an intertidal silty clay ('Scrobicularia' or 'OD' clay of authors) of high marsh environment, flooded only at high spring tides. This clay is overlain by a forest layer which, with the underlying clay, has been proved in boreholes over a wide area from Borth to Ynyslas, and eastwards to near Ty Mawr [629 927] and possibly to Afon Ddu [658 951] (p.127). A forest bed is also present north of the estuary [585 985], at Clarach Bay [585 838], and north [577 797] of Allt-wen. By about 4700 BP (Wilks, 1979, p.33), the area became an ombrogenous peat bog. Peat formation has continued to the present day in the central part of the bog, although Godwin (1943, pp.224–225) records a temporary estuarine transgresssion, with deposition of clayey peat, into its northern part early in Pollen Zone VIII. The age of this transgression has been determined by radiocarbon analysis as 2900 ± 110 BP (Godwin and Willis, 1964, p.128); its estuarine nature is apparent from foraminiferal evidence (Adams and Haynes, 1965).

The development of the coastal barrier has never been studied in detail, though it has been suggested (Wilks, 1979, p.33) that the development of the spit north of Ynyslas had started by about 5150 BP. At the present day it comprises a storm beach from Borth northwards to Twyni Bach [605 943] with a sand spit extending farther to the north. Extensive sand dunes are present north of Foel-ynys [607 930]. The northward growth of the beach was undoubtedly enhanced by the diversion in 1824 of Afon Leri from its original course just behind the beach, with its mouth near Foel-ynys, to its present canalised course. Details of the northward growth of the beach since 1887 have been published (Yates, 1968). The sand spit and the dune system have also grown in recent years. There has also been an eastward migration of the barrier, possibly by up to 1 km (Wilks, 1979, p.33), and it has overridden the forest layer and intertidal clay now exposed on the foreshore between Borth and Foel-ynys [607 930].

Recent sedimentation in the Dovey Estuary has been studied in detail (Haynes and Dobson, 1969). Since the earlier work on the Dovey salt marshes (Yapp and others, 1916, 1917) there has been a rapid spread of *Spartina townsendii* s.l. over the low marsh environment; it has also invaded the sand flats over a wide area on the southern side of the estuary.

Other areas

South of Upper Borth coastal deposits are limited to Storm Beach Gravels and Marine Beach Deposits between Allt-wen [575 797] and Constitution Hill [583 825], and at Clarach Bay and Wallog.

Immediately west of Aberdovey the beach is backed by a line of sand dunes up to 12 m high; north of these there is a low tract of Blown Sand, with peaty patches in depressions. Where the coast turns north at the mouth of the estuary, a continuous rampart of storm beach shingle lies seaward of the dunes and extends beyond the margin of the district.

GENERAL RELATIONSHIPS

The difference between Morainic Drift in the lower valleys, mainly lying in the south-west, and in the heads of four north-eastern valleys is noted below (pp.118). The former is considered to be older than the latter, but each characterises a different part of the valley system so that their relationship can be evaluated only indirectly.

The Garthgwynion ditch (Figure 51) shows that Morainic Drift of the lower valleys is overlain by nivation gravels (Head), which almost certainly date from the latest Devensian and are in turn overlain by peat at least as old as 7515 ± 100 BP (Welin, Engstrand and Vaczy, 1974, p.102). In another part of the ditch the peat overlies, and appears to pass up from, a limnic deposit of diatomite which is dated at 11430 ± 160 BP. There is no obvious sign of a hiatus between the diatomite and the overlying peat, though the boundaries are not well displayed.

In Hengwm, Morainic Drift in the head of the valley, overlies laminated glacial-lake clays devoid of fauna and flora. The clays contain lenticles of fine nivation gravel, which is taken to be evidence of a cold climate and which possibly dates from a time when nivation gravel was forming extensively on the local hill slopes. The lake-clays are thus younger than the Morainic Drift of the lower valley type but underlie the Morainic Drift of these valley heads. The latter

Morainic Drift, therefore, may be the product of a late glacial resurgence and approximately of the age of the diatomite.

A correlation of these events can now be attempted with those implicit in the work of Heyworth and others (in press). From boreholes at Clarach, they record peats, gravels and clays in a valley bottom location behind the marine storm beach which clearly indicate that ponding occurred here under conditions of lower sea-levels. Limnic peats and silts were encountered to depths of 10 m or so, and from their base an age of 11489 ± 106 BP was recorded. Above, no age was determined until that of 9708 ± 85 BP. The former age is comparable with that of the Garthgwynion peat/diatomite junction.

At Clarach the peat overlies fluviatile gravels containing wood dated 13303 ± 127 BP, suggesting the gravels are at least that old. These gravels seem to have been deposited by the river system after the retreat of the valley glacier, a glacier which earlier had been responsible for the kame-like Morainic Drift of the lower reaches of the valley. This gravel overlies cold-climate, laminated and barren clay. These clays are considered to be proglacial lake clays, deposited in the Clarach valley, itself freed from Welsh ice but still dammed seawards by Irish Sea ice. The valley bottom topography was uneven, its surface being that of the Morainic Drift just released from a glacier of Welsh ice. The cobbly gravel overlain by the clays may, therefore, be Morainic Drift.

SUMMARY OF LATE DEVENSIAN TO RECENT EVENTS

During the main Devensian glaciation an ice-cap developed over the Welsh uplands, and ice flowed radially outwards from the Plynlimon area. The flow of ice, tended to be channelled by the major valleys, which may have been overdeepened already during at least one earlier glaciation. Westward it spread out in a series of piedmont glaciers on the floor of Cardigan Bay. The sarns, ridges of glacial drift which extend from the coast into the Bay, mark the limits of each of the main piedmont glaciers. Thus ice from the Aran mountains, carrying debris of 'felsite' and other rocks from that area, flowed down the Dovey valley and spread southwards as far as Sarn Cynfelyn, west of Wallog [592 857]. This southerly limit coincides approximately with the limit of 'felsite' erratics (Jones and Pugh, 1935a, fig. 38). These authors also suggest, on the evidence of glacial striae, that this northern Welsh ice may have displaced local Central Wales ice which originally flowed westwards towards the Dovey estuary. Both inland and offshore the glaciers deposited extensive areas of till and Morainic Drift. At a later stage in the glaciation, ice derived from larger but more distant ice caps to the north moved southwards down the Irish Sea (the Main Irish Sea Glaciation), truncated the western ends of the weaker Welsh piedmont glaciers and deposited till which overlies that derived from Wales (Garrard and Dobson, 1974, pp.40–41, figs. 2 and 3).

Towards the end of the Devensian the glaciers fed from Central Wales retreated and ablated, depositing Morainic Drift in the valleys. These glaciers were ablating and in retreat earlier than those fed from the ice-cap of North Wales and the northern Irish Sea. Consequently (Jones and Pugh, 1935a, p.289 and fig. 38 and in discussion of Jones, 1965, pp.279–280), westward flowing streams between Glandyfi [697 970] and Llandre [626 869] derived from the melting Central Wales ice were blocked by North Wales ice to the west and were diverted southwards over the interfluves before reaching the sea at Aberystwyth or Clarach. During this diversion the meltwaters cut marginal channels through the interfluves. Varved clays from a borehole at Clarach may have been deposited in lakes impounded early in this episode. Varved clays in the Hengwm valley [780 950], north of Plynlimon, suggest that the northward-flowing streams were also impounded, but evidence of overflow to the west is not obvious. Alternatively they may have resulted from more local ice damming. As the North Wales ice also retreated, further meltwater channels were cut in the floor of the western part of Cardigan Bay (Garrard and Dobson, 1974). Morainic Drift present at the heads of some valleys and overlying the laminated clays may be associated with a temporary partial reassertion of glacial conditions.

In late-Devensian times (Pollen Zones I–III) extensive periglaciation took place, including the development of stratified screes (nivation gravel) and terraces of soliflucted till (Watson, 1968, 1970; Potts, 1971). The glacial cirques at Llyn Llygad Rheidol [792 875] and near Rhiw-gam [786 946; 792 940], together with some nivation hollows, are also very late Devensian.

As the ice disappeared the sea-level rose, and this rise continued at a decreasing rate through the Holocene (Wilks, 1977, 1979). Sedimentation in the Dovey estuary and the Cors Fochno (Borth Bog) area appears to have kept pace with the rise from about 8000 BP to nearly 6000 BP (Haynes and Dobson, 1969). At about this date, and associated with the northward extension of a coastal barrier from Upper Borth [608 890] (p.117) the rate of sedimentation overtook the rate of rise in sea-level in the area now occupied by Borth Bog, and the intertidal sediments gave way to terrestrial conditions with the development of a forest, now exposed as submerged forest on Borth beach, and eventually an ombrogenous peat bog. At Clarach, freshwater marsh conditions appear to have prevailed for most of the time since the latter part of the late-glacial (p.118), probably due to local ponding of Afon Clarach.

CHAPTER 8

Quaternary—details

BOULDER CLAY

Aberystwyth – Upper Borth

Patchy Boulder Clay is present on the high ground east and NE c.
Aberystwyth particularly around Waun Fawr where a road cutting
[5997 8175 – 6000 8189] exposed 1.2 m of strongly weathered
matrix-rich Boulder Clay with rounded clasts up to 20 cm diameter.
The matrix is buff-weathering with orange staining and white patches after rotted mudstone chips. Granule to cobble size clast-rich
Boulder Clay occupies the hollow between Plas Cwmcynfelin and
Comins Coch and covers the hill-slope towards Capel Dewi. A
stream section [6155 8241], SSW of Dorglwyd, shows 1.2 m of
brown weathered Boulder Clay on 1.5 m of stratified drift consisting
mainly of locally derived pebble-sized mudstone clasts resting on
solid.

North of the Clarach valley, Boulder Clay is present in the broad
hollow between Ty'r Abbi and Rhyd-meirionydd and thence
eastwards towards Wileirog and Castell Gwallter. Six metres of stiff
buff to grey clast-rich clay with angular and rounded clasts of pebble
to cobble size were seen in a stream section [6017 8544 – 6030 8538].

Capel Bangor – Goginan

Patches of bluish grey clast-rich clay, locally at least 1.8 m thick, occur around Cwm-mwythig and Llwyniorwerth.

Two 6-m-deep gullies [700 810] south of Blaen-dyffryn-isaf [700
812] give poor sections in orange-brown weathering grey clay with
poorly sorted and poorly bedded mudstone debris. The vague stratification suggests an element of resorting, possibly by solifluction.

Penrhyncoch – Talybont

Boulder Clay is present on the south side of Afon Stewy at Salem
and extends westwards along the northern side to Bow Street. The
clay is locally very clast-rich and in places almost a clayey gravel.
East of Bow Street, a stream section [6283 8497] shows up to 4.6 m
of brown weathered clay with a few boulders, some greater than a
metre in diameter. Boulder Clay fills the broad hollow east of
Penycefn. In a stream [6518 8609] SSW of Elgar a 6 m bank shows
brown unbedded clay with abundant, unsorted mudstone and sandstone pebbles up to 30 cm in diameter. Similar drift covers the slope
between the two small streams south of Cynnull-mawr [654 874],
and is incised by the head of a glacial drainage channel at Ty'n-y-
ffynnon.

Afon Leri and tributaries (upstream from Talybont)

On the southern side of Afon Leri, Boulder Clay occupies most of
the basin of Nant y Gwyddil [670 884 – 670 868] and extends
southwards over a col to Mynydd Gorddu mine [669 861]. There
the drift is probably thin, and there are many sandstone erratics on
the surface. Similar remanié erratics are abundant on the drift-free
high ground to the west of the Boulder Clay. Along the steep northern part of its course [670 878 – 670 884], Nant y Gwyddil is incised up to 8 m in brown very sticky clay with some large boulders.
Higher up the Leri valley Boulder Clay covers the flattish ground
west, south and east of Bontgoch. It is also present along the valleys
of Nant Lŵyd [686 880 – 707 861] and Nant Perfedd [690 875 – 700
873]. West of the bridge at Bontgoch, a river cliff [6811 8622] shows
6 m of unbedded stony clay with boulders up to 30 cm in diameter.

Much of the northern side of the Leri valley, from 1 to 2 km
upstream of Talybont, has a cover of Boulder Clay. A section at
Tyrhelig [668 890] showed about 2 m of brown stony clay with a
mixture of boulders, pebbles and angular debris of mudstone and
sandstone. The drift fills hollows in the ridge to the north, and
covers an extensive area on the southern side of Afon Ceulan from 1
to 2.5 km east of Talybont church. There are many large sandstone
erratics, some nearly 4 m across, particularly on the higher parts of
the slope.

Afon Clettwr

Boulder Clay is widespread in the upper part of the Clettwr valley,
particularly on the southern side of the river [677 914 – 706 913].
East of Caer-arglwyddes, it is largely covered by peat. The river is
deeply incised into the drift, in excess of 20 m to the WNW of Caer-
arglwyddes [693 917]. The Boulder Clay gives rise to broad sloping
terrace-like features, not unlike the solifluction terraces in the upper
Rheidol valley (Watson, 1968, 1970), although the threefold division of the Rheidol drifts (Watson, 1965b, pp.22 – 23) is not apparent. The lithology of the Boulder Clay is variable, but in general
there is an upstream transition from stony clay to clayey gravel.
Large angular blocks of local sandstone are common in the top 2 m.
The upstream transition can be demonstrated by three sections
[6872 9189; 6929 9182; 6959 8182] which show respectively 9.1 m of
clay with very poorly sorted fragments, pebbles and large blocks of
mudstone, sandstone and quartz; 4.5 to 6 m of unsorted very stony
clay or clayey gravel, coarse in the basal metre; and 4.5 m of grey to
buff roughly bedded and poorly sorted gravel with a clay matrix.

Afon Rheidol and tributaries (south of Dinas dam)

The valley of Afon Rheidol has a mature aspect from its source to
the Dinas dam [745 821] (Jones and Pugh, 1935a, p.294). South of
the dam the mature valley continues as a relict, but the river is incised into it in a southward-deepening gorge which includes the well
known and spectacular 'incised meanders' between Ponterwyd and
Parson's Bridge. It has been suggested (Challinor, 1933; Jones and
Pugh, 1935a) that these meanders result from local glacial diver-
sions rather than from incision on the mature valley floor, as the
original and more direct incised channel, now filled with Boulder
Clay, can be traced on the eastern side of each of the two meanders
(Figure 52). However, the meander pattern could at least in part
have developed on the drift-covered valley floor and have persisted
when downcutting resumed (Anderson, 1963). It has also been suggested (Jones and Pugh, 1935a) that the gorge south of the Dinas
dam at Craignant was due to similar glacial diversion and that the
old river channel, now filled with Boulder Clay, lay to the east. This
has since been verified by boreholes sunk for the Rheidol
Hydroelectric Project (Adams, 1961; Anderson, 1963) which showed the presence of several earlier channels beneath the Boulder
Clay. It has been proposed (Watson, 1968, p.41; 1970, p.129) that
the low ridge in the till immediately south of the Dinas Reservoir is
a moraine, and that this caused diversion of the river into the gorge.
It is clear from the presence of the drift-filled channels both north
and south of Ponterwyd that the formation of the gorge from Dinas
to Devil's Bridge, whether due to river capture (Jones and Pugh,
1935a) or tributary takeover (Challinor, 1947, 1969), must largely
have post-dated the last glacial episode.

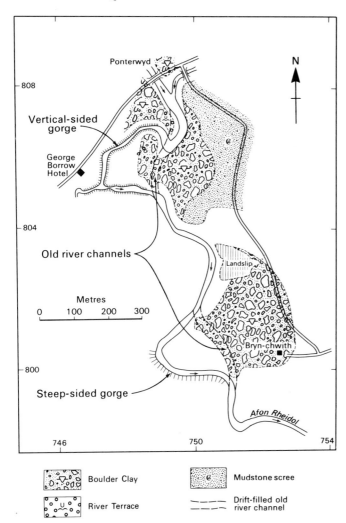

Figure 52 'Incised meanders', Ponterwyd

North of Ponterwyd the valley is extensively covered with Boulder Clay, especially to the east of the present river; the maximum thickness of drift proved in the Rheidol Hydroelectric Project boreholes was 35.4 m [7486 8192]. A 10 to 12 m bank [750 821] at the southern end of the Dinas Reservoir shows sections in slipped bluish grey unbedded silty clay with unsorted angular to rounded mudstone and sandstone debris, commonly striated, mostly no more than 15 cm in diameter.

Afon Rheidol and tributaries (north of Dinas dam)

Much of the tributary valley between the Dinas Reservoir and Llyn Syfydrin [723 847] and northwards to Bwlchystyllen [730 862], is floored with bluish grey stony clay of unknown thickness, commonly overlain by peat. From the Dinas Reservoir northwards much of the Boulder Clay forms 'solifluction terraces' (Watson 1968, fig.III.2); there are, however, widespread deposits beyond the limits of these 'terraces'. A tripartite division of the Boulder Clay (Watson 1965b, p.22; 1968, p.41) has only rarely been seen, and it has not been possible to determine the geometry of Watson's middle (gravel) member.

Between Nant Ceiro [751 821] and about 0.6 km north of Hirnant, a broad sloping 'terrace' lies on the eastern bank of the Rheidol; about 15 m of Boulder Clay overlie solid rock [7498 8277] SE of Nant-yr-cae-rhedyn. A borehole at the Dinas Power Station

site [7469 8301] proved a Boulder Clay filled valley with a depth of at least 16.8 m (bottom at 240 m above OD) (Anderson, 1963, p.45). A peat-covered 'solifluction terrace' is developed along the southern side of the Peithnant [776 836 – 760 849] and to a lesser extent on the northern side. A section in the stream [7707 8414] SE of Blaen-Peithnant shows 3.7 to 4.6 m of bluish grey clay with ill-sorted fragments and boulders to 0.9 m diameter overlying 1.5 m of bluish grey poorly sorted, bedded pebble to cobble gravel. The Boulder Clay extends across the low col at Lle'r-neuaddau [759 849] and along the eastern bank of the Rheidol to the Nant-y-môch dam [755 862] where 12.5 m was recorded (Anderson, 1963, p.44).

Boulder Clay, largely peat-covered, fills the higher parts of the Nant-y-moch and Maesnant valleys. It is at least 15 m thick in places along Nant-y-moch [e.g. 774 865] and large sandstone erratics (up to 3.7 m × 2.7 m × 1.2 m) are common in places.

MORAINIC DRIFT

Rheidol valley

Much coarse terraced gravel occurs in the lower Rheidol valley and has been regarded (Jones and Pugh, 1935a) as fluvial. However, the highest occurrence, which is singularly high in comparison with the remainder, we regard as a kame terrace of Morainic Drift. It is composed of bedded clayey gravels and sands, chaotic and collapsed in places, and containing pods of till. West of Lovesgrove, erosional benches occur in this deposit at various levels.

Aberystwyth Castle stands on a small flat between the Rheidol and the sea, 1 m of clayey pebble-cobble gravel being exposed [5795 8145] overlying Aberystwyth Grits. A flat at Penparcau consists of pebble gravel. It extends through the col between the Rheidol and Ystwyth valleys, suggesting that meltwater may have flowed at some time from one valley to the other. Morainic Drift forms small terraces along the northern side of the Rheidol valley from Aberystwyth to Llanbadarn Fawr. A valley running to the east of Buarth-mawr hill [588 816], between the Rheidol valley and the old course of the Rheidol that entered the sea on the north side of Aberystwyth, is possibly floored by Morainic Drift. Between Frongôg and Lovesgrove, clayey gravels occupy irregular sloping terraces and pass imperceptibly northwards into till. It would appear that meltwater flowed southwards into the Rheidol valley through the col at Lovesgrove while ice still occupied the Clarach and Rheidol valleys.

On the southern edge of this area, along a 7 – 9 m bank adjacent to the Rheidol alluvium, the variability of the deposit is demonstrated by the contrast between two sections [6160 8088; 6197 8113] which show bedded cobble to pebble gravel with a granule matrix, and a temporary one [6183 8103] between the other two, which showed 5.5 m of very clayey pebble to cobble gravel or gravelly clay with a 15 cm sandy granule bed near the base (overlying the Aberystwyth Grits).

On the northern side of the valley, a Morainic Drift terrace is well developed east of Pentre-rhyd-yr-onen [6370 8110]. In places it is 6 to 7.5 m above the flood-plain level. The best sections occur in cuttings behind houses [6452 8039; 6483 8038]. The first exhibits 2.1 m of locally bouldery, clayey pebble-cobble gravel with scattered dark brown sand lenses. Sorting is poor, whilst collapse structures and vertical stones are common; the second exhibits 6 m of similar gravel with a few small rafts of till included within it.

East of Glan Rheidol, the bottom of the valley is occupied mainly by Morainic Drift terraces. On the northern side, a terrace lies on average 4.6 m above the flood-plain level and has a gently sloping surface; scattered mounds are cored by solid rock with drapes of gravel. On the southern side a terrace lies 7.6 m above flood-plain level, and remains of another occur 4.5 m higher. The lower is well exposed in the river bank [6880 7915], where up to 3 m of gravel with collapse structures are seen.

Peithyll valley

The Peithyll valley has gently sloping terraces of Morainic Drift on either side of the modern alluvial tract. Most of the terrace is in gravel. At Capel Dewi, Morainic Drift extends southernwards from the Peithyll valley through the Lovesgrove col to connect with similar deposits on the northern side of the Rheidol valley (p.120).

Stewy valley

A rock-gorge, cutting the north side of the W–E Stewy valley, at one time spewed detritus-laden glacial meltwater into the Stewy valley at Cwrt [646 846]. The overspill water came from the glacier-filled valley to the north, but there is no alluvial cone at Cwrt. The detritus must, therefore, have been fed on to or into valley ice, standing probably at about the 300 ft (91 m) contour, to contribute eventually to the Morainic Drift of the Gogerddan valley, a deposit which, significantly, does not extend much upstream of Cwrt, where collapse structures were noted in it.

Around Penrhyncoch, Morainic Drift occurs only on the southern side of the valley as a single gently sloping terrace and as a flat between the two convergent streams Stewy and Silo. Most sections expose buff to orange pebble cobble gravel with a variable sandy clay matrix. A stream section [6866 8436] exposed 2.7 m of slightly bouldery, pebble-cobble gravel with scattered pods of blue-grey clayey gravelly till.

Bow Street valley

Almost identical gorges occur on the northern side of this valley, at Bryngwyn-canol and Dolgau, just NE of Bow Street. They were spillways, from the western slopes of Banc Mynydd-gorddu and the Leri valley glacier SW of Talybont respectively, on to ice in the Bow Street valley. As in the Leri and Stewy valleys there are no remnant alluvial cones where they debouched, but a ribbon of Morainic Drift originates near them at Bryngwyn-mawr [647 861]. This extends through Bow Street to join with the Morainic Drift of the Stewy valley and Peithyll valley and to emerge at the coast through the Clarach valley, as presumably did the parent tongue of valley ice.

As far north as the Llandre gap, Morainic Drift terraces are mainly present on the eastern side of the valley, the western side being mantled by head; up to three different terrace heights are present. In a cutting [6234 8486] in the highest terrace, 6 m of slightly bouldery pebble-cobble gravel with a buff clayey coarse-grained granule-rich sand matrix are exposed. The Bow Street valley splits north of Pen-y-garn into the Llandre gap and the Dolau valley. The mudstone scree that fills the Llandre gap overlies Morainic Drift immediately north of Pen-y-garn at the southern end of the gap. North of Pen-y-garn in the Dolau valley the Morainic Drift forms flats interspersed with mounds on either side of the modern alluvial tract. Some of the mounds appear to be thick drapes of Morainic Drift over solid relief rather than constructional features. Immediately north of Pen-y-garn, one was exposed in a road cutting [6282 8565 – 6285 8575]; it showed a core of solid with 3 m of overlying gravel in the north, and 4.5 m of very clayey gravel verging to till overlying solid in the south. Between Pwll-glâs and Dolgau five glacial overflow channels form the northern part of the Dolau valley. They are floored by Head and pass out on to a boulder clay flat. A large mound [642 862], on the southern side of the stream immediately east of Dolau, consists mainly of gravel but a section at Bryngwyn [6402 8610] exposed (above solid) 2.1 m of pebble-cobble gravel with interbedded beds of angular and subrounded mudstone chips, passing laterally into buff unstratified gravelly till rich in mudstone fragments. Farther up the valley Morainic Drift continues between Cwm Cae and Elgar. Two south-westerly trending glacial drainage channels [646 864; 649 863],

although cutting through this deposit, are intimately associated with it. In composition, the deposit varies from a brown-weathering very stony clay, completely unbedded and unsorted, as in a 6 m bank [6497 8603], to a bedded poorly sorted gravel as in a stream bank [6457 8612] south of Cwm Cae.

Clarach Valley

The Morainic Drift is best seen in the cliffs on the northern side of Clarach Bay where up to 3.6 m of pebble-cobble gravel with scattered coarse-grained sand lenses are mostly unbedded and rest on blue till. Traced northwards the gravels thin and merge fairly rapidly into a buff Head that, at the first small headland north of the bay, overlies the blue till. The Head overlies the gravels on the northern side of the Clarach valley.

In the main Clarach valley, Head occupies the southern side of the valley virtually as far east as Llangorwen; elsewhere a terrace occurs along both sides of the valley. Locally, on the northern side, two terraces are present. Their surfaces do not however, conform with any possible down-valley thalwegs relatable to river deposits. A section in the upper terrace occurs at Nant-cellan-fâch [5959 8403]. Here a north–south orientated ridge of solid traverses almost the entire width of the terrace, and up to 3.6 m of horizontally bedded slightly bouldery pebble-cobble gravel, with a few very thin beds of grey gravel-rich sand, are in contact with the ridge on both sides. Boreholes behind the storm-beach barrier (Taylor, 1973, pp.278 and 282) revealed that fluviatile gravels, at least 1.5 m thick, underlie a sequence of peats and carbonaceous silts dated at 10 100 ± 250 BP near its base. More recent work (Heyworth and others, in press) has provided a date of 11 488 ± 106 BP at the base of the peats, while the gravels were shown to be up to 2.5 m thick. At the top they contain wood dated at 13 303 ± 127 BP. They overlie up to at least 7 m of lake clays and silts, the basal part of which appears to be abruptly replaced laterally by a body of silty gravel; the whole sequence rests on tough ill-sorted stony clay 'Head' or '?till'. Part of these lower gravels may be the Morainic Drift mapped in nearby valleys.

Wallog

At the northern end of the col between the Clarach and Wallog valleys a gently sloping bench is present mainly on the southern side of the Wallog valley at Rhosgellan. Streams [e.g. 5968 8542 – 5972 8528] demonstrate the bench to be formed mainly on orange pebble-cobble gravel, locally rudely stratified, but gravelly till is also present. It may be partly a kame terrace, developed when melt water from wasting ice in the Wallog valley and its tributaries was draining through the narrow gap of the Wallog valley-mouth to the sea.

The narrow Wallog valley west of the above col is also filled by Morainic Drift. In the valley bottom this forms a narrow terrace, mainly on the southern side of the valley. Some sections expose pebble-cobble gravel and, locally gravelly till. In the stream bank [5908 8578], 3 m of pebble-cobble gravel are exposed.

Westwards the valley opens on to a narrow, gently seaward-sloping platform, parallel to the present day coastline. The platform is made of a variably thick veneer of Head which overlies Morainic Drift consisting of buff variably clayey pebble-cobble gravel with local boulders. Stratification by grain size is common but generally crude, though lenticular beds of granule-rich sand up to 0.3 m thick are locally common. These gravels crop out for some half a kilometre in the sea cliffs on either side of the Wallog valley [5885 8540 to 5913 8612]; they rest mainly on Aberystwyth Grits, though at one locality immediately south of the valley they overlie an earlier Head which itself rests against a fossil cliff cut in the Aberystwyth Grits subparallel to the present sea cliffs.

In general this spread of gravel coincides with a rather broad valley, but its deepest part some 250 m to the north [8601 5909] is not on the line of the present Wallog valley. A 30-m-wide channel is here cut below the modern wave platform of Aberystwyth Grits, and is directed to the NW; it contains 12.2 m + of gravel. Elsewhere the thickest gravels overlying bedrock are 10.7 m; this thickness decreases gradually to the north and south. Sarn Cynfelyn appears to be unrelated to the gravels. The junction of the gravels with the overlying Head appears to be fairly sharp.

Leri/Cyneiniog valley

A major point of entry of detritus can be identified on the southern side of the valley at Cwmere. Here, a large non-glaciated gorge breaches the valley side, allowing entry of the present headwaters of the Leri. This gorge is cut in solid rock for much of the way to Bontgoch, and has a steep gradient where it enters the main W–E valley. The Morainic Drift of the main valley does not rise towards, and into, this gorge as would an alluvial cone or fan, but occupies the valley-side indentation to a level of c.120 m above OD as elsewhere nearby. It seems probable therefore, that the detritus transported through the gorge during the active meltwater period was deposited on to and into a glacier occupying the valley at Cwmere. The glacier then redistributed it down the valley as part of the Morainic Drift. The valley-bottom Drift ceases to be gravelly in the Cyneiniog valley upstream of Cwmere, and there it cannot be distinguished from Boulder Clay. This part of the valley has a very limited catchment, while the Bontgoch-Cwmere gorge (and the adjacent but lesser Cwmere river) tapped a catchment which still includes the Llyn Craigypistyll basin. In late-glacial times it may have carried water or ice overspill from the Upper Rheidol catchment via the Bwlchystyllen col [737 867].

Just downstream of Cwmere, grey stoneless clay is exposed below gravel, at the base of the river banks, and in the river bed where it bulges upwards [6811 8848]. The base of the clay is obscured below the river bed, and so the deposit may be no more than a clay lenticle within the Morainic Drift. On the other hand, it may be comparable with the glacial lake-clays at Clarach and Hengwm.

The Leri valley from Cwmere to Talybont has an infilling of Morainic Drift up to at least 15 m thick. The river has cut down rapidly through these deposits, forming steep unstable slopes up to 15 m high. The instability has resulted in landslips on both sides of the river from 1 to 1.7 km downstream from Cwmere [683 882]. The variable nature of the drift is demonstrated by two sections within the valley. The first is in the south bank of the river [6730 8852] where 3.7 m of fawn banded silty clay with pockets of pebble gravel up to 30 cm thick, and also dark grey layered sand in curved scours, overlie 7.6 m of fairly clean gravel with rounded clasts varying in size from coarse sand to 23 cm diameter cobbles; this gravel includes a 1.2 m bed of finer sand with pebbles up to 0.5 cm in diameter. The second section [6628 8882], 1.1 km farther downstream, shows 4.6 m of very poorly sorted and bedded boulder gravel with lenses of clayey gravel.

Morainic Drift forms a gently uneven area around and to the west of Glanfraid. Solid bedrock [6273 8778] protrudes through this deposit 100 m NE of Pen-y-wern, though higher land immediately SE is gravel covered. Afon Leri has eroded a channel through this deposit, and incised deeply (over 17 m) into the underlying Silurian rocks. Thus a variable thickness of gravelly Morainic Drift now rests perched in the top part of the valley side, and in places [633 879; 625 879] has become unstable to produced slurried and landslipped slopes. The depth and steepness of the erosion into solid rocks, with a broad (up to 100 m) flood plain at the bottom of the cut, probably points to confinement of the river, either subaerially or possibly subglacially, by ice.

In the railway cutting [6262 8723 to 6261 8720] rounded gravel is seen in isolated sections in the lower part of the cutting, whilst mudstone scree is seen in the upper part. It is considered very unlikely that the Morainic Drift extends much farther south along

the Llandre 'gap', which for the main part is filled by nivation gravel (p.127).

Talybont – Furnace

North-west of Talybont, Morainic Drift extends over a low col and gives rise to hummocky topography on the margin of Cors Fochno between Ynysycapel [645 905] and Tan-yr-allt [655 911]. The predominant clasts are local sandstones and siltstones, but vein quartz and volcanic rocks are also present.

Two gullies [6790 9404; 6777 9420–6790 9413] (Figure 37a, loc.8), east of Llwyn-gwyn, show up to 4.6 m of buff weathering grey very stony clay to clayey gravel with clasts of mudstone, sandstone, quartz and 'felsite' up to 60 cm in diameter. Grey mudstone is exposed below the drift in the second gully. Boreholes along a proposed realignment of the A487 west of Furnace and Eglwysfach proved Drift, mainly gravel, of varying thickness and nature. The rock-head topography is very irregular and was presumably formed by scour along the strike either by ice or subglacial water. The thickest proved Drift was in a borehole [6842 9542] (surface level 8.49 m above O D) adjacent to Afon Einion; the hole proved sandy and silty clay with pebbles (Alluvium) to 2.9 m, on sandy gravel with some cobbles and boulders (?Morainic Drift) to 16 m, on clayey silt to 17 m, on sand and gravel to 19 m, on cobbles and boulders with mudstone fragments to 23 m, on solid mudstone.

Plynlimon

Morainic Drift occurs in the head of the Rheidol valley, on the north side of Plynlimon above and below Llyn Llygad Rheidol. It is a pile of unascertained thickness, mostly of large boulders disposed on the steep slope above the lake. Below the lake the deposit appears to merge with a thick (at least 18 m) valley-bottom fill of bouldery clayey silt, unsorted and unlayered, and mapped as Boulder Clay. Upon the terrace-like surface of this Boulder Clay there are many large boulders similar to those forming the deposit above the lake, but whether they are an integral part of the Boulder Clay or a later addition is not known.

Rhiw-gam valley

A small area of Morainic Drift is present in the head of the minor valley west of Rhiw-gam (Figure 53) where it occupies the bottom of a steep-sided semicircular cwm (Plate 26). It consists of stony and bouldery clay forming an uneven mound c.200 m long and possessing two major curved steps, up to 6 m high, on the surface of its general down-valley northward gradient. Below this area, the valley bottom and sides are covered with stony clay mapped as Boulder Clay, but showing signs of down-slope movements in many places. The headwall of the cwm is largely covered by nivation gravels which appear to lap onto the moraine. Although smaller and at a lower altitude this deposit is comparable with cirque moraines in the Brecon Beacons (Lewis, 1970, p.162; Ellis-Gruffydd, 1977) and late-glacial readvance moraines in Snowdonia (Gray, 1982).

Cwm-byr

An area of Morainic Drift is dissected by a stream which exposes some 5–6 m of deposits in its banks; these consist of rounded gravels, rather clayey and silty, on yellow stoneless clays containing regular, thin interbeds of sand (Figure 53)..

Hengwm

Hengwm is much larger than Cwm-byr and contains a more extensive spread of valley-bottom fill (Figure 49). On its surface [7723 9368] there is a small transient lake, Llyn y Delyn, considered to be

Figure 53 Drift deposits in Cwm-byr and near Rhiw-gam

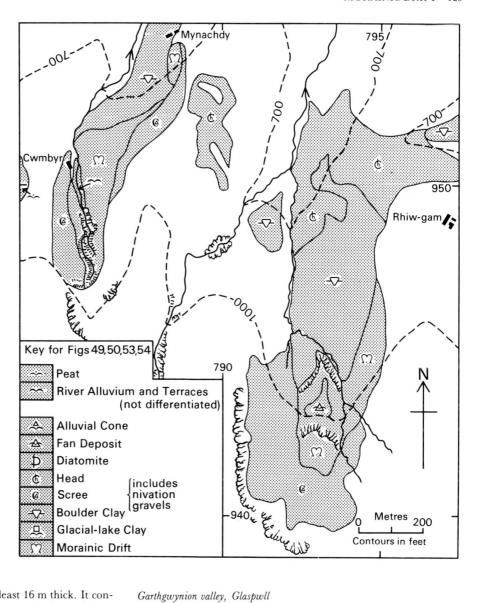

Key for Figs 49,50,53,54

〜	Peat
〜	River Alluvium and Terraces (not differentiated)
△	Alluvial Cone
⌂	Fan Deposit
Ð	Diatomite
ℭ	Head
℮	Scree } includes nivation gravels
▽	Boulder Clay
◻	Glacial-lake Clay
♏	Morainic Drift

a kettle hole. Morainic Drift is in places at least 16 m thick. It consists of boulders up to 0.3 m and rounded and subrounded to subangular gravel set in a silty matrix. The clasts are commonly orientated parallel with the surface, but in exposures near the head of the cwm they show imbrication with a 20° up-valley dip (Plate 27). In many places the deposit has foundered on an underlying body of laminated clays. In other places clay plugs have pierced the gravels, while pendulous load-lobes of gravel penetrate down into the clay. Near the head of the cwm, gravity faults and back-tilting have affected the gravels and enhance the appearance of clast imbrication.

Cwm Rhaiadr

Cwm Rhaiadr is the blind end of the Llyfnant valley and, like its neighbours to the east, possesses a small, but similar, lobe of Morainic Drift (Figure 54). This covers an area at the confluence of a small tributary valley from the SE, and consists of c.5 m of ill-sorted rounded gravel of sizes commonly up to 10 cm and some especially at the base of the deposit, up to 0.5 m. The stones are commonly orientated parallel with the surface, which is fairly even, declining away from the side of the valley (and the entry point of the tributary). There is little sign of bedding within the gravel but it rests on well bedded clayey and sandy deposits.

Garthgwynion valley, Glaspwll

Two dry valleys converge near Garthgwynion, continuing northwards as one. They are aligned at right angles to the drainage via the Llyfnant which has beheaded their pre-glacial catchments (Figure 50). The deposits include Morainic Drift of a form and composition like that in the valleys in the south-west. Presumably the valley remained blocked by ice after higher ground to the south had become ice-free. North Wales ice at the egress of this valley into the Dovey valley probably caused the blockage, so that drainage from the south spilled westwards along a depression following the line of the Llyfnant Fault. The level of a portion of the pre-glacial lower Llyfnant valley at Allt-ddu [715 973] suggests that the pre-glacial river here flowed eastward.

A ditch section near Garthgwynion is presented graphically (Figure 51). The deposits observed in the walls of the ditch A–B along certain lengths (measured from A towards B) are described below.

Plate 26   Two-tier moraine left by late-glacial corrie-glacier. [793 944]. Near Rhiw-gam. A14053

Plate 27 Gravelly Morainic Drift overlying glacial-lake clays. Hengwm [773 937]. A12599

Figure 54
Drift deposits of
Cwmrhaiadr

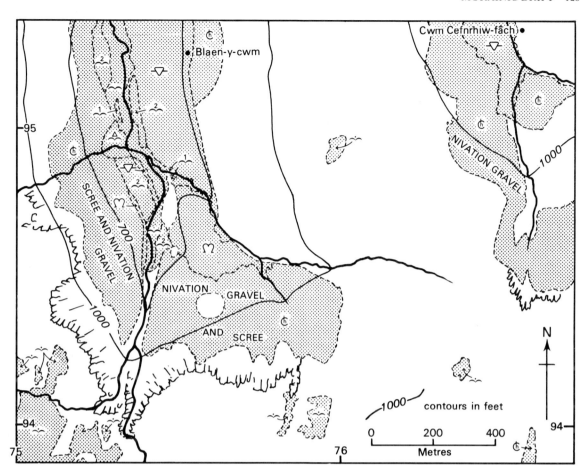

Distance from A m		Thickness m
0–32	Gravel, fine, of mudstone flakes, orientated towards valley bottom	0.69
50.3	Gravel, ill-sorted, coarse and cobbly	1.5
58.5	Silt, loamy, brown; abundant coarse angular stones becoming smaller and fewer eastward. Part of a solifluction fan	up to c.1.0
68.6	Clayey silt, fawn-grey and stoneless with incorporated plant debris	c.1.0
73.2	Clayey silt becoming peaty	
76.3	Silty peat	c.0.61
	Gravel, fine, cryoturbated	seen c.0.15
84.1	Silty peat	c.0.61
	Gravel, fine, cryoturbated	seen c.0.61
88.7	Silty peat. Sample from base	0.3 to 0.46
	Gravel, fine, cryoturbated	0.61 to 0.76
93.3	Gravel, clayey and silty (at top of bank–dug from ditch?)	0.46
	Peat, thin irregular layer, sample	c.0.07
	Gravel, fine, cryoturbated	seen

Distance m		Thickness m
97.8	Similar to 93.3 m. From here eastward peat layer becomes thicker, browner and softer	
114.3	Gravel, clayey and silty, thin	c.0.15
	Peat. Sample from base	0.77
	Gravel, fine, cryoturbated	seen c.0.15
114.3–155.4	Peat, brown, fibrous and wet with wood fragments	0.91–1.22

The following are descriptions of the deposits in the walls of ditch B–C along lengths measured from C towards B (Figure 51):

Distance from C m		Thickness m
0–43	Clay, grey, silty, stony with shaly fragments (?part of a solifluction fan on Morainic Drift)	c.1.53
c.46	Peat, brown, fibrous and wet. Silt, pale grey containing reedy vegetable fragments (Diatomite)	seen c.0.3
46–173.7	Peat, brown, fibrous and wet. Diatomite seen at base in SE	1.2–1.5

HEAD

The most extensive areas of Head are within 5 km of the coast from the Rheidol valley northwards to the southern margin of Cors Fochno. Up to 3 m of crudely stratified brown to light brown clayey sand with shale and sandstone clasts to 15 cm diameter is present on the hillside north and NW of Moriah [620 795], and stony clay covers the hillside at Penparcau. Farther north, a narrow apron of Head, overlying Morainic Drift, borders both sides of the Clarach valley between the railway line and the sea. These deposits are best seen in the coastal sections from Clarach Bay to Wallog.

Between Clarach Bay and Wallog a gently sloping platform, some 150 m wide, occurs at the top of the sea cliffs. Behind it is a steep back-feature or bevel interpreted by Wood (1959) here and elsewhere along the coast as a degraded marine cliff. The platform is covered by an apron of Head. The Head consists of a grey, but almost ubiquitously buff weathered, silty clay with numerous angular mudstone, Aberystwyth Grits and sandstone clasts, mainly to pebble size but locally to boulder size. Rounded pebbles including erratics are locally present. The percentage of clasts varies, and in the upper part of the platform the Head is commonly matrix-rich with only scattered granule-size material. The Head is on average 3 m but up to 4.5 m thick. For the most part it sits on frost shattered Aberystwyth Grits, but immediately on either side of the Wallog and Clarach valleys it rests on Morainic Drift. At Wallog and at Clarach, Morainic Drift is locally seen in the sea cliffs resting on an earlier Head deposit. In the cliffs on the southern side of Wallog [5890 8558], Head not only underlies and predates the Morainic Drift but is plastered against a vertical cliff cut in the Aberystwyth Grits. Up to 2 m of Head is seen, and consists of angular mudstone clasts up to a centimetre in diameter with scattered sandstone clasts to 0.15 m diameter in a buff clayey matrix. It is obviously very locally derived, and represents scree banked against a small cliff cut in the solid. It may have formed during an interstadial period immediately prior to the deposition of the Morainic Drift. Farther south at Clarach on the north side of the bay [5874 8410], Morainic Drift overlies a blue coloured Drift consisting of angular and rounded clasts to cobble size, mainly of local derivation, in a matrix of silty clay. It is more indurated than the Head that overlies the Morainic Drift. It may be a very locally derived till or a Head formed at the same time as the one at Wallog that pre-dates the Morainic Drift. In the northern part of Clarach Bay it is up to 5 m thick, and is overlain directly by the buff Head that postdates the Morainic Drift.

A similar, but narrower, strip of Head is present along the southern side of the valley at Gogerddan [630 836]; here it varies in composition from a clay with some granule and small pebble clasts [6283 8338], to a deposit of fragments of well-cleaved mudstones up to 2.5 cm diameter in a sparse clay matrix [6351 8359 – 6371 8365]. There is a broad area of Head on the western side of the valley between Bow Street and Aberceiro [626 860].

Farther inland, local and mainly small areas of Head are present on many of the valley sides. A typical 3 m section [7150 8124] south of Nantyrarian comprises angular mudstone fragments (mainly 5 to 25 mm diameter) and a few slabs of mudstone and sandstone up to 30 cm diameter in a greyish brown clay matrix. The Head is well bedded with a slight downhill dip in the direction of the hill-slope. Less typical is an area of Head on the northern side of Cwm Ceulan where a stream gully [6989 9049 – 6989 9034] shows up to 4.5 m of poorly bedded angular to subrounded fragments of dark grey mudstone in a poorly cemented slightly ferruginious sandy matrix.

SCREE

The Llandre gap between Abercerio and Llandre is filled by mudstone scree (Jones and Pugh, 1935a, p.291). The screes are composed of well sorted mudstone fragments a few mm to 2 cm in size. Bedding is defined by grain-size banding and by thin clay-rich seams. Most beds have virtually no matrix, but where it is present it consists of orange to buff silty clay. The bedding in the scree dips westwards on the eastern side of the valley: 16°SW [6260 8599] where 4.6 m is exposed; 18°SW [6267 8666] where 4.5 m are exposed; 5°W [6268 8694] where 2.1 m are exposed, and where it dips less steeply than the present valley sides. All this material must have been derived from the small hill immediately to the east of Llandre. On the gentle lower slopes of the western side of the Llandre valley there are no exposures of scree though the soil suggests it is present. As westward dips are recorded in the scree as far east as Aberceiro [6260 8599], it is likely that virtually all the scree was derived from the east. Although scree overlies Morainic Drift to the north and south of the Llandre gap (see page 121), the gap itself does not appear to contain Morainic Drift, and so it may not be an overflow channel as has been suggested (Jones and Pugh, 1935a). Elsewhere, however, as at Dol-gau [6414 8688] and at localities [6434 8730] near Fagwyr and SSW of Pen-y-cwm [6493 8516], mudstone screes occur on the sides of such channels and clearly post-date their incision.

Another good example of mudstone scree is in an old pit [7572 8062] NNW of Parcgwyn, where 4.5 m of angular unconsolidated dark grey mudstone were seen. The mudstone fragments have a maximum size of 5 cm but are mostly less than 2.5 cm across. The bedding parallels the hill-slope, except in the lower half of the sequence in the centre of the pit where it was disturbed by a fossil ice-wedge and shows other contortions (Watson, 1965a, p.48; 1965b, p.20).

Quite large screes occur at Bwlch-Hyddgen [771 934] on the western side of Hengwm and on either side of Pistyll y Llyn [754 944]. They are composed of cleavage fragments of mudstone and thin sandstone laminae still being fed by frost-shatter, insolation and other weathering processes from the crags above. Large parts are unvegetated and at Bwlch-Hyddgen, where conifer trees have been planted on it, it is clear the scree is still in motion for the tree spacings close downwards.

Block scree has been mapped in several localities. Examples are on the western slope of Plynlimon [788 866 – 789 871] and to the west of Banc Lletty-Evan-hen [714 851 – 7145 8545]. In both places it consists of unsorted angular blocks of sandstone, derived from west-facing crags of massive sandstone within the Pencerrigtewion Member. The greater part of the scree was probably laid down under periglacial conditions after the last glaciation, although present-day frost action is still contributing a small amount of debris.

GLACIAL SAND AND GRAVEL

Included under this heading are several small deposits of sand and gravel of uncertain origin. The most southerly of these forms a low mound [6575 8800] south of Argoed Fawr, and there is another smaller area 250 m to the east. A degraded pit [6599 8790] in this latter area shows 1.2 m of poorly sorted pebble gravel with a coarse sand matrix and interbedded seams of coarse sand.

Bod-yr-yrfa [7233 9874] stands on a small patch of Glacial Sand and Gravel. Fairly well sorted and fine-grained rounded gravel forms an elongate spread at the exit of a glacial spillway from the SW.

SUBMERGED FOREST

The well-known Borth submerged forest, of Holocene age, is exposed along a 200 m-wide strip, centred on the low water mark of medium tides, which extends for about 4.5 km between Upper

MARINE BEACH DEPOSITS AND TIDAL FLATS 127

Borth [606 889] and Twyni Mawr [603 933]. Exposures are very variable, depending on the amount of sand on the beach; at the time of survey (1965) they were limited to a small area at Upper Borth and a discontinuous 2 km strip extending southwards from Twyni Mawr. At that time the exposed sequence comprised about 0.3 m of bluish grey silty clay with roots overlain by up to 0.3 m of peat and peaty clay with plant debris and boles of trees (mainly *Pinus*).

The sequence through the forest bed and underlying intertidal clay has been studied by many workers, mainly from borehole evidence. The borings include a north–south series across Borth Bog (Godwin, 1943, p.223), another along Borth Sands (Adams and Haynes, 1965), two behind the storm beach at Borth (Haynes and Dobson, 1969), and a N–S line east of Ynyslas (Wilks, 1979). The latter paper summarises the borehole results and other work on the submerged forest. The boreholes have proved that the submerged forest and the forest bed under the bog are in continuity, and hence that the coastal barrier appears to have moved eastwards across the forest bed (p.117). They also show that the underlying interbedded clay has a remarkably flat surface at about 0.6 m below OD. A line of five boreholes [6580 9507–6583 9510] across the mouth of Afon Ddu, drilled by the Building Research Station in 1945, showed peat overlying soft grey clay on silty fine sand and silty clay (Appendix 1). The level of the upper surface of the clay is recorded as approximately 0.92 to 1.14 m below OD. Fragments of birchwood at the base of the peat in one of the holes [6580 9507] suggest that the forest bed may extend well up the Dovey Estuary.

The submerged forest at Clarach Bay [5866 8379] is of a similar age and height to that at Borth. However, a recent study of the underlying clay (Wilks 1977; 1979, p.33) has shown that this is terrestrial and not intertidal; thus the forest growth was not directly due to regression of the sea at Borth.

Small exposures of submerged forest [577 797] were seen north of Allt-wen and near the northern margin of the district [585 985]. Their extent and stratigraphical position is not known, but they appear similar to those at Borth and Clarach.

STORM GRAVEL BEACH DEPOSITS

The evolution of the storm beaches at Aberystwyth, between Allt-wen [577 796] and Constitution Hill [583 825], has been described in detail (So, 1974; Wood, 1978, 1980). Access of new material to the beaches by longshore drift from the south has almost ceased and the beaches are now undergoing erosion by attrition, compounded by the building of the Aberystwyth promenades along the tops of the beaches north of Aberystwyth harbour which has prevented their gradual landward migration. The most southerly beach [577 796–578 808] was broken by the mouth of the Ystwyth until that river was diverted into the harbour in the 18th Century. The northern beach [582 818–583 825] extends across a previous mouth of the Rheidol (p.129). Small storm beaches are also present across the valley mouths at Clarach Bay and Wallog.

The best developed storm beach extends northwards from Upper Borth to Twyni Bâch. It is composed of flattened rounded pebbles and cobbles of local greywacke with a small admixture of other rocks. In its southern part it is up to 6 m high and 80 m wide, and the village of Borth is largely built on it. North of the village, on the golf links, the gravels extend eastwards under a thin (less than 0.9 m) cover of Blown Sand approximately to the old course of Afon Leri. In this area, and especially from about 400 to 500 m NW of Aberlerry Farm, the gravel forms low ridges curving inland towards the NE, probably associated with varying positions of the mouth of Afon Leri before its canalisation (p.117). Northwards the beach gradually decreases in height and width, ending about 1.3 km north of Ynyslas. A gravel bank [613 941] on the eastern side of the sand spit may be a remnant of an earlier position of the beach, and gravel

is also reported [6113 9336] under 0.9 m of sand and 1.8 m of clay or silt about 0.7 km to the SSW.

North of the Dovey a narrow storm beach extends from the river mouth to beyond the northern margin of the district. The beach is about 20 m wide and 2.5 m high, but the gravels extend eastward for an unknown distance under Blown Sand. Human interference has reinforced this beach in places where it had previously been breached.

MARINE AND ESTUARINE ALLUVIUM

The marine and estuarine alluvium comprises deposits laid down between the mean high water level of spring tides and the highest level of spring tides, corresponding to the high marsh environment (Haynes and Dobson, 1969, p.243). Their greatest extent is on the southern side of the Dovey estuary from Twyni Bâch eastwards to Ynys-Edwin [678 963] and thence on both sides of the river to the northern margin of the district. On the southern side of the estuary only those areas of alluvium which lie north of the Cambrian Railway or within the artificial levées of Afon Leri and other northward flowing streams are now subject to tidal flooding and active deposition of sediment.

The deposits are mainly silts and clayey silts with a variable proportion of fine sand. Silt and clay is present under the blown sand east of Twyni Mawr, and occupies a broad area east of Afon Leri to Glanmorfa [638 932] and along the old meandering course of Afon Clettwr to Craig-y-Penrhyn [654 928]. Peaty clay and clayey peat down to 1 m above OD has been recorded in boreholes (Wilks, 1979) [621 928–621 936] east of the mouth of the Leri, and silt or silty clay is seen in many ditch sections with a maximum thickness of 2.4 m [6339 9350] NW of Glanmorfa. Southwards the alluvium thins to a feather edge on the peat of Cors Fochno. Similar deposits are present along the old course of the Afon Ddu WNW of Hen-Hafod [662 941], north of Ynys Greigiog [673 948], adjoining Afon Einion north of Furnace [685 952], and on both sides of the Dovey to the north of Glandyfi [696 970]. Estuarine alluvium is also present behind the blown sand area west of Rhownia [599 982].

MARINE BEACH DEPOSITS AND TIDAL FLATS

These deposits comprise the gravels, sands, silts and muds of the present day beaches and tidal flats below the level of the Storm Gravel Beach Deposits. Their greatest extent is in the Dovey estuary where they include the open sand flats and low marsh (Haynes and Dobson, 1969). The unvegetated open sand flats, which consists of fine- to medium-grained sand, are covered at all high tides. They are traversed by the shifting channels of the River Dovey and its tributaries whose positions in the last 80 years has been documented (Haynes and Dobson, 1969, fig.12). The lower limit of the low marsh is taken at the approximate lower limit of *Spartina townsendii* s.l., and the upper limit is usually marked by a low 'cliff' (up to 0.6 m high) approximately at the mean high water level of spring tides. Since 1916 (Yapp and others, 1916, 1917), *S.townsendii* has colonised most of the low marsh area and in many places has invaded areas which were then open sand flats. The sediments of the low marsh range from clayey silts to fine sand.

At Aberystwyth and Clarach Bay the beach material is mainly pebbly gravel with some sand at low tide. For 500 m south of Wallog [592 857] the foreshore comprises boulders, probably of glacial origin, with a thin veneer of more recent sand and gravel. Sarn Cynfelyn, which extends westwards from Wallog, is a boulder/cobble gravel bar some 1.2 to 1.8 m above low tide level; it is composed of local rocks and also exotic material including gneiss and greenschist. It has been interpreted (p.118) as marking the limit of a piedmont glacier which flowed down the Dovey valley into the Irish Sea during the Devensian. The beaches to the north and south of the Dovey, from Upper Borth [607 889] to the northern margin of the district, are composed almost entirely of fine- to medium-grained sand.

RIVER TERRACES

The upper part of the Myherin valley has a low gravel terrace about 1 m above the alluvium and a remnant of a probable terrace 2 to 3.5 m higher. A section [7920 7948] in the latter shows about 1.5 m of clayey pebble to cobble gravel. To the north, in the Castell valley, a terrace about 1 m above the alluvium extends for about 1 km upstream from the Dyffryn Castell Hotel [774 816].

Numerous small boulder gravel terraces, rarely exceeding 40 m in width, are present in the Rheidol gorge between Ponterwyd and the Dinas Dam, and upstream from the northern end of the Dinas Reservoir to the confluence with the Peithnant [7535 8440]. They range in height from 1.5 to 4.5 m above the river. Flights of river terraces occur below the Morainic Drift terraces in the valley east of Pwllcenawon to Capel Bangor. At least three levels occur at various heights above the modern flood plain level. The lowest terrace passes beneath modern flood plain deposits on either side of the valley at Pwllcenawon.

Small terraces are present in places along Afon Melindwr and Afon Leri, and there is a distinct terrace on Afon Ceulan above Talybont at about 3 m above the alluvium. About 600 m east of Penpompren Hall, 2.4 m of roughly bedded coarse gravel are exposed [6670 8989].

Further very small terraces, at two levels, follow the course of the headwaters of the Llyfnant below Pistyll y Llyn [753 942]. They terminate downstream at a rock ledge which crosses the bed of the river at Cwm-rhaiadr-fâch [7625 9597].

ALLUVIAL CONES OR FANS

The largest fan is at Tre'r-ddôl [660 922] where Afon Clettwr emerges from the hills on to the low ground of Cors Fochno (Borth Bog). The fan has a gentle convex form and is about 0.85 km wide and 0.7 km across. Roughly bedded boulder gravel is seen in the banks of the Clettwr [6583 9227].

Several fans are present on either side of Afon Castell upstream from the Castell mine [773 813]. The largest is around Llysarthur [7865 8250] with a section [7860 8260] near its margin showing 2.7 m of gravelly silt interbedded with fine to coarse gravel. On the southern side of the Castell valley a section [7755 8135] in Nant Meirch showed 3.7 m of roughly bedded imbricated poorly sorted coarse gravel.

There are several well developed fans in the upper reaches of the Myherin valley, with 4.5 m of roughly bedded poorly sorted boulder gravel seen at the confluence [7864 7940] of Nant y Creiau and Afon Myherin.

ALLUVIUM

Only the westernmost kilometre of the alluvial plain of Afon Ystwyth falls within this district. The river originally flowed out to sea through a gap in the storm beach, but it was diverted north-wards into Aberystwyth Harbour in the late 18th century (Wood, 1978, p.63). Up to 1.8 m of tabular-bedded sand and gravel is exposed in the river banks [590 793] SW of Crugiau.

The broad alluvial plain of Afon Rheidol is up to 0.8 km wide; it extends from near Capel Bangor westwards to Aberystwyth and the development of meanders has been studied (Lewin, 1978).

Boreholes (Appendix 1) in the alluvium show that the valley was overdeepened during the glacial period. Several [c.638 808] north of Pwllcenawon proved a variable succession, with sand and gravel for 9 to 12 m on gravel with some clay to a depth of 30 m. The lowest river terrace passes beneath the alluvium just to the east; this suggests that the upper 9 to 12 m of gravel include both alluvium and terrace resting on Morainic Drift or Boulder Clay.

Three groups of boreholes in the alluvium, respectively about 400 m south [c.5945 8115] of the National Library of Wales, some 600 m east [589 813] of Trefechan Bridge, and in a car park [585 815–586 813] adjacent to Aberystwyth railway station, all show sequences of clay and silt (maximum thickness 9.4 m) including beds of peat and organic matter up to 3.8 m thick overlying sand and gravel with some clay beds. These sands and gravels rest on Boulder Clay (proved to 22.4 m) at the first location, solid mudstone (at 11.9 m) at the second, and were not penetrated (at 22.3 m) at the third. The Rheidol originally flowed out to sea through the northern part of the town to the NNW of the railway station, and the silts, clays and peats in the upper parts of the boreholes were probably laid down along this earlier course after it was abandoned in favour of the present one through Trefechan. A temporary section [5840 8194], showed 0.9 m of flat-bedded coarse yellow sand with granule laminae and beds of gastropods.

The broad alluvial flat of Afon Clarach extends westwards from Plas Gogerddan [630 836] to Llangorwen [603 878]. The alluvium narrows abruptly to 30 m at Llangorwen church where the river passes through a ridge of Morainic Drift, then widens locally to 250 m between Llangorwen and the sea. Sections [6238 8379 – 6216 8377] in the banks of the river west of Gogerddan show 1.2 to 1.5 m of pebble and cobble gravel, and 1.8 m of silt, sand and gravel is seen [5958 8381 – 5913 8380] west of Llangorwen church. Boreholes through the alluvium immediately behind the storm beach at Clarach Bay (Wilks, 1977) show a sequence of clays and peats, mainly of freshwater marsh environments, to a depth of about 5.5 m (c.2.5 m below OD), with radiocarbon dates ranging from 11 220 ± 1600 BP to 1140 ± 80 BP.

An almost continuous stretch of floodplain alluvium extends downstream along Afon Leri from north [634 880] of Glanfraid to near Aberlerry Farm. Upstream it is present only in a few narrow and discontinuous strips. From Glanfraid to Glan-Lerry [617 886] it comprises up to 1.5 m of sand and silt with beds and pockets of gravel, especially towards the base. North-west of Glan-Lerry it widens from about 150 m to about 1 km, before narrowing again around the western side of the solid 'island' of Ynys-fergi. In this broad alluvial belt there is up to 1.2 m of silt exposed in places, with a peaty layer in the lower part and some lenses of gravel. Locally pine tree remains are present at the base of the silt which rests on peat [6103 8890] or on sand and silt [6149 8889]; samples of wood at the latter locality gave radiocarbon dates of 1097 ± 45 and 1215 ± 50 BP (Harkness and Wilson, 1979, p.255). North of Borth a narrow strip of clay, partly obscured by recent Blown Sand, marks the old (pre-1824) course of the Leri behind the storm beach; it disappears under blown sand about 400 m NW of Aberlerry Farm.

BLOWN SAND

There are two main areas of Blown Sand, to the north and south of the mouth of the River Dovey.

South of the Dovey there is a narrow strip of Blown Sand behind the storm beach from Borth [6085 8959] to Aberlerry Farm [609 910], widening northwards to about 1 km north of Foel-ynys [607 929] and terminating in the dunes of Twyni Bâch. Medium- to coarse-grained sand which emerges from beneath the old Afon Leri alluvium [612 899] just north of St. Matthew's Church, Borth, may be a remnant of an earlier deposit of Blown Sand.

Between Borth and the road junction [608 924] south of Foel-ynys, the Blown Sand is bisected by the alluvium of the old channel of the Leri. West of this channel there is a thin cover of sand, mainly on storm beach gravel, while to the east the sand thins out to a feather-edge on the peat of Cors Fochno. North of Ynyslas Turn [608 924] an active dune system is present behind the storm beach. Northwards to about 500 m north of Foel-ynys, the dune system [606 934] is up to 4.5 m high and 100 m wide; farther north it widens to the whole width of the Blown Sand outcrop, and the highest dunes behind the beach reach a height of 10.5 m. Between the dunes and Afon Leri, NE of Foel-ynys, there is a flat area of Blown Sand, originating partly from sand blown over the top of the dunes and partly from that blown down the estuary. A pit [6113 9336] in this area showed 0.9 m of sand on 1.8 m of clay and silt on coarse gravel.

Evidence from old maps indicates that the dune system has grown considerably since about 1900, particularly between and to the west of Twyni Mawr and Twyni Bâch and to the north of Twyni Bâch.

The coast from Aberdyfi north-westwards to the northern margin of the district is fringed by sand dunes. They occupy a belt c.120 m wide in the south and 40–70 m wide in the north. In the south they attain a height of 12 m. A flat tract behind the sand hills is also made largely from Blown Sand and there are lines of low, relict older sandhills [595 970 to 590 980].

MADE GROUND

There are many, mainly small, tips from old mineral workings and quarries. Most of the material in the few larger and more accessible tips, as at Old Goginan [689 816] and Cwmerfyn [696 829], has been removed for use as hard-core or for road-making.

LANDSLIP

There are no major landslips within the district, but small scale slips within drift deposits are not uncommon. Several small landslips have occurred in Morainic Drift from 1 to 1.7 km downstream from Cwmere [683 882] where this has been deeply incised by the Leri, and minor slips are common along the steep fronts of Boulder Clay 'terraces' (p.112), especially where these are affected by active fluvial erosion.

Foundering of Morainic Drift (gravel) on a bed of glacial-lake clay has occurred in the valley around Hengwm-cyfeiliog [781 948], and Boulder Clay on the left bank of Nant Ceniarth has slipped and soliflucted towards the river in several places [772 968; 773 974; 771 977].

Similar remobilised Boulder Clay is present on the western bank of Afon Dulas between Coed Pant-y-bedw [794 973] and Dolcaradog (p.116).

CHAPTER 9

Mineral products

METALLIFEROUS MINERALS

The Aberystwyth district includes the north-western part of the Central Wales Mining Field. There is a long history of mining, dating from Roman, and possibly pre-Roman times (Lewis, 1967, pp.22–25). Production was at its peak from 1850 to 1870, after which it fell rapidly due to near exhaustion of the known reserves of many of the mines compounded by a fall in prices due to the importation of foreign ore. A few mines remained open into the present century, but all are now closed (Ball and Nutt, 1976). The total output for the Central Wales Mining Field (Dunham, 1943–44) was 486 296 tonnes of lead concentrates (1845–1938), and 153 193 tonnes of zinc concentrates (1854–1938); a total of 15 153 tonnes of copper ore has been recorded for the period 1845–1969 (Foster-Smith, 1978, 1979). These figures are not a true reflection of the richness of the mining field as much lead and copper ore was won before 1845, and large quantities of zinc ore (sphalerite) were left underground or dumped before that mineral became of economic importance in the mid-nineteenth century. The lead ore (galena) also contains a proportion of silver ranging from about 80 to 1250 ppm of lead metal. The richest mines for silver in the district were between Goginan and Daren, and some of the early workings were primarily for silver rather than lead. The most comprehensive account of the mining field was given by Jones (1922) at a time when some mines were still working and others had but recently closed. Brief descriptions of the mines including many of the smaller and less successful trials have been published more recently (Foster-Smith, 1978, 1979).

The main primary ore minerals are galena (lead sulphide, with a variable silver content), sphalerite (zinc sulphide) and, to a lesser extent, chalcopyrite (copper/iron sulphide). Two types of iron sulphide, pyrite and, less commonly, marcasite, also occur in the ore bodies but neither is workable as an ore. Pyrite is widely distributed throughout the country rocks of the district. A sample of pyrite (MR 34023) from a quarry [7609 9976] near Forge contained 0.13 ppm of gold and 30 ppm of silver. The most abundant gangue mineral is clear or bluish white quartz; ankerite is locally the chief gangue mineral, and in a few places calcite or dolomite is present. Quartz was precipitated throughout the ore-forming process and the generalised paragenetic sequence of the other minerals is normally pyrite-ankerite-chalcopyrite-sphalerite-arsenopyrite-galena (Raybould, 1974, fig.11). At certain mines, notably at the western end of the Castell and Camdwr lodes, the paragenetic positions of galena and sphalerite are reversed. Secondary minerals are present only locally and in small quantity, mainly near the surface (Jones, 1922, p.180); the commonest are hydrozincite, gypsum and limonite. There appears to have been no secondary enrichment of the ores.

Manganese minerals have been noted from a number of localities. On the Camdwr lode manganese ores associated with pinkish and red-stained shales have been recorded (Jones, 1922, pp.27,144) between Llawrcwmbach [708 854] and Llettyhen [694 848] mines, and low grade psilomelane (p.26) from the eastern side of Drosgol. Pink-stained shales have also been recorded (p.40) on the Dyfngwm-Dylife lode near Carn Gwilym [792 908], and to the SW near Hyddgen. The occurrence of manganese ores on these two lodes was given (Jones, 1922, p.41) as one reason for suggesting that the lodes are in continuity. However, the present survey indicates that the two lodes are not connected; the Dyfngwm-Dylife lode persists south-westwards through Hyddgen mine [783 907] before dying out, and the Camdwr lode swings to W–E through Banc Lluestnewydd [792 896] to terminate about a mile farther east. Samples (Lab. Nos. 4880 A–F.) Bain and others, 1968) from these two lodes were analysed by the Mineralogy and Petrology Research Group of BGS to elucidate the association of red-stained country rocks with the presence of manganese minerals in the lodes. Secondary manganese oxides were recorded from two localities [7770 9016; 7641 8833], including lithiophorite at the second locality. Red colouration of sandstone at this second locality, and of a nodule [7627 8817], is due to the presence of iron oxides, principally goethite. It was concluded that the manganese oxides were a late addition to the lodes, and may be due to percolation of subaerial water using the same channels as the primary sulphide mineralisation but genetically unconnected with it. The red-stained rocks appear to have been produced by the action of sulphuric acid (from weathered sulphides) on ferromagnesian minerals in the country rock. The association of red-stained rocks with manganese minerals thus appears to be largely fortuitous.

The ore minerals occur predominantly along ENE-trending normal faults and breccia zones. There are a few short NW-trending lodes in the northern part of the district. There are also some important easterly or ENE-trending faults (e.g. the Ystwyth Fault) which are not mineralised and which in some places cut the lodes.

The relationship between the mineralisation and the stratigraphy and structure of the adjacent country rocks has been briefly described (Jones, 1922, pp.181–182). Four major lithostratigraphical divisions were recognised; the Van Formation (Ordovician), overlain by the Gwestyn (Cwmere Formation), Frongoch (Derwenlas to Borth Mudstones formations inclusive) and Cwmystwyth (Aberystwyth Grits Formation and Cwmystwyth Grits) formations. Most of the major mines were in the Frongoch Formation, and very few of the ore bodies in that formation appear to extend downwards into the pyritous mudstones of the Gwestyn Formation. Mines in that formation were few and of low output, although it should be noted that the proportion of copper ore at intersect with the Gwestyn is considerably higher than at intersect with the beds above or below. There were few significant mines in the Van Formation in the Aberystwyth district. Westwards, the lodes cease to be productive beyond a line running approximately SSE from Aberdyfi. It is uncertain if this results from stratigraphical control, though the present survey has shown that in this district the upward limit of mineralisation approximates to the boundary between the Devil's Bridge and Borth Mudstones formations. It has also been pointed out (Jones, 1922, p.182) that the ore bodies are commonly related to anticlinal axes in the country rock, and that individual ore-shoots within them plunge steeply to the west; this structural relationship has been amplified by Hughes (1969, p.6). Enrichment of lodes also occurred in some places where they bend or branch (Jones, 1922, p.13).

The age of the mineralisation has long been the subject of some controversy. Clearly it post-dates the Cwmystwyth 'formation', the youngest rocks in which it is known to be present, but there is no evidence to provide an absolute upper stratigraphical age limit. Finlayson (1910, p.287) considered that the transverse ore-bearing faults were later than the Caledonian folding and strike-faulting, and that the mineralisation was coincident with the formation of these transverse faults; he related it to a major widespread ore-forming episode during the Hercynian orogeny. Other authors have also argued for a post-Carboniferous date for the mineralisation (Archer, 1959; and Hughes, 1959). Jones (1922, pp.177–178), on the other hand, considered that the transverse faults were formed in association with the folding, though at a late state; since the ore-bearing faults are cut by some major unmineralised faults (p.130) unlikely to be later than Hercynian, he concluded that the mineralisation was post end-Silurian to pre end-Carboniferous. The mechanism of hydraulic fracturing has been put forward (Phillips, 1972) to account for the development of breccia zones as extensions to normal faults. Both he and Raybould (1974, 1976) consider that the fluids causing this brecciation also carried the ore minerals, and Raybould (1974, p.18) tentatively ascribes this brecciation to the latter stages of the Caledonian fold cycle.

Attempts at direct age dating of the mineralisation by isotopic methods has led to conflicting results. An average lead isotope model age of 430 ± 40 Ma (early Silurian) has been determined (Moorbath 1962, p.325) for samples from the Harlech Dome and Plynlimon areas, but later work (Ineson and Mitchell, 1975, p.14) on K-Ar isotopic ages produced a mean age of 356 ± 7 Ma (late Devonian or early Carboniferous) for samples from the Plynlimon area. The age relationship between the mineral lodes and other, irregular, bodies of sulphide-barren quartz veining in the country rock is also significant. The barren vein-quartz is commonly associated with strike-faults and is of a different type from that associated with the ore minerals. Three samples of mudstone within these quartz veins [6015 9630 (2 specimens); 5971 9703] produced a mean age of 384 ± 9 Ma, some 30 Ma older than the mean age for the sulphide mineralisation (Ineson and Mitchell, 1975, pp.13–14). This supports evidence that the transverse faults displace faults and folds with which the vein quartz is associated.

The origin of the mineralising fluids is also uncertain. It was first suggested (Finlayson, 1910, p.228; Jones, 1922, pp.178–179) that the ores were primarily derived from magmatic sources, though it was mentioned that no connection could be made between the lodes and any igneous rocks.

More recently, Raybould (1974, p.118), in a study of the paragenesis of the veins of mid-Wales, concluded that the hydrothermal solutions originated as interstitial brines which leached the metals from the country rock and transported them as bisulphide complexes. The ores were then deposited during periods of pressure release as the faults were extended by hydraulic fracture (Phillips, 1972, pp.352–353).

No new lodes or ore bodies have been discovered during the present survey. The Camdwr Fault has been mapped westwards to a point [629 813] south of Lovesgrove, but its extension westward from Bronfloyd Mine [660 834] does not appear to be mineralised. In the Clettwr valley, 1.5 km east of Tre'r-ddôl, there is a small mine (Clettwr) not mentioned by other authors. It was being worked in 1940, and at that time a lower adit [6765 9173] had been driven NNE from the stream to intersect two narrow E–W trending lodes approximately 2 m apart about 50 m from the portal. The lodes, which carried galena, sphalerite and chalcopyrite in a gangue of quartz and calcite had a southerly dip of about 70° and had been followed for some 50 m to the east. An upper adit 20 m upstream had intersected a third parallel quartz-galena lode about 10 m from the portal with a northward dip of about 70°. The mine is situated in the topmost beds of the Bryn-glâs Formation.

The prospects for future development of the mining field have been discussed (Jones, 1922; Hughes, 1959; Ball and Nutt, 1976); all agree that no major lodes remain to be discovered in the district, and this has been confirmed by the present survey. It is evident that all the ore bodies of any consequence which crop out at the surface have been discovered and worked; consequently any future prospects must lie in ore bodies which have not been uncovered by erosion or which are concealed beneath drift or peat. Concealed ore bodies might be present in at least three locations. Firstly, they may occur at relatively shallow depths along the known lodes between the old mines. Although there are numerous trial shafts and trenches along most of the lodes between the known ore bodies, very few seem to have been more than a few metres in depth. Additionally there was very little underground exploration beyond the confines of each individual ore body. Such shallow occurrences could possibly be located by a combination of geophysical, geochemical and structural methods. Secondly, there is the possibility that ore bodies exist below old mine workings. Most mines were abandoned as soon as the ore body being worked died out; in particular, no mines in the Frongoch Formation (Jones, 1922) ever penetrated through the underlying Gwestyn Formation into the possibly productive Van Formation below. Such deeper ore bodies would be difficult to locate and expensive to work. Lastly (Ball and Nutt, 1976), the productive 'flats' in the sandstones of the Van mine near Llanidloes were not discovered until mining had almost ceased in the area, and consequently there was no exploration for similar occurrences elsewhere. These deposits did not occur along a lode but in vertical fractures in gently dipping massive Ordovician sandstones. It must be emphasised that apart from the Van 'flats' all the workable minerals so far discovered in the mining field occur in vein deposits of high grade, but limited extent and tonnage. There is no evidence of any extensive low grade disseminated mineralisation which could be exploited by large scale opencast or underground mining.

BUILDING STONE AND SLATE

There are now no working quarries within the district, but in the past almost all sufficiently hard and resistant rock types have been used locally for building and road-making. Small quantities of poor quality roofing slates were obtained from various mudstone horizons, but most slate was quarried in the form of slabs.

Mudstones near the top of the Nant-y-Môch Formation (the Narrow Vein of Pugh (1923)) were quarried in Cwm Einion [6993 9440] for slate slabs (p.19).

Sandstones of the Pencerrigtewion Member were worked at the Hafan Quarry, Carn Owen [732 881] for paving setts and other purposes during the last century. The only record of output is of 1690 tons 4 cwt (1717.3 tonnes) for the year 25 June 1897 – 25 June 1898 (Wade, 1976, p.16). A quarry nearby was opened to provide constructional material for the Rheidol Hydroelectric Project between 1959 and 1962 (Anderson, 1963, p.39). These sandstones have also been quarried by the roadside in Cwm Ceulan [688 901].

The Mottled Mudstone Member has been worked for slate slabs and rough building stone in several localities. The largest quarry was that of the Cardiganshire Slate Company [698 959], about 500 m south of Cymerau, which provided cleavage slabs. The high pyrite content of the paler beds within the rock prevented the production of sawn slabs (Jones and Pugh 1916, p.352). There are two more quarries [691 945; 6915 9463] at this horizon at Tyn y Garth. Farther south there are several smaller quarries along the outcrop for some 500 m SW of the Ogof Fault [687 937]. The Mottled Mudstone Member was also worked at a number of places around the Plynlimon inlier, notably SW [745 838] of Aber-Peithnant and just north [794 828] of Cwmergyr.

Mudstones in the lower part of the Derwenlas Formation have been quarried for local buildings just south [684 881] of Cwmere where the cleavage and bedding are coincident, and also east [656 911] of Tan-yr-allt. The arenaceous beds in the upper part of the formation have also been utilised as, for example, on the northern sides of the Cyneiniog valley and the Leri valley west of Werndeg; also south [757 800] of Parcgwyn.

There are many small quarries in the basal arenaceous beds of the Devil's Bridge Formation, the largest being west [7405 8090] of Ponterwyd (p.92) where these beds total about 15 m with individual sandstone beds up to 30 cm thick, and also at the western end of Cwm Ceulan [6663 8976] (p.88), where the upper part of the Cwmsymlog formation was also worked. Thin sandstones in the main part of the formation were locally used for poor quality slabs as on the ridge [714 328] south of Llyn Blaenmelindwr.

The Borth Mudstones are softer and more easily weathered than most of the other rocks in the district, and consequently have been used for buildings only in areas where other stone was not close at hand, as at Ynyslas, where there are several small quarries, at Llancynfelyn [644 922] and at Penhelig where stone was mined.

The sandstones of the Aberystwyth Grits have been widely used as a building stone especially around Aberystwyth itself. Most of the older buildings in the town are built from the Grits which were quarried at the northern edge of the town and east [588 815] of the railway station; notable among the other quarries were those at Allt-wen [577 795], west [611 795] of Moriah, and at Allt-glais [597 835]. Mudstones from these quarries were sometimes used for poor quality roofing slates, but most slate used in the town was brought in from elsewhere. More recently, quarry waste and some in-situ rock from Constitution Hill [583 826] has been used to replenish the beach and prevent storm damage at Victoria Terrace, Aberystwyth (So, 1974; Wood, 1978, 1980).

SAND AND GRAVEL

There are no major deposits of sand and gravel in the district. In the past, small pits have been opened in a variety of deposits for purely local use.

The most widespread gravel deposits are those of the Morainic Drift, but they have been little utilised, probably because of their unsorted and heterogeneous nature. The clay content of the matrix is very variable and clast size ranges from granules to boulders.

Storm beach gravels have been worked [600 925] north of Borth, and also north of the Dovey estuary. They are pebble to cobble gravels with predominantly greywacke clasts and little matrix.

Head gravel has been dug in a number of places, as for example south [757 806] of Parcgwyn (p.127) and at Ponterwyd [752 806]. As it consists entirely of angular mudstone fragments it can be used only as a poor quality fill or hard-core.

One of the main sources of hard-core has been mine-waste from the larger of the old tips. Those at Old Goginan have been entirely removed, together with much material from Cwmsymlog, Cwmerfyn and elsewhere. Much of this waste has been used for surfacing tracks, and the high lead and zinc content of some of it can cause local pollution problems.

PEAT

The peat of Cors Fochno (Borth Bog) was at one time sold locally for fuel and digging continued into the early part of this century (Lewis, 1969, p.59). Many of the other peat deposits in the district have been dug for small scale local use.

REFERENCES

ADAMS, T. D. 1961. Buried valleys of the Upper Rheidol (Cardiganshire). *Geol. Mag.*, Vol.98, 406–408.

— 1963. The geology of the Dinas Cwm-Rheidol hydroelectric tunnel. *Geol. Mag.*, Vol.100, 371–378.

— and HAYNES, J. 1965. Foraminifera in Holocene marsh cycles at Borth, Cardiganshire (Wales). *Palaeontology*, Vol.8, 27–38.

ANDERSON, J. G. C. 1963. The geology of the Rheidol hydroelectric project—Cardiganshire. *Proc. S. Wales Inst. Eng.*, Vol.78, 35–45.

ANKETELL, J. M. and LOVELL, J. P. B. 1976. Upper Llandoverian Grogal Sandstones and Aberystwyth Grits in the New Quay area, Central Wales: a possible upwards transition from contourites to turbidites. *Geol. J.*, Vol.11, 101–108.

ARCHER, A. A. 1959. The distribution of non-ferrous ores in the Lower Palaeozoic rocks of North Wales. 259–276 in *The future of non-ferrous mining in Great Britain and Ireland. A symposium.* 614pp. (London: Institute of Mining and Metallurgy.)

BAIN, J. A., McKISSOCK, G. M. and LIVINGSTONE, A. 1968. IGS Mineralogy Unit Rep. No.1. [Unpublished].

BAKER, S. J. 1981. The graptolite biostratigraphy of a Llandovery outlier near Llanystumdwy, Gwynedd, North Wales. *Geol. Mag.*, Vol.118, 355–365.

BALL, T. K. and NUTT, M. J. C. 1976. Preliminary mineral reconnaissance of Central Wales. *Rep. Inst. Geol. Sci.*, No.75/14. 12pp.

BATES, D. E. B. 1982. Aberystwyth Grits. 81–90 in *Geological excursions in Dyfed, south-west Wales.* BASSETT, M. G. (editor). 327pp. (Cardiff: National Museum of Wales.)

BICK, D. E. 1974–1978. *The old metal mines of Mid-Wales. Parts 1–5.* (Newent, Glos: The Pound House.)

BLUNDELL, D. J., GRIFFITHS, D. H. and KING, R. F. 1969. Geophysical investigations of buried river valleys around Cardigan Bay. *Geol. J.*, Vol.6, 161–180.

BOUMA, A. H. 1962. *Sedimentology of some flysch deposits—A graphic approach to facies interpretation.* 168pp. (Amsterdam/New York: Elsevier.)

BOWEN, D. Q. 1973. The Pleistocene history of Wales and the borderland. *Geol. J.*, Vol.8, 207–224.

— 1974. The Quaternary of Wales. 373–426 in *The Upper Palaeozoic and Post-Palaeozoic Rocks of Wales.* OWEN, T. R. (editor). 426pp. (Cardiff: University of Wales Press.)

BRIDGES, P. H. 1975. The transgression of a hard substrate shelf: the Llandovery (Lower Silurian) of the Welsh Borderland. *J. Sediment. Petrol.*, Vol.45, 79–94.

CAVE, R. 1967. Pl.1, fig.1 in INSTITUTE OF GEOLOGICAL SCIENCES. *Annual report for 1966.* 197pp. (London: HMSO.)

— 1971. P.13 and pl.1 in INSTITUTE OF GEOLOGICAL SCIENCES. *Annual report for 1970.* 199pp. (London: Institute of Geological Sciences.)

— 1974. P.20 and pl.1, fig.2 in INSTITUTE OF GEOLOGICAL SCIENCES. *Annual report for 1973.* 220pp (London: Institute of Geological Sciences.)

— 1975a. Glan-fred Borehole, Dol-y-bont. P.7 in INSTITUTE OF GEOLOGICAL SCIENCES. IGS Boreholes 1974. *Rep. Inst. Geol. Sci.*, No. 75/7, 26pp.

— 1975b. P.26 in INSTITUTE OF GEOLOGICAL SCIENCES. *Annual report for 1974.* 234pp. (London: Institute of Geological Sciences).

— 1976. P.24 in INSTITUTE OF GEOLOGICAL SCIENCES. *Annual report for 1975.* 239pp (London: Institute of Geological Sciences.)

— 1978a. A reconnaissance of the bases of the Aberystwyth Grits/Rhuddnant Grits. Internal Report, Institute of Geological Sciences. 5pp.

— 1978b. Report of field excusion, 6th March 1977. P.251 in *Conference report. Deformation of soft sediments.* FITCHES, W. R. and MALTMAN, A. J. *J. Geol. Soc. London,* Vol.135, 245–251.

— 1979. Sedimentary environments of the basinal Llandovery of mid-Wales. 517–526 in *The Caledonides of the British Isles—reviewed.* HARRIS, A. L., HOLLAND, C. H. and LEAKE, B. E. (editors). *Spec. Publ. Geol. Soc. London,* No.8. 768pp (Edinburgh: Scottish Academic Press.)

— and HAINS, B. A. 1967. Pp.65–66 in INSTITUTE OF GEOLOGICAL SCIENCES. *Annual report for 1966.* 197pp. (London: HMSO.)

— — 1968. P.79 in INSTITUTE OF GEOLOGICAL SCIENCES. *Annual report for 1967.* 198pp. (London: HMSO.)

CHALLINOR, J. 1928a. Curious rock-marks from Transbaikalia. *Geol. Mag.*, Vol.65, 241–244.

— 1928b. A shelly band in graptolitic shales. *Geol. Mag.*, Vol.65, 364–368.

— 1929. Further curious rock-marks from the Cardigan coast. *Geol. Mag.*, Vol.66, 354–356.

— 1933. The 'Incised Meanders' near Pont-erwyd, Cardiganshire. *Geol. Mag.*, Vol.70, 90–92.

— 1947. The physiography of the two localities Ponterwyd and Devil's Bridge. 139–140 in Report of the Centenary meeting. *Archaeologica Cambrensis*, Vol.99 [for 1946]. [Amended and printed privately as 'River capture or tributary take-over? The question arising at Devil's Bridge'. 1976.]

— 1949. The origin of certain rock structures near Aberystwyth. *Proc. Geol. Assoc.*, Vol.60, 48–53.

— 1969. A review of geological research in Cardiganshire, 1842–1967. *Welsh Geol. Q.*, Vol.4, 3–37.

— 1978. Curious markings on bedding surfaces. *Geol. Mag.*, Vol.115, 383.

— and WILLIAMS, K. E. 1926. On some curious marks on a rock surface. *Geol. Mag.*, Vol.63, 341–343.

CRIMES, T. P. and CROSSLEY, J. D. 1980. Inter-turbidite bottom current orientation from trace fossils with an example from the Silurian flysch of Wales. *J. Sediment. Petrol.*, Vol.50, 821–830.

CUMMINS, W. A. 1959. The Lower Ludlow Grits in Wales. *Liverpool Manchester Geol. J.*, Vol.2, 168–179.

DAVIES, K. A. 1933. The geology of the country between Abergwesyn (Breconshire) and Pumpsaint (Carmarthenshire). *Q. J. Geol. Soc. London*, Vol.89, 172–201.

DAVIES, W. 1980. On slaty cleavage and structural development in Lower Palaeozoic Wales. *University College of Wales, Aberystwyth, Department of Geology Publications*, No.8, 26pp.

— and CAVE, R. 1976. Folding and cleavage determined during sedimentation. *Sediment. Geol.*, Vol.15, 89–133.

DENAEYER, M. E. 1948. Les gisements de cone-in-cone de France et de Grande-Bretagne (deuxième partie). *Bull. Soc. Belg. Geol. Palaeontol. Hydrol.*, Vol.56, [for 1947], 382–412.

DEWEY, J. F. 1982. Plate tectonics and the evolution of the British Isles. *J. Geol. Soc. London*, Vol.139, 371–412.

DUNHAM, K. C. 1943–44. The production of galena and associated minerals in the northern Pennines; with comparative statistics for Great Britain. *Trans. Inst. Min. Metall. London*, Vol.53, 181–214.

ELLIS-GRUFFYDD, I. D. 1977. Late Devensian glaciation in the Upper Usk basin. *Cambria*, Vol.4, 46–55.

FINLAYSON, A. M. 1910. The metallogeny of the British Isles. *Q. J. Geol. Soc. London*, Vol.66, 281–298.

FITCHES, W. R. 1972. Polyphase deformation structures in the Welsh Caledonides near Aberystwyth. *Geol. Mag.*, Vol.109, 149–155.

FLEUTY, M. J. 1964. The description of folds. *Proc. Geol. Assoc.*, Vol.75, 461–492.

FOSTER-SMITH, J. R. 1978. The mines of Montgomery and Radnorshire. *Br. Min.*, No.10, 41pp.

— 1979. The mines of Cardiganshire. *Br. Min.*, No.12, 99pp.

GARRARD, R. A. and DOBSON, M. R. 1974. The nature and maximum extent of glacial sediments off the west coast of Wales. *Mar. Geol.*, Vol.16, 31–44.

GODWIN, H. 1943. Coastal peat beds of the British Isles and North Sea. *J. Ecol.* Vol.31, 199–247.

— and WILLIS, E. H. 1964. Cambridge University Natural Radiocarbon Measurements VI. *Radiocarbon*, Vol.6, 116–137.

GRAY, J. M. 1982. The last glaciers (Loch Lomond Advance) in Snowdonia, N. Wales. *Geol. J.*, Vol.17, 111–133.

GREIG, D. C., WRIGHT, J. E., HAINS, B. A. and MITCHELL, G. H. 1968. Geology of the country around Church Stretton, Craven Arms, Wenlock Edge and Brown Clee. *Mem. Geol. Surv. G.B.*, Sheet 166, 379pp.

GRIFFITHS, D. H. and GIBB, R. A. 1965. Bouger gravity anomalies in Wales. *Geol. J.*, Vol.4, 335–342.

HARKNESS, D. D. and WILSON, H. W. 1974. Scottish Universities Research and Reactor Centre, Radiocarbon measurements II. *Radiocarbon*, Vol.16, 238–251.

— — 1979. Scottish Universities Research and Reactor Centre, Radiocarbon measurements III. *Radiocarbon*, Vol.21, 203–256.

HAYNES, J. and DOBSON, M. R. 1969. Physiography, foraminifera and sedimentation in the Dovey estuary, Wales. *Geol. J.*, Vol.6, 217–256.

HENDRICKS, E. M. L. 1926. The Bala-Silurian succession in the Llangrannog district (South Cardiganshire). *Geol. Mag.*, Vol. 63, 121–139.

HEYWORTH, A. and KIDSON, D. 1982. Sea-level changes in southwest England and Wales. *Proc. Geol. Assoc.*, Vol.93, 91–111.

HEYWORTH, A., KIDSON, D. and WILKS, P. J. (in press). Late-glacial and Holocene sediments at Clarach Bay, near Aberystwyth. *J. Ecol.*

HUGHES, W. J. 1959. The non-ferrous mining possibilities of central Wales. 277–294 in *The future of non-ferrous mining in Great Britain and Ireland. A symposium.* 614pp. (London: Institution of Mining and Metallurgy.)

HUNT, R. 1848. Notices of the history of the lead mines of Cardiganshire. *Mem. Geol. Surv. G.B.*, Vol.2, part 2, 635–654.

INESON, P. R. and MITCHELL, J. G. 1975. K-Ar isotopic age determinations from some Welsh mineral localities. *Trans. Inst. Min. Metall. London*, Sect.B: *Appl. Earth Sci.*, Vol.84, 7–16; discussion, 120–121.

JAMES, D. M. D. 1971a. The Nant-y-môch Formation, Plynlimon inlier, west central Wales. *J. Geol. Soc. London*, Vol.127, 177–181.

— 1971b. Petrography of the Plynlimon Group, west central Wales. *Sediment. Geol.*, Vol.6, 255–270.

— 1972. Sedimentation across an intra-basinal slope: the Garnedd-wen Formation (Ashgillian), west central Wales. *Sediment. Geol.*, Vol.7, 291–307.

JEHU, R. M. 1926. The geology of the district around Towyn and Abergynolwyn (Merioneth). *Q. J. Geol. Soc. London*, Vol.82, 465–489.

JONES, O. T. 1909. The Hartfell-Valentian succession around Plynlimon and Pont Erwyd (North Cardiganshire). *Q. J. Geol. Soc. London*, Vol.65, 463–537.

— 1922. Lead and zinc. The mining district of north Cardiganshire and west Montgomeryshire. *Mem. Geol. Surv. Spec. Min. Res. G.B.*, Vol.20, 207pp.

— 1938. Anniversary Address [On the evolution of a geosyncline]. *Q. J. Geol. Soc. London*, Vol.94, lx–cx.

— 1965. The glacial and post-glacial history of the lower Teifi Valley. *Q. J. Geol. Soc. London*, Vol.121, 247–281.

— and PUGH, W. J. 1916. The geology of the district around Machynlleth and the Llyfnant Valley. *Q. J. Geol. Soc. London*, Vol.71 [for 1915], 343–385.

— — 1935a. The geology of the districts around Machynlleth and Aberystwyth. *Proc. Geol. Assoc.*, Vol.46, 247–300.

— — 1935b. Summer field meeting to the Aberystwyrth district. *Proc. Geol. Assoc.*, Vol.46, 413–428.

JONES, W. D. V. 1945. The Valentian succession around Llanidloes, Montgomeryshire. *Q. J. Geol. Soc. London*, Vol.100 [for 1944], 309–332.

KEEPING, W. 1878. Notes on the geology of the neighbourhood of Aberystwyth. *Geol. Mag.*, Vol.15, 532–547.

— 1881. The geology of central Wales. With an appendix on some new species of Cladophora by C. Lapworth. *Q. J. Geol. Soc. London*, Vol.37, 141–177.

KELLING, G. and WOOLLANDS, M. A. 1969. The stratigraphy and sedimentation of the Llandoverian rocks of the Rhayader district. 255–282 in *The Precambrian and Lower Palaeozoic rocks of Wales.* WOOD, A. (editor). 461pp. (Cardiff: University of Wales Press.)

KUENEN, PH. H. 1953. Graded bedding, with observations on Lower Palaeozoic rocks of Britain. *Verhandel. Koninkl. Ned. Akad. Wetenschap.*, *Afdel. Natuurk.*, Sect.1, Vol.20, 2–47.

LAPWORTH, C. 1878. The Moffat Series. *Q. J. Geol. Soc. London*, Vol.34, 240–346.

— 1879. On the tripartite classification of the Lower Palaeozoic rocks. *Geol. Mag.*, Vol.16, 1–15.

LEPPARD, R. K. 1978. Convolute laminations in the turbidites of the Aberystwyth Grits. 248 in *Conference report. Deformation of soft sediments.* FITCHES, W. R. and MALTMAN, A. J. *J. Geol. Soc. London*, Vol.135, 245–251.

LEWIN, J. 1978. Meander development and flood plain sedimentation: a case study from mid-Wales. *Geol. J.*, Vol.13, 25–36.

LEWIS, C. A. 1970. The upper Wye and Usk regions. 147–173 in *The glaciations of Wales and adjoining regions*. LEWIS, C. A. (editor). 378pp. (London: Longman.)

LEWIS, H. P. 1946. Bedding-faults and related minor structures in the Upper Valentian rocks near Aberystwyth. *Geol. Mag.*, Vol.83, 151–161.

LEWIS, W. J. 1967. *Lead mining in Wales*. 377pp. (Cardiff: University of Wales Press.)

— 1969. *Cardiganshire historical atlas*. 88pp. (Aberystwyth: Cymdeithas Lyfrau Ceredigion Gyf.)

LOVELL, J. P. B. 1970. The palaeogeographical significance of lateral variations in the ratio of sandstone to shale and other features of the Aberystwyth Grits. *Geol. Mag.*, Vol.107, 147–158.

MARR, J. E. 1883. *The classification of the Cambrian and Silurian rocks*. 147pp. (Cambridge: Deighton, Bell & Co.)

— and NICHOLSON, H. A. 1888. The Stockdale Shales. *Q. J. Geol. Soc. London*, Vol. 44, 654–732.

MITCHELL, G. F. 1960. The Pleistocene history of the Irish Sea. *Adv. Sci. London*, Vol.17, 313–325.

— 1972. The Pleistocene history of the Irish Sea: second approximation. *Sci. Proc. R. Dublin Soc. A*, Vol.4, 181–199.

MOORBATH, A. 1962. Lead isotope abundance studies on mineral occurrences in the British Isles and their geological significance. *Philos. Trans. R. Soc. London*, Ser. A. Vol.254, 295–360.

MURCHISON, R. I. 1839. *The Silurian System, founded on geological researches in the counties of Salop, Hereford, Radnor, Montgomery, Caermarthen, Brecon, Pembroke, Monmouth, Gloucester, Worcester and Stafford; with descriptions of the coalfields and overlying formations*. 768pp. (London: John Murray.)

MUTTI, E. 1977. Distinctive thin-bedded turbidite facies and related depositional environments in the Eocene Hecho Group (South-central Pyrenees, Spain). *Sedimentology*, Vol.24, 107–132.

— and RICCI LUCCHI, F. 1972. Le torbiditi dell Apennino settentrionale: introzione all anatisi di facies. *Mem. Soc. Geol. Italy*, Vol.11, 161–199. [English translation in *International Geol. Rev,*, Vol.20, 125–166]

NETTLE, J. T. 1964. Fabric analysis of a deformed vein. *Geol. Mag.*, Vol.101, 220–227.

NICHOLSON, R. 1966. The problem of origin, deformation and recrystallization of calcite-quartz bodies. *Geol. J.*, Vol.5, 117–126.

— 1978. Folding and pressure solution in a laminated calcite-quartz vein from the Silurian slates of the Llangollen region of N. Wales. *Geol. Mag.*, Vol.115, 47–54.

NORMARK, W. R., PIPER, D. J. W. and HESS, G. R. 1979. Distributary channels, sand lobes and mesotopography of Navy Submarine Fan, California Borderland, with applications to ancient fan sediments. *Sedimentology*, Vol.26, 749–774.

NOTHOLT, A. J. G. and HIGHLEY, D. E. 1979. *Raw Materials Research and Development Dossiers, 4. Phosphate*, 234pp. (Commission of the European Communities.)

PENN, I. E. and EVANS, C. D. R. 1976. The Middle Jurassic (mainly Bathonian) of Cardigan Bay and its palaeogeographical significance. *Rep. Inst. Geol. Sci.*, No. 76/6 8pp.

PERRIN, R. M. S. 1971. *The clay mineralogy of British sediments*. 247pp. (London: Mineralogical Society.)

PHILLIPS, W. J. 1972. Hydraulic fracturing and mineralisation. *J. Geol. Soc. London*, Vol.128, 337–359.

POTTS, A. S. 1971. Fossil cryonival features in central Wales. *Geografiska Annaler*, Vol.53A, 39–51.

POWELL, D. W. 1956. Gravity and magnetic anomalies in North Wales; with an appendix on the magnetic anomalies over the Lleyn Peninsula, by D. H. GRIFFITHS and R. F. KING. *Q. J. Geol. Soc. London*, Vol.111 [for 1955], 375–397.

PRICE, N. J. 1962. The tectonics of the Aberystwyth Grits. *Geol. Mag.*, Vol.99, 542–557.

PUGH, W. J. 1923. The geology of the district around Corris and Aberllefenni (Merionethshire). *Q. J. Geol. Soc. London*, Vol.79, 508–545.

— 1929. The geology of the district between Llanymawddwy and Llanuwchllyn (Merioneth). *Q. J. Geol. Soc. London*, Vol.85, 242–306.

RAMSAY, A. C. 1866. The geology of North Wales. *Mem. Geol. Surv. G.B.*, Vol.3.

— 1881. The geology of North Wales. *Mem. Geol. Surv. G.B.*, Vol.3 (2nd edition). 611pp.

RAYBOULD, J. G. 1974. Ore textures, paragenesis and zoning in the lead-zinc veins of mid-Wales. *Trans. Inst. Min. Metall. Sect. B: Appl. Earth Sci.*, Vol.83, 112–119; discussion *in* Vol.84 (1975), 67–69.

— 1976. The influence of pre-existing planes of weakness in rocks on the localisation of vein-type ore deposits. *Econ. Geol.*, Vol.71, 636–641.

RICH, J. L. 1950. Flow markings, groovings and intra-stratal crumplings as criteria for recognition of slope deposits, with illustrations from Silurian rocks of Wales. *Bull. Am. Assoc. Petrol. Geol.*, Vol.34, 717–741.

RICHARDSON, S. W. and OXBURGH, E. R. 1978. Heat flow, radiogenic heat production and crustal temperatures in England and Wales. *J. Geol. Soc. London*, Vol.135, 323–337.

RICKARDS, R. B. 1964. The graptolitic mudstone and associated facies in the Silurian strata of the Howgill Fells. *Geol. Mag.*, Vol.101, 435–451.

— 1976. The sequence of graptolite zones in the British Isles. *Geol. J.*, Vol.11, 153–188.

ROBERTS, R. O. 1929. The geology of the district around Abbey-cwmhir (Radnorshire). *Q. J. Geol. Soc. London*, Vol.85, 651–676.

RUDEFORTH, C. C. 1970. Soils of north Cardiganshire. *Mem. Soil Surv. G.B.*, Sheets 163 and 178. 153pp.

RUPKE, N. A. 1978. Deep clastic seas. 372–415 in *Sedimentary environments and facies*. READING, H. G. (editor). 557pp. (Oxford: Blackwell.)

SEDGWICK, A. 1852. On the classification and nomenclature of the Lower Palaeozoic rocks of England and Wales. *Q. J. Geol. Soc. London*, Vol.8, 136–168.

SMYTH, W. W. 1848. On the mining district of Cardiganshire and Montgomeryshire. *Mem. Geol. Surv. G.B.*, Vol.2, part 2, 655–684.

So, C. L. 1974 Some coast changes around Aberystwyth and Tanybwlch, Wales. *Trans. Inst. Br. Geogr.*, Vol.62, 143–153.

STRONG, G. E. 1979. An Upper Llandoverian turbidite sequence of the Bwlch-glas area, near Aberystwyth, Dyfed. *Geol. J.*, Vol.14, 99–106.

SUDBURY, M. 1958. Triangulate Monograptids from the *Monograptus gregarius* Zone (Lower Llandovery) of the Rheidol Gorge (Cardiganshire). *Philos. Trans. R. Soc. London*, Ser. B, Vol.241, 485–555.

SYNGE, F. M. 1963. A correlation between the drifts of south-east Ireland and those of west Wales. *Ir. Geogr.*, Vol.4 [for 1959–1963] 360–366.

TAYLOR, J. A. 1973. Chronometers and chronicles. *Prog. Geogr.*, Vol.5, 250–334.

— and TUCKER, R. B. 1970. The peat deposits of Wales: an inventory and interpretation. *Proc. 3rd Int. Peat Congress* (Quebec 1968), 163–173.

THOMAS, D. 1972. Diatomaceous deposits in Snowdonia. *Rep. Inst. Geol. Sci.*, No.72/5. 8pp.

WADE, E. A. 1976. *The Plynlimon and Hafan Tramway.* 64pp. (London: Gemini Publishing Co.)

WALKER, R. G. 1967. Turbidite sedimentary structures and their relationship to proximal and distal depositional environments. *J. Sediment. Petrol.*, Vol.37, 25–43.

WARREN, P. T., HARRISON, R. K., WILSON. H. E., SMITH, E. G. and NUTT, M. J. C. 1970. Tectonic ripples and associated minor structures in the Silurian rocks of Denbighshire, North Wales. *Geol. Mag.*, Vol.107, 51–60.

WATERS, R. A. 1974. P.20 *in* INSTITUTE OF GEOLOGICAL SCIENCES. *Annual report for 1973.* 220pp. (London: Institute of Geological Sciences.)

WATSON, E. 1965a. Periglacial structures in the Aberystwyth region of central Wales. *Proc. Geol. Assoc.*, Vol.76, 443–462.

— 1965b. Grèzes liteés ou éboulis ordonnés tardiglaciaires dans la région d'Aberystwyth, au centre du Pays de Galles. *Bull. Assoc. Geogr. Fr.*, Nos.338–339, 16–25.

— 1968. The periglacial landscape of the Aberystwyth region. 35–49 in *Geography at Aberystwyth: Essays written on the occasion of the Departmental Jubilee 1917–18 to 1967–68.* BOWEN, E. G., CARTER, H. and TAYLOR, J. A. (editors). 276pp. (Cardiff: University of Wales Press.)

— 1970. The Cardigan Bay area. 125–145 in *The glaciations of Wales and adjoining regions.* LEWIS, C. A. (editor). 378pp. (London: Longman.)

WEDD, C. A., SMITH, B. and WILLS, L. J. 1927. The geology of the country around Wrexham. Part 1. Lower Palaeozoic and Lower Carboniferous rocks. *Mem. Geol. Surv. G.B.*, Sheet 121, 179pp.

WELIN, E., ENGSTRAND, L. and VACZY, S. 1974. Institute of Geological Sciences, Radiocarbon dates V. *Radiocarbon*, Vol.16, 95–104.

WILKS, P. J. 1977. *Flandrian sea-level change in the Cardigan Bay area.* Unpublished PhD thesis, University of Wales.

— 1979. Mid-Holocene sea-level and sedimentation interactions in the Dovey estuary area, Wales. *Palaeogeography, Palaeoclimatology, Palaeoecology*, Vol.26, 17–36.

WOOD, A. 1959. The erosional history of the cliffs around Aberystwyth. *Liverpool Manchester Geol. J.*, Vol.2, 271–287.

— 1978. Coast erosion at Aberystwyth; the geological and human factors involved. *Geol. J.*, Vol.13, 61–72.

— 1980. Prevention of beach erosion at Aberystwyth: a success story. *Geol. J.*, Vol.15, 135–36.

— and SMITH, A. J. 1959. The sedimentation and sedimentary history of the Aberystwyth Grits (Upper Llandoverian). *Q. J. Geol. Soc. London*, Vol.114 [for 1958], 163–195.

WOODLAND, A. W. (editor). 1971. The Llanbedr (Mochras Farm) Borehole. *Rep. Inst. Geol. Sci.*, No.71/18. 115pp.

YAPP, R. H., JOHNS, D. and JONES, O. T. 1916. The salt marshes of the Dovey estuary. Part 1. Introductory. *J. Ecol.*, Vol.4, 27–42.

— — — 1917. The salt marshes of the Dovey estuary. Part II. The salt marshes. *J. Ecol.*, Vol.5, 65–103.

YATES, R. A. 1968. Surveying techniques in coastal geomorphology. 129–142 in *Geography at Aberystwyth: Essays written on the occasion of the Departmental Jubilee 1917–18 to 1967–68.* BOWEN, E. G., CARTER, H. and TAYLOR, J. A. (editors). 276pp. (Cardiff: University of Wales Press.)

ZIEGLER, A. M., COCKS, L. R. M. and McKERROW, W. S. 1968. The Llandovery transgression of the Welsh Borderland. *Palaeontology*, Vol.11, 736–782.

— McKERROW, W. S., BURNE, R. V. and BAKER, P. E. 1969. Correlation and environmental setting of the Skomer Volcanic Group, Pembrokeshire. *Proc. Geol. Assoc.*, Vol.80, 409–439.

APPENDIX 1

List of boreholes

This appendix lists, by six-inch maps, the main borehole records for the district. Copies of these records can be obtained from the Wales office of the Survey at a fixed tariff. Each entry in the list shows first the permanent record number and location of the borehole and then its stratigraphical range.

SN 58 SE/1–4   Aberystwyth Telephone Exchange
Site investigation boreholes proving Drift on Aberystwyth Grits Formation [588 814].

SN 58 SE/5–10   Aberystwyth Municipal Car Park
Site investigation boreholes in Drift [5848 8149–5864 8135].

SN 58 SE/11   Aberystwyth, North Road Flats
Site investigation borehole in Drift [5863 8195].

SN 58 SE/12–14   Aberystwyth, Crown Buildings
Site investigation borehole proving Drift on Aberystwyth Grits Formation [5869 8181].

SN 58 SE/20–21   Aberystwyth Telephone Exchange
Site investigation boreholes in Drift [588 814].

SN 58 SE/22–26   Aberystwyth Swimming Pool
Site investigation boreholes in Drift [595 812].

SN 58 SE/27–47   Aberystwyth, A44 Eastern Approach Road
Site investigation boreholes in Drift [586 816 – 598 806].

SN 67 NE/1–9   Cwm Rheidol Dam
Site investigation boreholes proving Drift and Devil's Bridge Formation [694 794].

SN 68 NW/1   Glanfraid, Llandre
Glan-fred (IGS) Borehole. Borth Mudstones, Devil's Bridge and Cwmsymlog formations. Full details published in Cave (1975a) [6305 8812].

SN 68 SW/1   6   Rheidol Valley, west of Capel Bangor
Boreholes in Drift [637 808].

SN 69 NE/1–18   Furnace Bridge, A487 improvement
Site investigation boreholes proving Drift and Devil's Bridge Formation [6819 9500–6882 9608].

SN 69 NE/25–29   Southern side of Dovey Estuary, Afon Ddu
Site investigation boreholes in Drift [658 951].

SN 69 NE/30–33   Pont Melin-y-garreg, A487 improvement
Site investigation boreholes proving Drift and Devil's Bridge Formation [689 962].

SN 69 SW/1–15   Borth and Dovey Estuary
Site investigation boreholes in Drift on the margins of Cors Fochno
[614 911–617 935, 647 942].

SN 69 SE/1–5   Furnace Bridge, A487 improvement
Site investigation boreholes proving Drift and Devil's Bridge Formation [6796 9461–6812 9491].

SN 77 NW/2   Cwm Rheidol Power Station
Site investigation boreholes proving Drift and Devil's Bridge Formation [708 793].

SN 77 NW/3   Bwa-drain, near Ystumtuen
Borehole in Devil's Bridge Formation [7146 7972].

SB 78 NE/1   Nant-y-môch Dam
Site investigation boreholes proving Drift and Drosgol Formation [784 862].

SN 78 SW/1   One kilometre E of Disgwylfa Fâch
Borehole in Bryn-glâs Formation [7453 8392].

SN 78 SW/2–9   Dinas Power Station
Site investigation boreholes proving Drift, and Cwmere and Bryn-glâs formations [747 830].

SN 78 SW/10–39   Dinas Dam
Site investigation boreholes proving Drift and Cwmere Formation [7438 8230–7495 8186].

APPENDIX 2

List of Geological Survey photographs

Copies of these photographs (mainly taken by C. J. Jeffery) are deposited in the libraries of the British Geological Survey at Keyworth, Nottingham, NG12 5GG and Bryn Eithyn Hall, Llanfarian, Aberystwyth, Dyfed, SY23 4BY. They all belong to Series A and may be supplied as black and white prints or lantern slides, and (except Nos. 387 to 6557) as colour prints or 2 × 2 in colour transparencies, all at a fixed tariff.

387 Fall of Afon Llywernog into the Rheidol Gorge, Ponterwyd.

388 Mudstones and thin flags, near Tynyffordd, Ponterwyd.

389 Incised meanders of the Afon Rheidol, south of Ponterwyd.

390 As 389.

6556 Cormorant Rock, a sea stack on the Aberystwyth Grits shore platform, Aberystwyth.

6557 Cliff of banded boulder clay, near Aberystwyth.

11861 Incised meanders of the Afon Rheidol, south of Ponterwyd.

12583 Aberdovey [5990 9654] Inverted strata, upper Llandovery Series.

12584 Aberdovey [5990 9654] Rotated cleavage, upper Llandovery Series.

12585 Cwm Safn-ast [6008 9706] Inverted strata, Borth Mudstones Formation.

12586 Aberdovey [6102 9601] Sideritic concretions, Devil's Bridge Formation.

12587 Aberdovey [6124 9600] Banded mudstone, Cwmsymlog Formation.

12588 Aberdovey [6127 9611] Asymetrical anticline, Borth Mudstones Formation.

12589 Ynys Penmaen [6860 9780] Tectonic ripples, Devil's Bridge Formation.

12590 Eglwysfach [6924 9579] Asymmetrical anticline, Devil's Bridge Formation.

12591 Bwlcheinion [691 944] Feature formed by Cwmere Formation.

12592 Craig Caerhedyn [7104 9677] Interbedded mudstones and sandstones, Nant-y-Môch Formation.

12593 Foel Uchaf [800 910] Hill peat.

12594 Foel Uchaf [800 910] Decaying hill peat.

12595 As 12594.

12596 Hengwm, Uwchygarreg [777 944] Valley-bottom drift and scree.

12597 Nant Taren-fedw-ddu [782 936] Late or post-glacial gorge.

12598 Creigiau Bwlch-Hyddgen, Uwchygarreg [770 935] Scree.

12599 Afon Hengwm [7730 9367] Morainic drift.

12600 Talbontdrain, Uwchygarreg [7790 9555] Nivation gravel (scree).

12601 Pistyll y Llyn, Cwm-rhaiadr [755 945] Blind glaciated valley.

12602 Nant y Gog, Cwm-rhaiadr [755 948] Boulder clay and morainic drift.

12603 Nant y Gog, Cwm-rhaiadr [7554 9482] Boulder clay and morainic drift.

12604 Craig y Dullfan, Nant-y-môch Reservoir [7721 8843] Thinly bedded turbidites, Nant-y-Môch Formation.

12605 Carn Owen [7298 8792] Sedimentary pillows.

12606 Cwmere [6822 8861] Folding and cleavage, Bryn-glâs Formation.

12607 Tyrhelig, Talybont [6720 8850] Valley bottom drift in the Leri valley.

12693 Dollwen, Goginan [6870 8128] Goginan mine tips.

12694 Cwmsymlog [6813 8393] Head of Cwmsymlog Valley.

12695 Pen-bont Rhydybeddau [6808 8394] Asymmetrical anticline, Devil's Bridge Formation.

12696 Pen-bont Rhydybeddau [6803 8396] Fold-pair, Devil's Bridge Formation.

12697 Pen-bont Rhydybeddau [6791 8311] Opencast, Great Daren (Old Daren) Mine.

12698 Capel Bangor [6654 8067] Turbidites, Devil's Bridge Formation.

12699 Capel Bangor [6593 8031] Upright anticline, Devil's Bridge Formation.

12700 Capel Bangor [6594 8031] Inverted strata and thrust, Devil's Bridge Formation.

12701 Garth Penrhyncoch [6543 8407] Cleavage, Borth Mudstones Formation.

12702 Garth Penryhyncoch [6510 8410] Folding, Borth Mudstones Formation.

12703 Capel Bangor [6570 7955] Gravel deposits, Afon Rheidol.

12704 Capel Bangor [6570 7955] Gravel deposits, Afon Rheidol.

12928 Ynyslas [6070 9290] Sand dunes and storm beach.

12929 Ynyslas [6070 9294] Sand dunes.

12930 Ynyslas [6070 9290] Dovey estuary

12931 Nant-y-môch Reservoir [738 875] View to Plynlimon and Y Garn.

12932 Bwlchystyllen [7311 8643] View along the Camdwr Fault.

12933 Craigypistyll [7088 8500] Camdwr Fault.

12934 Blaendyffryn, Goginan [7057 8118] View of the Melindwr valley.

12935 Ponterwyd [7438 8169] Dinas dam and gorge of the Afon Rheidol.

12936 Ponterwyd [749 802] Incised meanders of the Afon Rheidol.

12937 Hen-Hafod, near Tre'r-ddôl [6663 9442] An 'island' of Devil's Bridge Formation surrounded by recent sediments of the Dovey Estuary.

12938 Talybont [6620 8879] Syncline, Devil's Bridge Formation.

12939 Nant-y-môch [7576 8605] Nant-y-môch dam and reservoir.

12940 Nant-y-môch [7580 8598] Afon Rheidol below Nant-y-môch dam.

12942 Eisteddfa Gurig [7972 8406] 'Exhumed surface' top of Bryn-glâs Formation.

12943 Plynlimon [7882 8632] Features in Drosgol Formation.

12944 Plynlimon [7897 8694] View from summit towards Drosgol and Cardigan Bay

12945 Dyffryn Castell [7744 8110] Castell valley

12946 Dyffryn Castell [7755 8177] Castell Mine

12947 Moriah [6115 7955] Gently inclined fold with long limbs inverted, Aberystwyth Grits Formation

12948 Moriah [6115 7955] Flute casts, Aberystwyth Grits Formation

12949 Moriah [6115 7955] Prod-marks, Aberystwyth Grits Formation

12950 Moriah [6115 7960] View of Aberystwyth and the Rheidol valley

12951 Bow Street [6248 8420] View along Clarach valley

12952 Llandre [6260 8599] Mudstone scree

12953 Wallog [5920 8640] Harp Rock Type turbidite, Aberystwyth Grits Formation

12954 Wallog [5920 8640] Turbidites, Aberystwyth Grits Formation

12955 Borth [5960 8765] Craig y Delyn (Harp Rock), Aberystwyth Grits Formation

12956 Wallog [5931 8671] Complex syncline, Aberystwyth Grits Formation

12957 Wallog [5908 8540] Sarn Cynfelyn

12958 Wallog [5885 8539] Fold-pair, Aberystwyth Grits Formation

12959 Wallog [5866 8486] Folding and thrusting, Aberystwyth Grits Formation

12960 Wallog [5876 8518] Anticline with thrust and faults, Aberystwyth Grits Formation

12961 Clarach [5856 8437] Fold-pair, Aberystwyth Grits Formation

12962 Clarach [5852 8461] Cone-in-cone nodules, Aberystwyth Grits Formation

12963 Aberystwyth [5828 8262] Turbidites, Aberystwyth Grits Formation

12964 Aberystwyth [5834 8297] Tectonic structures, Aberystwyth Grits Formation

12965 Allt-wen, Aberystwyth [5760 7940] Intense folding and faulting, Aberystwyth Grits Formation

14053 Aberhosan [7925 9425] Corrie with glacial moraine

14054 Upper Borth [5960 8765] Harp Rock, Aberystwyth Grits Formation (base)

14055 Bwlch Nant-yr-arian [717 812]
and   Folds in the Devil's Bridge Formation
14056

APPENDIX 3

a Localities with non-graptolitic fauna

(Numbers in round brackets are specimen numbers)

Derwenlas Formation

1 [7126 8554] Outcrop on Craigypistyll, at level of steep path to W of stream and W of scree, opposite bend in main steam at lower end of waterfall (DEX 9953–10 000 and YFF 4001–4007, not inclusive)
2 As for locality 1, but 3 m stratigraphically higher (YFF 4008–4041, not inclusive)
3 [7430 8187] Stream section, 1130 m at 328° from Ponterwyd chapel, 65 m downstream from road bridge [7426 8183] (DEX 6922–6923)
4 [7426 8183] Stream section as for locality 3, but 8 m down from bridge (DEX 6968)
5 [7119 8281] Outcrop 265 m at 54° from Carndolgau, near Cwmerfyn (DEX 9848–9862)
6 [7215 8838] Roadside section 400 m at 38° from Cyneiniog, near Bontgoch (DEX 7078–7095)

?Derwenlas or Cwmsymlog Formations

7 [6967 8635] Small exposure just WSW of track across Banc Bwlch-Rosser, 2070 m at 5° from Bryn-goleu farmhouse (FG 124–127)
8 [6967 8635] Trackside 2075 m at 5° from Bryn-goleu farmhouse, S side of track across Banc Bwlch-Rosser (FG 128–129)

Cwmsymlog Formation

9 [6864 8414] Trackside section at junction of forest roads, 975 m at 33° from Cwmdarren, Pen-bont Rhydybeddau (DEX 7263)
10 [6848 9478] Road section, 571 m at 287° from Tyn y Garth (DEX 9249–9254)
11 [6804 8178] Small exposure on side of knoll, 402 m at 73° from Bro-dawel (DEX 7281)

Devil's Bridge Formation

12 [6977 7967] Stream section in Coed Ty-llwyd, 1120 m at 81° from Ty'n-y-nant Farm, about 5 km NW of Devil's Bridge (JW 1803–1824)
13 [6993 7922] Trackside adjacent to S side of Cwm Rheidol Reservoir, 1300 m at 102° from Ty'n-y-nant Farm, about 5 km NW of Devil's Bridge (JW 1825–1828)
14 [6995 7909] Cutting on the Devil's Bridge Railway, S of Gothic Mine, 1360 m at 107° from Ty'n-y-nant Farm, about 5 km NW of Devil's Bridge (JW 1829–1832)
15 [7761 7951] Forest track 823 m at 348° from Dolwen, SE of Ponterwyd (DEX 7334–7405)
16 [7938 7987] Bank of old leat near hill summit 1989 m at 54° from Dolwen (DEX 7603–7613)
17 [7994 8112] Stream section at waterfall, near outlet of small reservoir, N of Blaen-Myherin, Devil's Bridge, 2532 m at 101° from the corner of Dyffryn Castell, Ponterwyd (YFF 7091–7154)
18 [7977 8006] Section in Nant Rhedol, 2830 m at 123½° from the E corner of Dyffryn Castell, Ponterwyd (DEX 9434–9435, loose specimen)

Dolwen Mudstones Formation

19 [7948 7920] Cliff by stream at clearing in forest 1780 m at 74½° from Dolwen (YFF 4357–4358)
20 [7983 7931–7984 7923] Stream and adjacent crags 2140 m at 74° from the N corner of Dolwen (YFF 4377–4398)

b List of fossils collected (excluding graptolites).

Numbers refer to localities listed in a

BRYOZOA
bryozoan—indeterminate   1, 2, 5, 7, 15, 17, 18

BRACHIOPODA
Aegiria grayi (Davidson)   6, 12, 15, 17
A.?   1, 5, 6, 14, 15, 16
atrypoid   1(?), 9, 17
Coolinia applanata (Salter)   15, 19 (cf.)
C. pecten (Linnaeus)   17
C. sp.   8(?), 15
Dicoelosia alticavata (Whittard and Barker)   5 (cf.), 12 (cf.), 15 (cf.), 17
D. sp.   15
enteletacean—indeterminate   17
Eocoelia?   2, 12, 13
Eoplectodonta penkillensis (Reed)   15 (cf.), 17, 18
E. sp.   1, 15, 18
Glassia sp.   15(?), 17
Howellella?   1
Hyattidina?   1, 13
Leangella scissa (Davidson)   15 (cf.), 17
L. segmentum (Lindström)   5, 15 (cf.)
L. sp.   1, 2, 5, 12(?), 15, 16, 18
Leptostrophia?—juv. 15
Mendacella?   13
Pentamerus sp.   13(?), 18
Pholidostrophia?   17
plectambonitid—indeterminate 12
Resserella?   5
rhynchonelloid?   12
Skenidioides lewisii (Davidson)   1(?), 15, 17, 20 (cf.)
spiriferoid   15(?), 17
Visbyella pygmaea (Whittard and Barker)   1(?), 15(?), 17, 18
V?   6, 16
Ygerodiscus?   17
brachiopod fragments—indeterminate 4, 7, 14
horny brachiopod fragment—indeterminate 15

MOLLUSCA
gastropod fragment—indeterminate 17

TRILOBITA
Calymene?   17
odontopleurid—indeterminate 15

ARTHROPODA—*incertae sedia*
Discinocaris cf. *browniana* Woodward 3

OSTRACODA
beyrichiid—indeterminate 4

CRINOIDEA
crinoid columnals 5, 6, 7, 12, 14, 15, 16, 18, 20

MISCELLANEA
conodont? 11
shell fragments—indeterminate 10

INDEX OF FOSSILS

No distinction is made here between a positively determined species and variants or examples doubtfully referred to it (i.e. with qualifications aff., cf. or ?, etc).

Fossils identifiable at generic level only (e.g. *Acanthograptus sp.*) are listed after the named species.

Acanthograptus sp. 60, 75
Aegiria grayi (Davidson) 91, 93, 140
Aegiria sp. 140
Algal debris 43, 44
Arthropoda 140
Atavograptus atavus (Jones) 59, 61, 63, 64, 68, 73, 74
Atrypoid 140

Benthic fossils 45, 51
Benthos 44
Beyrichiid 140
Biserial graptolites 73
Brachiopoda 140
Brachiopods 6, 45, 50, 57, 97, 140
Bryozoa 140
Bryozoan 140
Burrows 42, 49, 57, 58, 84

Calymene sp. 140
Carbonaceous debris 43, 48
Cephalograptus cometa cometa (Geinitz) 59, 61, 63
Cephalograptus cometa extrema Bouček & Přibyl 59, 61, 76
Cephalograptus cometa (Geinitz) s.l. 69, 77, 82
Chondrites sp. 42, 49, 65, 84, 86
Climacograptus alternis Packham 59, 77
Climacograptus innotatus Nicholson 59
Climacograptus medius Törnquist 59, 64, 65
Climacograptus miserabilis Elles & Wood 59, 65, 68
Climacograptus nebula (Toghill & Strachan) 61, 92
Climacograptus normalis Lapworth 59, 61, 63, 65, 68, 74
Climacograptus rectangularis (McCoy) 59, 63, 74
Climacograptus scalaris (Hisinger) s.l. 59, 74, 75, 76, 77
Climacograptus simplex Rickards 60, 61, 63, 82, 85
Climacograptus sp. 52, 53, 66, 67, 73, 74, 77, 78, 85, 86, 90
Conodont 140
Coolinia applanata (Salter) 140
Coolinia pecten (Linnaeus) 140
Coolinia sp. 140
Coronograptus cyphus (Lapworth) 59, 61, 63, 64, 68, 73, 74

Coronograptus gregarius (Lapworth) 59, 63, 70, 73, 74, 77
Crinoid columnals 58, 93, 140
Crinoidea 140
Crinoid fragments 50
Cystograptus penna (Hopkinson) 59
Cystograptus vesiculosus (Nicholson) 59, 63, 74
Cystograptus sp. 73

Dendrograptus sp. 60, 75
Dicellograptus anceps (Nicholson) 16
Dicoelosia alticavata (Whittard & Barker) 91, 93, 140
Dicoelosia sp. 140
Dictyodora sp. 49
Dictyonema sp. 52, 60
Diplograptus magnus H. Lapworth 59, 61, 63, 69, 70, 73, 74, 75, 77, 78, 79
Diplograptus modestus parvulus (H. Lapworth) 59
Diplograptus modestus Lapworth s.l. 59, 65
Diplograptus sp. 68, 73
Discinocaris browniana Woodward 69, 140
Diversograptus ramosus Manck 53, 60
Diversograptus rectus Manck 60
Diversograptus runcinatus (Lapworth) 52, 53, 60, 86, 90, 91, 92, 93, 94, 95
Diversograptus sp. 77

Enteletacean 140
Eocoelia sp. 57, 93, 140
Eoplectodonta penkillensis (Reed) 91, 140
Eoplectodonta sp. 140

Fauna 19, 113, 117
Faunal remains 6
Flora 113, 117

Gastropod fragment 140
Gastropods 129
Glassia sp. 140
Glyptograptus elegans Packham 60, 82
Glyptograptus incertus Elles & Wood 60, 63, 85
Glyptograptus persculptus (Salter) 6, 37, 40, 41, 59, 61, 63, 64, 65, 66, 67, 68
Glyptograptus persculptus (Salter)-thin variant 59
Glyptograptus serratus Elles & Wood s.l. 60
Glyptograptus sinuatus Nicholson 59, 73, 74, 78
Glyptograptus tamariscus fastigans Haberfelner 52, 60, 92
Glyptograptus tamariscus linearis (Perner) 60, 69, 73
Glytograptus tamariscus (Nicholson) s.l. 59, 61, 63, 73, 74, 77, 78
Glyptograptus sp. 52, 53, 67, 73, 74, 76, 77, 78, 82, 85, 86, 90, 91, 93, 95
Glyptograptus (Pseudoglyptograptus) vas. Bulman & Rickards 59, 61, 63, 73, 74, 79
Glyptograptus (Pseudoglyptograptus) sp. 73, 74
Graptolites 2, 3, 16, 19, 42, 44, 45, 50, 51, 53, 57, 58, 61, 62, 63, 65, 66, 67, 68, 69, 70, 73, 74, 75, 76, 77, 79, 80,

82, 83, 85, 86, 87, 90, 91, 92, 94, 95, 96, 97, 133, 140
Graptolitic debris 43, 44, 67
Graptolite faunas 2
Graptolite fragments 16

Horny brachiopod fragments 140
Howellella sp. 140
Hyattidina sp. 93. 140

Ichnofauna 57

Lagarograptus acinaces (Törnquist) 59, 61
Lagarograptus tenuis (Portlock) 53, 60, 61, 63, 85, 86, 89
Leangella scissa (Davidson) 91, 140
Leangella segmentum (Lindström) 140
Leangella sp. 93, 140
Leptostrophia—juv. 140
Linguloid brachiopods 6

Macrofossils 2
Marine organisms 44
Mendacella sp. 93, 140
Mollusca 140
Monoclimacis crenularis (Lapworth) 60, 61, 74, 76, 77, 82, 87
Monoclimacis galaensis (Lapworth) 53, 60, 93
Monoclimacis sp. 91
Monograptus acus Elles & Wood 53, 60
Monograptus argenteus (Nicholson) 59, 61, 63, 73, 74, 77, 82
Monograptus austerus Törnquist 59
Monograptus barrandei (Suess) 52, 53, 60, 94
Monograptus capis Hutt 52, 59, 61, 69, 73, 74, 82, 85
Monograptus cerastus Hutt 59, 61, 73
Monograptus clingani (Carruthers) 60, 61, 63, 69, 74, 82
Monograptus communis communis Lapworth 59, 61, 74
Monograptus communis obtusus Rickards 60
Monograptus communis rostratus Elles & Wood 60, 69, 73
Monograptus communis Lapworth s.l. 53, 59, 63, 69, 73, 74, 77, 78, 82, 85, 86, 90
Monograptus convolutus (Hisinger) 48, 60, 61, 63, 69, 70, 71, 73, 74, 75, 76, 77, 80, 82, 87
Monograptus decipiens Törnquist 60, 61, 63, 69, 73, 76, 77, 82, 84
Monograptus delicatulus Elles & Wood 60, 69, 74, 82, 85
Monograptus denticulatus Törnquist 59, 61, 69, 73, 74, 75, 76, 80, 82
Monograptus dextrorsus Linnarsson 61
Monograptus difformis Törnquist 59, 73
Monograptus distans (Portlock) 60, 77, 83, 90
Monograptus elongatus Törnquist 60
Monograptus exiguus (Nicholson) 52, 53, 60, 61, 95
Monograptus gemmatus (Barrande) 52, 53, 60, 69, 86, 93

Monograptus halli (Barrande)   51, 52, 53, 57, 60, 63, 82, 85, 86, 90, 91, 92, 93

Monograptus intermedius (Carruthers)   60

Monograptus involutus Lapworth   52, 59, 63, 69, 71, 73, 85, 90, 91

Monograptus knockensis Elles & Wood   53, 60, 85

Monograptus limatulus Törnquist   60, 61, 63, 69, 70, 71, 73, 76, 77, 80, 82

Monograptus lobiferus (McCoy)   48, 52, 53, 59, 61, 63, 69, 70, 71, 73, 74, 75, 76, 77, 80, 82, 83, 85, 86, 87, 91

Monograptus marri Perner   53, 60, 63, 90, 91, 92, 94

Monograptus millepeda (McCoy)   59, 61

Monograptus nobilis Törnquist   59

Monograptus nodifer Törnquist   61

Monograptus petilus Hutt   52, 60

Monograptus planus (Barrande)   60, 86, 87, 93, 94

Monograptus pragensis (Přibyl)   82

Monograptus pragensis ruzickai (Přibyl)   60, 61, 63. 73, 75, 85, 93

Monograptus proteus (Barrande)   52, 53, 60, 61, 63, 76, 82, 86, 91, 92, 93, 94

Monograptus pseudobecki Bouček & Přibyl   52, 53, 60, 92, 93, 94, 95

Monograptus pseudoplanus Sudbury   59, 61, 70, 73

Monograptus revolutus Kurck s.l.   59, 61, 63, 69, 73, 74, 78

Monograptus rheidolensis Jones   61

Monograptus rickardsi rickardsi Hutt   53, 60, 92

Monograptus sedgwickii (Portlock)   48, 51, 52, 53, 57, 60, 61, 63, 70, 73, 74, 75, 76, 77, 82, 83, 84, 85, 86, 87, 88, 90, 91

Monograptus spiralis (Geinitz)   60, 61, 90

Monograptus toernquisti brevis Sudbury   59, 61

Monograptus toernquisti Sudbury s.l.   90

Monograptus triangulatus fimbriatus (Nicholson)   59, 61, 63, 69, 70, 73, 74, 77, 78, 79, 82

Monograptus triangulatus major Elles & Wood   59, 73, 74, 78

Monograptus triangulatus separatus Sudbury   59, 61, 73, 74, 78

Monograptus triangulatus triangulatus (Harkness)   59, 61, 63, 70, 73, 74, 78

Monograptus triangulatus (Harkness) s.l.   59, 73, 74, 82

Monograptus tullbergi Bouček   60, 91

Monograptus turriculatus (Barrande)   51, 52, 53, 61, 63, 86, 92, 94, 95

Monograptus undulatus Elles & Wood   60, 69, 73, 82, 85, 91

Monograptus veles (Richter)   61

Monograptus sp.   73, 87

Nereites sp.   49

Odontopleurid   140

Organic debris   4

Orthograptus bellulus (Törnquist)   59, 78

Orthograptus cyperoides (Törnquist)   60, 69, 75

Orthograptus insectiformis (Nicholson)   60, 74

Orthograptus mutabilis Elles & Wood   59, 73

Orthograptus truncatus abbreviatus Elles & Wood   19

Orthograptus truncatus intermedius Elles & Wood   19

Orthograptus truncatus Lapworth s.l.   16, 19, 61

Orthograptus sp.   60, 67, 68, 78

Ostracoda   140

Palaeodictyon sp.   52, 57, 58, 97

Parakidograptus acuminatus (Nicholson) s.l.   59, 61

Pentamerus sp.   140

Pernograptids   73

Petalograptus altissimus Elles & Wood   60

Petalograptus kurcki Rickards   60, 61, 86, 91

Petalograptus minor Elles   59, 73

Petalograptus ovatoelongatus (Kurck)   59, 63, 69, 70, 73, 74, 75, 77, 80

Petalograptus ovatus (Barrande)   60

Petalograptus palmeus (Barrande) s.l.   52, 53, 60, 61

Petalograptus tenuis (Barrande)   52, 60, 61, 63

Petalograptus wilsoni Hutt   60, 92, 94

Petalograptus spp.   52, 53, 77, 82, 85, 86

Pholidostrophia sp.   140

Plectambonitid   140

Pribylograptus argutus (Lapworth) s.l.   59, 70

Pribylograptus incommodus (Törnquist)   59, 73, 77

Pribylograptus leptotheca (Lapworth)   59, 61, 63, 69, 73, 74, 78, 82

Pribylograptus sandersoni (Lapworth)   59, 63

Pristiograptus concinnus (Lapworth)   59, 69, 71, 73, 74, 82

Pristiograptus fragilis (Rickards) s.l.   59, 73

Pristiograptus jaculum (Lapworth)   60, 76, 77, 84, 85, 90

Pristiograptus nudus (Lapworth)   52, 53, 60, 63, 90, 92, 94

Pristiograptus regularis regularis (Törnquist)   53, 59, 63

Pristiograptus regularis solidus Přibyl   53, 60, 86, 91, 92, 95

Pristiograptus regularis (Törnquist) s.l.   52, 53, 59, 61, 69, 71, 73, 74, 75, 76, 77, 78, 80, 82, 83, 84, 85, 86, 87, 89, 90, 91, 93, 94, 95

Pristiograptus variabilis (Perner)   61, 84, 93

Pristiograptus sp.   73, 77

Pseudoclimacograptus (Clinoclimacograptus) retroversus Bulman & Rickards   59, 61, 63, 69, 71, 73, 74, 76, 77, 82, 85, 87

Pseudoclimacograptus (Metaclimacograptus) hughesi (Nicholson)   59, 61, 68, 73, 74, 75, 78

Pseudoclimacograptus (Metaclimacograptus) undulatus (Kurck)   59, 61, 63, 69, 71, 73, 74, 75, 76, 77, 80, 82, 83, 84, 85, 86, 87, 91

Pseudoclimacograptus (Metaclimacograptus) sp.   74, 77

Pseudoclimacograptus sp.   16, 52

Pseudoplegmatograptus obesus (Lapworth)   60, 61, 85, 93

Rastrites approximatus geinitzi Törnquist   60, 73

Rastrites distans spengillensis Rickards   52, 61

Rastrites distans Lapworth   52, 53, 60

Rastrites fugax Barrande   60, 73, 85, 86, 90, 93

Rastrites hybridus Lapworth   60, 73, 77, 82, 85

Rastrites linnaei Barrande   53, 60, 61, 85, 86, 90

Rastrites longispinus Perner   59, 61, 63, 70, 73, 74, 78, 80, 90

Rastrites maximus Carruthers   50, 51, 52, 53, 61, 63, 85, 86, 90, 93

Rastrites peregrinus Barrande   59, 69, 73

Rastrites spina (Richter)   60, 73

Rastrites sp.   52, 61, 73, 74, 77, 82, 94

Rastritid   82

Resserella sp.   140

Retiolites perlatus Nicholson   60, 76

'*Retiolites*' *sp.*   69

Retiolitid   75

Rhaphidograptus toernquisti (Elles & Wood)   52, 59, 61, 63, 64, 67, 70, 73, 74, 75, 77, 78, 82, 87

Rhynchonelloid   140

Scolecodont   75

Shell fragments   50, 57, 58, 69, 91, 93, 98, 140

Shelly band   133

Shelly faunas   2

Skenidioides lewisii (Davidson)   91, 140

Spiriferoid   140

Squamodictyon sp.   52, 57, 58

Trace fossils   9, 19, 49, 52, 57, 94, 97, 133

Triangulate monograptids   3, 64, 74, 75, 76, 77, 82, 85, 135

Trilobita   140

Trilobite fragments   50

Visbyella pygmaea (Whittard & Barker)   91, 140

Visbyella sp.   140

Worm tracks   56

Ygerodiscus sp.   140

GENERAL INDEX

Aberceiro 126
Aberdovey 45, 46, 51, 91, 95, 96, 100, 109, 111, 112, 117, 129, 131
Aberlerry Farm 127, 128
Aber-Peithnant 132
Abertafol Halt 81, 82
Aberystwyth 1, 3, 45, 46, 53, 57, 58, 97, 103, 104, 108, 116, 118–120, 127–129, 137
 Castle 110, 120
 Grits 2, 3, 50–52, 55, 56, 57, 58, 87, 95, 97, 103, 108, 109, 111, 120, 121, 122, 126, 130, 132
  graptolite faunas 53
  Formation 50, 50–58, 96, 97
  section 96
 Harbour 128
Accretionary prism 2
Aeromagnetic anomaly map 104
Afon (River) See individual river names
Allerød 113
Alltddu 105, 123
Allt-glais 132
Alltgochymynydd 6, 38, 75, 105, 106
 Inlier 38
Alltgweiddyn 84
Allt-wen 57, 97, 108, 109, 117, 127, 132
Alluvial fans 115, 116, 122, 128
 floodplains 115, 128, 129
Anglers' Retreat 86
Ankerite 130
Anticlines,
 Carn March Arthur 99
 Coed Dipws 38, 65, 84
 Moel y Garn 99
 Moel-y-llyn 99
 Pemprys–Tarren Tyn-y-maen 36
 Pen y Graig-ddu 70
 Towy 11, 99, 100
Apatite 4, 44, 49, 82
Aran Mountains 118
Arch, The 49
Argoed Fawr 126
Arsenopyrite 130
Ashgill Series 2, 6
Atavograptus atavus Zone 61, 64

Ball and pillow structures 21, 24, 29, 35, 38, 40, 56
Banc Llechwedd-mawr 11, 26, 37
Banc Lletty-Evan-hen 6, 11, 29, 99, 126
 Inlier 29, 38
Banc Lluestnewydd 16, 21, 130
Banc Mynydd-gorddu 121
Banc Ty-Newydd 87
Baryte 2, 103
Basal Sandstone 91
Bedd Taliesin 88
Bedform 6, 10, 29

Bentonite 82
Biotite 82
Bioturbation 4, 16, 42, 44, 47, 49, 73, 86
Blaen Myherin Mudstones 58
 Formation 58
Blaen-Ceulan 46, 71, 75, 87
Blaen-dyffryn-isaf 119
Blaeneinion 65, 88
Blaen-geuffordd 51
Blaen-Peithnant 120
Block scree 127
Blown Sand 117, 127, 128, 129
Bod-yr-yrfa 126
Bontgoch 42, 119, 122
Bontgoch–Cwmere gorge 122
Boreholes 45, 113, 117–119, 121, 122, 127–129
 Glan-fred 46, 49–51, 88, 104, 109
 Rheidol Hydroelectric Project 120
Borth 50–53, 57, 58, 88, 96, 108, 109, 117, 126, 128, 129, 132
 Bog 51, 95, 113, 116–118, 127, 132
 Mudstones 5, 50–52, 57, 58, 87, 95, 102, 108, 132
 Mudstones Formation 49–53, 88, 95, 96, 130, 131
  stratigraphy 54
 Sands 127
Bottom structures 9, 23, 40
Boudins 99
Bouguer gravity anomaly 104
Boulder Clay 112, 113, 116, 119–122, 129
Bounce casts 55
Bow Street 113, 119, 121, 126
Box folds 103
Brachiopods 6, 45, 50, 57, 97
Braich Garw 94, 106
Braichycelyn 95
 Lodge 90, 95
Breccia zones 100, 105, 130, 131
Brecon Beacons 122
Bronfloyd Mine 131
Brwyno
 Overthrust 8, 11, 18, 30, 35, 40, 65, 100, 106
 River 19, 36
 valley 30
Bryn Brith 48
Bryn Mawr 49
Brynbras 45, 83, 91
Bryn-glâs 64
Bryn-glâs Formation 6, 10, 11, 14, 15, 23, 24, 26, 29, 31, 32, 35, 36, 37, 38–42, 44, 47, 64, 65, 67, 68, 100, 103, 131
 Group 14, 37
 Hill 36
 Mudstones 14
Bryngwyn 121
Bryngwyn-canol 121
Bryngwyn-mawr 121
Bryniau Rhyddion 71, 85
Brynybeddau 16
Buarth-mawr hill 120
Building Research Station 127

Building stone 132
Builth 99
Bwa-drain 119
Bwlch Corog 11
Bwlch Hyddgen 94, 126
Bwlch Nant-yr-arian 92, 105
Bwlch yr Adwy 48, 84
Bwlcheinion 11, 19, 31, 32, 35, 40
Bwlch-glas 6, 38, 43, 65, 71, 75, 87, 93, 99, 105
 Inlier 38
Bwlchystyllen 36, 37, 120, 122

Cader Idris 104
Caer-arglwyddes 119
Calcareous nodules 68, 83, 89, 94
Calcite 106, 130, 131
Caledonian Orogeny 99, 108, 131
Cambrian 3, 4, 99
 Railway 127
 Shelf sea 2
Camdwr Fault 36, 38, 64, 68, 105, 131
Camdwr lode 130
Capel Bangor 95, 115, 119, 128
Capel Dewi 119, 121
Caradoc 99
Carbonate mineralisation 100, 101, 106
Carboniferous 131
Cardigan Bay 1, 2, 111, 112, 118
Cardiganshire 3
 Slate Company 132
 Slate Quarry 5, 66, 67
Careg Milfran 51, 95, 108, 109
Carn Gwilym 130
Carn March Arthur 99
 Anticline 99
Carn Owen 6, 11, 15, 27, 28, 29 35, 36, 39, 65, 75, 99, 105, 106, 132
 Inlier 10, 26–29, 38
 map 27
 Pericline 11, 15, 29, 99, 100
 section 28
Carndolgau 83
Carregcadwgan 66
Carregifan 14
Cascade Overthrust 100
Castell, Afon 36, 128
 Fault 91
Castell Gwallter 119
 lode 130
 mine 128
 valley 128
Cefn Coch 65
Cefn-cynhafal 116
Cefn-gweirog 35, 40
Cefn-gwyn 87
Cefnhendre Farm 97
Cefnmaesmawr 36
Cefnyresgair 36, 37, 99
Central Wales 104, 112, 118
 glaciation 118
 Mining Field 130
Cerig Blaen-Clettwr-fawr 36
Cerig Gwinion Grits 43
Cerrigyrhafan 45
Ceulan, Afon 88, 119, 128
 valley 87, 89

Chalcopyrite 105, 130, 131
Channel fill deposits 11, 21, 22, 26, 29, 30, 35, 36, 49, 75, 87
  section *35*
Chlorite 10, 35, 57, 82, 106, 113
Cil-olwg 95
Cirques 118, 122
Clarach 57, 58, 108, 112, 113, 116, 118, 121, 122, 127
  Afon 115, 118, 128
  Bay 56, 57, 97, 109, 117, 121, 126–129
  valley 97, 118–121, 126
Clay minerals 100, 113
Cleavage 3, 5, 6, 9, 15, 21, 37, 41, 46, 47, 49, 65–68, 75, 78–80, 84, 86, 87, 90, 91, 94–96, 99–103, 106–110, 126, 127, 132
  double 6
  fracture 3, 56, 95
  slaty 82
Clettwr
  Afon 35, 40, 65, 66, 119, 127
  mine 131
  valley 119, 131
Climatic variations 113
Clogau Shales 4
Coed Dipws Anticline 38, 65, 84
Coed Pant-y-bedw 116, 129
Coed Ty-llwyd 93
Coed y Fedw 90
  sketch map *79*
Coed y Gofer 45
Comins Coch 119
Cone-in-cone-concretions 51, 54, 57, 95–97
Conglomerates 11, 22, 30, 35
Constitution Hill 97, 108, 117, 127, 132
Contour currents 58
Contourites 54, 58
Coronograptus cyphus Zone 42, 44, 45, 61, 64, 65, 67, 68, 73, 74
Coronwen 19
Corries 113
Cors Fochno 51, 95, 116–118, 122, 126, 127, 128, 129, 132
Craig Caerhedyn 8, 19, 30
  section *33*
Craig y Delyn 50, 95, 96
Craig y Dullfan 16–18
  subfacies 8, 16, 17, *18*
Craig y Fedw 21, 23
Craig y Gâth 35
Craig-y-Penrhyn 127
Craig yr Eglwys 24
Craig yr Hesg 93
Craignant 119
Craignant-mawr 69
Craigypistyll 47–49, 64, 70, 84, 93, 105
  Dam 29
  sketch map 65
Craigyrwylfa 95, 109
Creigiau Bwlch Hyddgen 94
Cretaceous 111
Cribyresgair 37
Crinoids 50, 58, 93
Crip y Frân 37

Cross-lamination 23, 26, 35, 36, 38, 44, 45, 51, 58, 64, 75, 76, 89, 90, 93, 95
Crugiau 128
Cryoturbation 126
Current rippling 23, 24, 26, 48, 49, 89
Current scour casts 9
Cusp and dome structures 100
Cusp and furrow structures 100, 101, *102*, 103
Cwm Cae 121
Cwm Ceulan 14, 35, 47, 65, 79, 80, 87, 106, 126, 132
Cwm Einion 30, 45, 132
Cwm Rhaiadr 86, 87, 113, 123,
Cwm Sylwi Fault 100
Cwm-byr 48, 76, 113, 122, *123*
Cwmere 6, 14, 39, 40, 65, 87, 88, 91, 99, 105, 122, 129, 132
  Formation 2, 4, 6, 37, 38, 40, 42, 43, 44, 45, 61, 64–69, 73, 74, 78, 80, 99, 100, 130
  Group 42
  Inliers 38
  River 122
  sketch map *78*
Cwmerfyn 49, 132
Cwmergyr 132
Cwm-mwythig 119
Cwm-rhaiadr-fâch 128
Cwms 122, 123
Cwmsymlog 45–47, 82, 83, 132
  Formation 46, 47, *48*, 49, 50, 70, 71, 75–77, 79, 81, 82–94, 100, 102, 113, 132
  sections *72, 88*
Cwmystwyth Formation 130, 131
  Grits 58, 130
Cwrt 121
Cymerau 65, 67, 89, 94, 132
Cyneiniog
  Afon 38, 75, 79, 87, 93, 94, 112
  valley 122, 132
Cynnull-mawr 119

Daren 130
Ddu, Afon 40, 67, 117, 127, 128
Derwen Group 44, 47
Derwenlas 45, 80, 89
  Formation 2, 15, 44, *45*, 46–48, 58, 64, 65, 68, 69, *70*, 71–82, 85–87, 90, 100, 105, 130, 132
  sections *69, 72, 80, 81, 88*
Devensian 2, 3, 111–113, 117, 118, 128
Devil's Bridge 49, 119
  Formation 5, *46*, 47, 49, *50*, 51, 52, 58, 75–77, 79, 81, 83–95, 100, 103, 105, 107, 131, 132
  sections *88*
  stratigraphy *54*
  Group 49, 58
Diagenesis 103, 107
Diatomite 113, 115, 117, 118, 125
Diatoms 113
Dicellograptus anceps Zone 2, 6
Dinas 119
  Reservoir 64, 119, 120, 128
Diplograptus magnus Zone 45, 61, 69, 70, 73–75, 77–79, 82

Disgwylfa 40
Disgwylfa Fâch 45, 64, 83, 88
Disturbed Beds 11, 36
Dolau 95
  valley 121
Dolcaradog 116, 129
Dolgau 121, 126
Dolomite 100, 103, 106, 130
Dolrhyddlan 85
Dolwen Mudstones 49, 58, 91
  Formation 58, 97, 98
Dolybont 95
Dorglwyd 119
Dovey
  Estuary 1, 3, 11, 43, 44, 46, 68, 90, 95, 99, 108, 117, 118, 127, 128, 132
  Junction 117
  River 1, 6, 45, 49, 51, 81, 115, 127, 128, 129
  salt marshes 117
  Valley 3, 113, 118, 123
Downtonian 99
Drosgol *frontispiece*, 11, 16, 18, 21, 22, 24–26, 130
  Formation 6, 9–16, *19*, 20, *21*, 22, 23–25, 26, 27–29, *30–32*, 33, *34*, 35–37, 40, 42, 100
  channel margin *33*
  location map *20*
  Grits 3, 6, 9, 11
  section *23*
Drum Peithnant 16
Dryas 113
Drybedd 24
Dulas, Afon 129
Dyfed 58
Dyffryn Castell Hotel 106, 115, 128
Dyfngwm–Dylife lode 130
Dyll Faen 24, 36
Dynyn 19
  River 19
  valley 30

Earthquakes 103
Eglwysfach 110, 122
Einion, River 19, 77, 89, 122, 127
Eisteddfa Beds 8
Eisteddfa Gurig 36, 37, 64, 68, 82, 83, 91
Elgar 119, 121
Epidote 82
Erglodd 80, 90
Erratics 118, 119, 126
Esgair 37
Esgair Foel-ddu 40
Esgair Fraith 45, 47
Esgair-hir 105
Estuarine deposits 117, 127

Fagwr-fawr 91
Fagwyr 126
Fainc-ddu Isaf 16, 18
  Sandstone 8, 16, 17
Fault, strike 17, 19, 45, 65, 100, 105, 106, 110, 131
Faults 2, 46
  Camdwr 36, 64, 68, 105, 131

Castell 91
Cwm Sylwi 100
Glandyfi 91, 93
Hafan 26, 28, 70, 84
Llyfnant 11, 19, 30, 35, 36, 100,
    105, 126
Ogof 18, 30, 35, 40, 105, 132
Pennal 100
Wern-deg 106
Ystwyth 105, 130
Faulting 66, 82, 85, 87, 91, 95–97, 99,
    100, 102, 105, 106, 108, 113, 130
Ferromagnesian minerals 130
Ferruginous nodules 94
Flandrian 3
Flute casts 17–19, 23, 26, 40, 52, 55,
    56, 64, 76, *89*, 90, 93–97
Fluvioglacial gravel 112, 120, 121
Foel Einion 11
Foel Fawr 94
Foel Goch 8, 65
    Outlier 65
Foel Uchaf 22
Foel-ynys 117, 128
Foraminifera 117
Forge 103, 107, 130
    Overthrust 100, 105
Formations, facies *54*
Fridd Cae-crŷdd
    geological sketch *67*
Fridd Cwmere 87, 93
Fron-gôch 95
    Formation 130, 131
Frongôg 120
Fronlas 40, 80
Furnace 110, 122, 127

Galena 130, 131
Gangue minerals 130, 131
Garn Wen 11, 35
Garth Penrhyncoch 95
Garthgwynion 113, 116–118
    valley 123
Gastropods 129
Gelli Goch Overthrust 100
Geophysics 99–111
Geothermal measurements 104
Glaciation 111, 112, 118, 122
    Ordovician 2
    Pleistocene 2
Glan-fred Borehole 46, 49–51, 88, 104,
    109
    graptolites 52
Glan-Lerry 115, 128
Glan Rheidol 120
Glandyfi 94, 110, 118, 127
    Fault 91, 93
    Tract 99, 100, 103, *105, 106, 107,*
    108, 109
Glanfraid 88, 122, 128
Glanmorfa 127
Glaspwll 113, 123
    drift deposits *115, 116*
Glide planes 99
Glyptograptus persculptus Band 44, 64, 65,
    67, 68
Glyptograptus persculptus Zone 2, 6, 8, 42,
    61, 64

Gneiss 127
Goethite 130
Gogarth 99
Gogerddan 115, 126, 128
    valley 121, 126
Goginan 45, 46, 49, 69, 91, 93, 119,
    130
Graben 102, 110
Graded bedding 10, 29
Graptolite zones 2, 6, 41, 44, 61
    Silurian *41, 59–61*
Graptolites 3, 6, 16, 19, 41–46, 48, 50,
    51, 56–58, 61, 65–71, 74, 75, 77,
    79–87, 89–92, 94–97
    Silurian 62, 63
Gravel 3
Greenschist 6, 82, 127
Grogal Sandstones 51, 58, 102
Groove casts 23, 55, 97
Gwar-cwm 79
Gwarin, Afon 24
Gwenffrwd-uchaf 64
Gwenffrwd Formation 130, 131
Gweunbwll 75
Gypsum 130

Hafan Fault 26, 28, 70, 84
    Incline 28
    Quarry 132
Harlech 111
    Dome 131
Harp Rock 50–53, 57, 95, 96
    beds 51, 54, 55, 95, 97
    debris-flow deposits 96
    sandstones 95
    turbidites 57
Head 112, 113, 117, 121, 122, 126,
    132
Heat flow 104
Hemipelagite 4, 43, 44, 47, 48, 75, 86,
    95, 103
Hen-Hafod 127
Hengwm 102, 113, 117, 122, 126
    Afon 24
    drift deposits *114*
    valley 118
Hengwm-cyfeiliog 113, 129
Henllys 95
Hercynian Orogeny 131
Hieroglyphs 103
Hiraeth 67
Hirnant 22, 24, 120
Holocene 95, 117, 118, 126
Horsts 102, 110
Hyddgen 36, 37, 86, 116, 130
    Afon 116
    mine 130
Hydraulic fracturing 131
Hydro-electric schemes 1
Hydroplastic deformation 101
Hydrozincite 130

Iapetus Ocean 99
Ice movements 2
Illite 113
Inliers,
    Alltgochymynydd 38
    Banc Lletty-Evan-hen 29, 38

Bwlch-glas 38
Carn Owen 11, 26–29, 38
Cwmere 38
Llyn Ieuan *83*
Machynlleth 6, *7*, 8, 10, 11, 14, 18,
    19, 30–36, *39*, 40, 43, 44, 49, 99,
    100, 105, 106
Pen y Castell 75
Penrhyngerwyn 40
Plynlimon 1, 6, *7*, 8, 10, 11, 14,
    16–18, 20–26, 36, 37, 49, 99, 105
Inter-channel deposits 11, 23, 35–37,
    40, 48
Irish Sea 112, 118, 128
    Boulder Clay 112
    ice 118
    landmass 52

Kame terraces 3, 112, 118, 120, 121
Kettle holes 123

Lacustrine clays 102, 112, 113, 117,
    118, 121, 122, 129
Lagarograptus acinaces Zone 44, 61, 68
Lake deposits 2
Lake District 61
Laminated Muddy Siltstones 4
Late Glacial 113
Lead 130, 132
Leri, Afon 93–95, 112, 113, 115, 117,
    119, 122, 127, 128, 129
    gorge 38
    valley 49, 75, 79, 94, 112, 113, 119,
    121, 122, 132
Limonite 67, 130
Listric faults *110*
Lithiophorite 130
Llanbadarn Fawr 120
Llancynfelyn 110, 132
Llandovery Series 2, 3, 41, 42, 44, 47,
    48, 50, 58, 73, 82, 87, 99, 111
    facies *54*
Llandre 95, 118, 126
    Gap 121, 122, 126
    valley 127
Llangorwen 121, 128
Llangranog 14, 53, 57
Llanidloes 15, 131
Llanwrin 107
Llanymawddwy 44
Llawrcwmbach mine 130
Llechwedd Cwm-byr 87
Llechwedd Gwineu 71, 84
Llechwedd Llŵyd 40, 65
Lle'r-neuaddau 24, 36, 120
Llettyhen 130
Lleyn peninsula 46
Lluestgota, Afon 45, 46, 65, 71, 84, 86,
    89, 116
Lluest-newydd 24
Lluest y Graig 23
Llwyn-gwyn 122
Llwyniorwerth 119
Llyfnant
    Fault 11, 19, 30, 35, 36, 100, 105, 123
    River 123, 128
    valley 34, 36, 67, 123, 126
Llyn Barfog 100, 115

Llyn Blaenmelindwr   82, 83, 132
Llyn Conach   94
Llyn Craigypistyll   29, 38, 64, 122
Llyn Ieuan Inlier   83
Llyn Llygad Rheidol   21, 22, 118, 122
  location diagram   24
Llyn Pen-rhaiadr   86
Llyn Syfydrin   69, 120
Llyn y Delyn   122
Llysarthur   49, 128
Loaded scour pockets   18
Locality map   1
Longshore drift   117, 127
Lovesgrove   120, 121, 131
Lower Leri Valley   113
Lower Quarry   92

Machynlleth   43, 58, 113
  Inlier   6, 8, 9, 10, 11, 18, 19,
    30–36, 39, 40, 43, 44, 49, 99, 100,
    105, 106
    Ordovician sequences   7
Made ground   129
Maesnant   16, 17
  subfacies   8, 16
  valley   120
Main Irish Sea Glaciation   118
Manganese minerals   130
Marcasite   130
Marine Beach deposits   117, 127
Marine deposits   117, 127
Mass-flow deposits   10, 11, 15, 21, 22,
    29, 35
Mass-movement   2, 15, 29, 47
Meanders   119, 120
Meirionnydd   112, 113
Melanges   6, 36
Melindwr, Afon   70, 128
  valley   75
Melin-y-cwm   45
Melt-water   3, 112, 118, 120–122
Mesozoic   111
Metalliferous lodes   3, 130, 131
Metalliferous minerals   130
Metalliferous-Slate Group   50
Microfaulting   84, 87
Mineral products   130–132
Mineralisation   2, 4, 130
Mining   1, 3, 87, 91, 96, 119,
    128–130, 132
Moel Cyneiniog   71, 85
Moel Fferm   45, 76, 77, 87
Moel Golomen   45, 48, 84, 87
Moel Hyrddod   11, 30, 34, 36
Moel y Garn   11, 35, 99
  Anticline   35, 38, 99
Moel-y-Llyn   11, 35, 36, 40, 77, 99
  Anticline   36, 99
Moelglomen   94
  Mine   87
Monograptids   64, 73–77, 85
Monograptus argenteus Zone   42
M. convolutus Zone   45, 61, 68–71, 73,
    74, 76, 77, 79, 80, 82, 83
M. crispus Zone   57, 58
M. leptotheca Band   71
  Zone   42, 69, 73, 74, 77, 82

M. rheidolensis Zone   44
Monograptus sedgwicki shales   46, 47, 71,
    75, 77, 81–83, 85, 89, 90
  Zone   47, 48, 57, 61, 63, 71, 75, 76,
    82–86, 88, 91, 93
M. triangulatus Zone   44, 45, 61, 70, 73,
    74, 78, 82
M. turriculatus Zone   2, 42, 47, 48, 50,
    51, 57, 58, 61, 91–95
Montgomeryshire   3, 4
Montmorillonite   2
Moraine   124
Morainic Drift   112, 113, 117, 118,
    120–123, 124, 125–129, 132
Morben Quarry   89
Moriah   126, 132
Mottled Beds   42
Mottled Mudstone   42, 44, 64, 65, 67,
    68
Mottled Mudstone Member   4, 42, 43,
    61, 64, 67, 132
Mudstone scree   113, 121, 122, 127
Muscovite   82, 113
Myherin, Afon   58, 91, 98, 128
  Group   49, 58
  valley   128
Mynydd Cae-du   36
Mynydd Gorddu mine   119
Mynydd y Llyn   68
Mynydd yr Ychen   91

Nant Bwlch-glâs   75
Nant Ceiro   36, 120
Nant-cellan-fach   121
Nant Ceniarth   129
Nant Fuches-gau   37, 64
Nant Fuches-wen   36, 68, 91
Nant Lŵyd   94, 119
Nant Maesnant-fach   16
Nant Meirch   68, 83, 91, 128
Nant Perfedd   99, 119
Nant Rhyddlan   65
Nant Rhŷs   49, 91
Nant Silo   95
Nant y Creiau   28
Nant y Gwyddil   119
Nant y Llyn   18, 21
Nant-y-môch   17
  Dam   16, 20, 21, 120
  Formation   4, 6, 8, 9, 10, 16–18, 19,
    30, 43, 100, 103, 132
  Reservoir   21, 24, 36
  valley   112, 116, 120
Nant-y-nôd   65
Nant-yr-cae-rhedyn   120
Nantglyn Flags   4
Nantperfedd   46, 49
Nantyrarian   126
Narrow Vein   9, 19, 132
National Library of Wales   128
Navy Submarine Channel   47
  Fan   47, 49
Neuadd-yr-ynys   94
New Quay   57, 58
Newquay   54
Nivation gravel   113, 117, 118, 122
Nod Glas   99

North Wales glaciation   118, 123
Northern Ireland   112

Ochr Lygnant   37
Ogof Fault   18, 30, 35, 40, 105, 132
Ogof Morris   31, 35, 40
Old Goginan   93, 129, 132
Older Dryas   113
Ordovician   3, 4, 6–41, 43, 44, 99, 104,
    110, 130, 131
  Glaciation   2
  Sedimentation   2
Ordovician–Silurian boundary   6, 8
Outlier,
  Foel Goch   65
Overbank deposits   11, 22
Overthrusts   105, 106

Palaeochannels   84
Palaeocurrents   17, 52, 58, 93–96
Palaeogeography, Hirnantian   14
Pant Eidal Wood   95
Paragenetic sequence   130
Parakidograptus acuminatus Zone   44, 61,
    64
Parc   11
Parcgwyn   126, 132
Parson's Bridge   119
Peat   115–121, 126–129, 131, 132
Peithnant, River   24, 120, 128
Peithyll   97
  Afon   95
  valley   121
Pembrokeshire   104
Pemprys   11, 36
Pemprys–Tarren Tyn-y-maen Anticline
    36
Pen Cerig   36
Pen Dinas   79
Pen y Castell Inlier   75
Pen y Graig-ddu   46, 47, 82, 83, 90–92
  Anticline   70
  geological sketch map   71
Pen-bont Rhydybeddau   46, 84
Pen-y-cwm   126
Pen-y-garn   121
Pen-y-graig   94
Pen-y-wern   122
Penbryn   93
Pencarreg-gopa   11
Pencerrigtewion   10, 22, 23, 38
  Member   6, 10–14, 20, 22, 26, 29,
    30, 35, 36, 39, 40, 126, 132
  Facies distribution   13, map 25
  sedimentational model   12
Penhelig   96, 132
Pennal Fault   100
Penparcau   120, 126
Penpompren Hall   128
Penrhyncoch   95, 119, 121
Penrhyngerwyn   6, 43, 110
  Inlier   40
Pentre-rhyd-yr-onen   120
Penycefn   119
Periclines   99, 100
Periglacial action   111, 112
Pernerograptids   73

Phosphatic concretions 5, 8, 42, 49, 90, 91, 95, 96, 103
Piedmont glaciers 118, 128
Pistyll y Llyn 126, 128
Plas Cwmcynfelin 119
Plas Gogerddan 128
Plate collisions 103
Plate tectonics 2
Pleistocene 2, 111, 113
Plynlimon 1, 6, 22, 24, 30, 38, 39, 112, 118, 122, 126, 131
  dome 53, 58
  Inlier 1, 6, 8, 9, 10, 11, 16–18, 20–26, 36, 37, 49, 99, 105, 132
  Ordovician sequences *7*
  ridge 22
Pollen Zones 116, 117
Pont Cwm-pandy 66
Pont Llyfnant 89
Ponterwyd 42–46, 49, 58, 61, 64, 68, 71, 79, 82, 119, 120, 128, 132
Pore-water 99, 103, 107, 109
Precambrian 104
Pribylograptus leptotheca Band 68
  Zone 45, 61, 73
Přídolí 2
Pseudonodules 15, 21, 26, 29, 30, 38
Psilomelane 130
Pumice 82
Pumlumon Fach 21
Pwll-glâs 121
Pwllcenawon 128
Pyrite 8–10, 16, 18, 22, 42–44, 46, 47, 57, 65–68, 84, 86, 87, 90, 92, 94, 97, 99, 106, 130, 132
Pyrolusite 84

Quarries 11, 19, 22, 26, 27, 29, 37, 38, 64, 66–68, 79, 80, 87, 89–97, 110, 129, 130, 132
Quartz mineralisation 2, 21, 82, 130, 131
Quartz veining 100, 105, 108, 109, 112, 113, 122
Quaternary 2, 111–129
Quaternary deposits, distribution *111*

Radiocarbon dating 117, 129
Radioisotopic dating 131
Ramping 103
Rastrites Band 85
Rastrites maximus Subzone 47, 50, 51, 57, 61, 63, 85–87, 90, 91, 93
Recent 118
Retiolitid 75
Rhayader 4, 43–45, 49
  canyon 48
  Pale Shales 48
Rheidol
  Afon 1, 16, 20, 24, 36, 64, 112, 115, 119, 120, 127, 128
  basin 112
  Gorge 3, 45, 64, 128
  Hydroelectric Project 119, 132
  valley 3, 45, 112, 113, 115, 119–122
Rhiw-gam 45, 47, 74, 75, 86, 113, 118, 122

valley 122, 126
Rhosgellan 121
Rhownia 127
Rhuddnant Grits 3, 58
Rhyd-meirionydd 119
Rhydyfelin 112
Rhythmites 71, 75
River terraces 112, 115, 116, 128
Roofing slates 5, 132
Rutile 82

St Matthew's Church 128
Salem 119
Sand and gravel 112, 113, 126–129, 132
Sand dykes 102
Sarn Cynfelyn 118, 122, 127
Sarns 118
Scolecodont 75
Scour casts 90
Scrobicularia clay 117
Sea level changes 2, 3, 44, 46, 48, 50, 99, 118, 127
Sedimentary structures 10, 11, 23, 29
Sedimentation
  Ordovician 2
  Silurian 2
Sericite 10
Shelf seas 2
Shelve 99
Siambr Traws-fynydd 74
Siderite 55, 57, 95, 106
Siliceous concretions 91, 96
Silo, Afon 121
Silurian 3, 4, 6, 8, 29, 38, 41–99, 103, 104, 122
  Sedimentation 2
Silver 130
Slate 132
Slide-planes, bedding-parallel 106, *107–110*
Slumping 10, 14, 15, 26, 29, 35, 103
Snowdonia National Park 1, 123
Sole markings 44, 53, 55, 58, 97
Solifluction 112, 113, 118–120, 126, 129
South Moelglomen Mine 87
Southern Uplands 2
Sphalerite 130, 131
Sponges 73
Staylittle 110
Stewy, Afon 45, 75, 84, 119, 121
  valley 121
Storm Beach Gravels 117, 128–129, 132
Structure 2, 99–103, *104*, 105–111
Structural events, sequence 103
Structural features *101*
Submarine fans 2, 39, *47*, 48, 49
Submerged forest 116, 126
Subzones,
  Rastrites maximus 47, 50, 51, 57, 61, 63, 85–87, 90, 91, 93
Sulphides 2, 48, 100, 103, 105, 110, 130, 131
Sulphuric acid 130
Syfydrin 68, 69

Taliesin 67, 94, 110
Talybont 6, 48, 49, 88, 94, 95, 110, 112, 113, 119, 121, 122, 128
Tan-yr-allt 122
Tarenig, Afon 36, 64
Tarren Neuadd-lŵyd 35
Tarren Tyn-y-maen 11, 36, 105
Tectonics 2
Tertiary 111
Thalwegs 112, 121
Tidal flats 127
Till 2, 112, 113, 118–121, 126
Titanium-oxides 82
Tool-mark casts 9, 103
Towy Anticline 11, 99, 100
Trace fossils 9, 49, 57, 94, 97
Tre-boeth 87
Trefechan 128
  Bridge 128
Trefeddian 100, 109
Trefri 96
Tre'r-ddôl 128, 131
Trilobites 50
Tuff 82
Turbidites 2–4, 8, 10, 18, 22, 23, 38–40, 43–46, 48–53, 54, 55–58, 65, 75, 79, 80, 82, 85, 87, 88, 94–97, 103, 106
  encroachment *46*
  lithology *4*
  terminology *4*
Turbidity currents 2–4, 9, 37, 38, 42, 44–46, 48, 52, 58
Twyni Bâch 117, 127–129
Twyni Mawr 127–129
Twywn 104
Ty Bwlch Hyddgen 86
Ty Mawr 96, 117
Tyddyn-y-briddel 45
Ty'n-y-ffynnon 119
Tyn y Garth 80, 89, 90, 132
  sketch map *79*
Tynffordd 68
Tynyffordd 91
Ty'r Abbi 119
Tyrhelig 87, 119
Tywyn 42

Upper Borth 51, 95, 117–119, 127, 128
Upper Rheidol 122

Valley glaciers 3, 112, 118, 121, 122
Van Formation 130, 131
Van mine 131
Varves 113, 118
Victoria Terrace 132
Volcanic activity 2
Volcanic activity 2
Volcanic ash 54, 55, 57, 82, 97

Waen Badell 36
Wallog 52, 56–58, 96, 97, 108, 117, 118, 121, 126–127
  valley 121, 122, 126, 127
Waun Fawr 119
Welsh Basin 2, 4, 99, 105

Welsh Borders 112
Welsh ice 118
Welshpool 99
Wenlock Series 2, 4, 58, 99
Wern-deg Fault 106
Werndeg 79, 132
Wileirog 119
Worm track casts 56
Wylfa 105

X-ray diffraction 57, 82

Y Chwareli 71, 82
 section 85
Y Faen gul 9
Y Glog 83
Ynys-Edwin 110, ·127
Ynys Eidiol 80
Ynys-fergi 117, 128

Ynys Greigiog 127
Ynys-Hir 67, 80, 90
Ynyslas 112, 117, 127, 128, 129, 132
Ynysycapel 122
Younger Dryas 113
Ystrad Meurig 45
 Grits 52
Ystumtuen 91
Ystwyth
 Afon 115, 127 – 129
 Fault 105, 130
 valley 112, 113, 120

Zinc 130, 132
Zones,
 Atavograptus atavus 61, 64
 Coronograptus cyphus 42, 44, 45, 61, 64, 65, 67, 68, 73, 74
 Dicellograptus anceps 2, 6

Diplograptus magnus 45, 61, 69, 70, 73 – 75, 77 – 79, 82
Glyptograptus persculptus 2, 6, 8, 42, 61, 64
Lagarograptus acinaces 44, 61, 68
Monograptus argenteus 42
M. convolutus 45, 61, 68 – 71, 73, 74, 76, 77, 80, 82, 83
M. crispus *57, 58*
M. leptotheca 42, 69, 73, 74, 77, 82
M. rheidolensis 44
M. sedgwickii 47, 48, 57, 61, 63, 71, 75, 76, 82 – 85, 88, 91, 93
M. triangulatus 44, 45, 61, 70, 71, 73, 74, 78, 82
M. turriculatus 2, 42, 47, 48, 50, 51, 57, 58, 61, 91 – 95
Parakidograptus acuminatus 44, 61, 64
Pribylograptus leptotheca 45, 61, 73

BRITISH GEOLOGICAL SURVEY

Keyworth, Nottingham NG12 5GG

Murchison House, West Mains Road,
Edinburgh EH9 3LA

The full range of Survey publications is available
through the Sales Desks at Keyworth and
Murchison House. Selected items are stocked by
the Geological Museum Bookshop, Exhibition
Road, London SW7 2DE; all other items may be
obtained through the BGS London Information
Office in the Geological Museum. All the books
are listed in HMSO's Sectional List 45. Maps are
listed in the BGS Map Catalogue and Ordnance
Survey's Trade Catalogue. They can be bought
from Ordnance Survey Agents as well as from
BGS.

*The British Geological Survey carries out the geological
survey of Great Britain and Northern Ireland (the latter as
an agency service for the government of Northern Ireland),
and of the surrounding continental shelf, as well as its
basic research projects. It also undertakes programmes of
British technical aid in geology in developing countries as
arranged by the Overseas Development Administration.*

*The British Geological Survey is a component body of the
Natural Environment Research Council.*

Maps and diagrams in this book use topography
based on Ordnance Survey mapping

HER MAJESTY'S STATIONERY OFFICE

HMSO publications are available from:

HMSO Publications Centre
(Mail and telephone orders)
PO Box 276, London SW8 5DT
Telephone orders (01) 622 3316
General enquiries (01) 211 5656
Queueing system in operation for both numbers

HMSO Bookshops
49 High Holborn, London WC1V 6HB
   (01) 211 5656 (Counter service only)
258 Broad Street, Birmingham B1 2HE
   (021) 643 3740
Southey House, 33 Wine Street, Bristol BS1 2BQ
   (0272) 264306
9 Princess Street, Manchester M60 8AS
   (061) 834 7201
80 Chichester Street, Belfast BT1 4JY
   (0232) 238451
71–73 Lothian Road, Edinburgh EH3 9AZ
   (031) 228 4181

HMSO's Accredited Agents
(see Yellow Pages)

And through good booksellers